ENGINEERING MANAGEMENT AND INDUSTRIAL ENGINEERING

PROCEEDINGS OF THE 2014 INTERNATIONAL CONFERENCE ON ENGINEERING MANAGEMENT AND INDUSTRIAL ENGINEERING (EMIE 2014), XIAMEN, CHINA, 16–17 OCTOBER 2014

Engineering Management and Industrial Engineering

Editor

A. Leung

Advanced Science and Industry Research Center, Hong Kong

CRC Press
Taylor & Francis Group
Boca Raton London New York Leiden

CRC Press is an imprint of the
Taylor & Francis Group, an **informa** business

A BALKEMA BOOK

Published by:
CRC Press/Balkema
P.O. Box 447, 2300 AK Leiden, The Netherlands
e-mail: Pub.NL@taylorandfrancis.com
www.crcpress.com – www.taylorandfrancis.com

ISBN 13: 978-1-138-02773-2 (hbk)
ISBN 13: 978-1-315-69056-8 (ebk)

Typeset by V Publishing Solutions Pvt Ltd., Chennai, India

Visit the Taylor & Francis Web site at
http://www.taylorandfrancis.com

and the CRC Press Web site at
http://www.crcpress.com

Engineering Management and Industrial Engineering – Leung (Ed.)
© 2015 Taylor & Francis Group, London, ISBN 978-1-138-02773-2

Table of contents

Engineering Management and Industrial Engineering – Leung (Ed.)
© 2015 Taylor & Francis Group, London, ISBN 978-1-138-02773-2

Preface

The 2014 International Conference on Engineering Management and Industrial Engineering (EMIE2014) has been held in Xiamen, China on October 16–17, 2014. As a major international convention in such fields, I am proud to say that EMIE2014 has both made satisfying contributions to related disciplines and served its participants well.

Beginning from the late twentieth century, engineering management and industrial engineering which both deal with the optimization of complex processes or systems, are mainly concerned with the development, improvement of integrated systems in industrial manufacture. It is easy to know from that definition that our knowledge in these fields can directly help the advance of management and manufacture. This is also the reason why we have decided to organize EMIE2014.

During the convention, innovative ideas and thoughts were presented by our participants; some hot topics in many fields like engineering education and training, industrial management and decision analysis were also intensively discussed. I am sure that these presentations will be of great research value to their listeners.

In our preparation for EMIE2014, we have received hundreds of related researches and studies. But in this book, only a small part of that amount is included after strict selection. Into 6 chapters these papers are categorized: 1) Modeling, Simulation and Engineering Application, 2) Manufacturing Systems and Industrial Design, 3) Information Processing and Engineering, 4) Management Technology and Engineering, 5) Engineering Education and Training, and 6) E-Business and E-Commerce.

As a member of the committee, I would like to specially thank those who have showed their great kindness to EMIE2014 in our preparation for it. I would like to specially express my thankfulness to the CRC Press for their publishing this book, to all our contributors and participants for their contribution and participation, and to all other people who have helped us including the reviewers and the organizers.

It is the effort of all these people's that has made EMIE2014 such a success and it is my sincere hope that we would have another chance to work together in the future.

EMIE2014 Committees

Engineering Management and Industrial Engineering – Leung (Ed.)
© 2015 Taylor & Francis Group, London, ISBN 978-1-138-02773-2

Organizing committee

GENERAL CHAIR

Prof. Mu-Song Chen, *Da-Yeh University, Taiwan*

PUBLICATION CHAIR

Prof. Jun Yeh, *Tallinn University of Technology, Estonia*

PROGRAM CHAIR

Dr. Ally Leung, *Advanced Science and Industry Research Center, Hong Kong*

INTERNATIONAL SCIENTIFIC COMMITTEE

Dr. Akbar Nikkhah, *University of Zanjan, Iran*
Dr. Bhaba R. Sarker, *Louisiana State University, USA*
Dr. Ali A. Ensafi, *Isfahan University of Technology, Iran*
Dr. N. Ethiraj, *Dr. M.G.R. University, India*
Dr. Morteza Enhessari, *Islamic Azad University, Iran*
Dr. Kevin Weller, *Science and Engineering Research Center, Hong Kong*
Dr. Q. Wen, *Huazhong University of Science and Technology, China*
Dr. M.M. Kim, *Chonbuk National University, Korea*
Dr. Ricardo Rodriguez, *Technological University of Ciudad Juarez, Mexico*
Dr. Faraj Alshmri, *Slamn bin Abdulaziz University, Saudi Arabia*
Dr. Q.B. Zeng, *Shenzhen University, China*
Dr. Z.Y. Jiang, *University of Wollongong, Australia*
Dr. V.K. Jain, *Indian Institute of Technology, India*
Dr. T.R. Vijayaram, *VIT University, India*

Modeling, simulation and engineering application

Engineering Management and Industrial Engineering – Leung (Ed.)
© 2015 Taylor & Francis Group, London, ISBN 978-1-138-02773-2

Construction quality control scheme for CFG pile of Wuhan–Guangzhou passenger dedicated line

Y.S. Liu
School of Mathematics, Lanzhou City University, Lanzhou, Gansu Province, China

H. Gao
Zhixing College of Northwest Normal University, Lanzhou, Gansu Province, China

ABSTRACT: In Wuhan-Guangzhou passenger dedicated line, there are large numbers of Cement Fly-ash Gravel (CFG) piles to improve the bearing capacity of subgrade and reduce the settlement of subgrade. This paper researched construction quality control of the CFG piles with many methods, such as concrete strength inspection, loading test, low strain integrity testing and core-drilling detection methods. Furthermore, we provided a new feasible control scheme.

1 INTRODUCTION

The length of the new Wuhan–Guangzhou passenger dedicated line is 874.406 km. The design speed is 350 km/h. According to a comprehensive analysis of the geological conditions and settlement estimation, settlement of test section (DK1228 + 500~DK1238 + 750) on Wuhan does not meet some sections of passenger dedicated railway track laying conditions non-residue on the settlement after construction, but Cement Fly-ash Gravel (CFG) pile and composite foundation were reinforced. CFG pile on Wuhan test section design uses 265,000 linear meters.

CFG pile construction quality is one of the key factors affecting construction quality of the Wuhan–Guangzhou passenger dedicated line, construction quality control program CFG pile of research is to ensure the quality of the key aspects of CFG pile.

1.1 *The basic situation of CFG pile*

CFG pile strengthening technology is based on the gravel pile foundation developed a processing technology, which is a high bond strength rigid pile made by mixing cement, fly-ash and gravel pile. Its principle is to make public bear from the external loads through the mattress layer between the two piles, thus to ensure maximum effectiveness of composite foundation.

In Wuhan engineering test section, the network of CFG piles is consist of piles, pile caps, cushion and geogrid. The space length between piles is 1.4~1.6 m, where the pile position presents a square arrangement and the diameter of the pile is 0.5 m. A 0.9 m (length) × 0.9 (width) × 0.3 (thickness) C15 concrete pile cap is set up on top of the pile block. Above the pile cap, there is a 50 cm (thickness) cushion which is consist of 20 cm gravel, 5 cm sand, 110 kN/m biaxial warp knitting geogrid, 5 cm sand and 20 cm gravel from bottom to top.segment network structure consisting of CFG piles, pile caps, cushion, geogrid structure, pile spacing 1.4~1.6 m, square layout, pile diameter 0.5 m, and pile located 0.9 m (length) × 0.9 (W) × 0.3 (thickness) C15 concrete pile cap. Cushion above the pile cap, thick 50 cm, set the form from the bottom of 20 cm (gravel) + 5 cm (sand) + geogrid (110 kN/m two-way geogrid warp) + 5 cm (sand) + 20 cm (gravel).

2 KEY CONSTRUCTION QUALITY CONTROL

2.1 *Concrete strength inspection*

CFG pile composite material strength design value of 15 MPa. CFG pile group carried out within a certain area sampling, a total of 347 sample group, the maximum compressive strength test block 28 days of 19.6 MPa, the minimum is 13.4 MPa, a total of about 92% more than the required strength test block design value.

2.2 *Loading test*

2.2.1 *The testing principle and method*

Site maintained pile loading tests is using a slow loading method, progressively loaded with manual pumps, maximum load capacity according to the

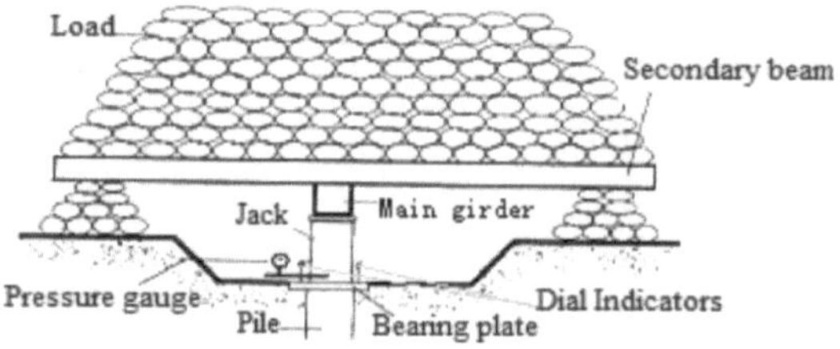

Figure 1. Loading test schematic diagram.

design value of vertical bearing capacity of two doubly contained.

Loading test is shown in Figure 1, the static pile loading test using Φ500 mm circular plate bearing plate, using the I-beam and springboard barricaded weights platform reaction force device and sand bags stacked.

Heap Pile Composite Foundation site loading test twice and according to the design value of the replacement ratio, consolidated bearing plate area. Unload twice the amount of their amount was loaded.

Pile composite foundation bearing plate loading test is using 1.5×1.5 m square plate, a bottom plate laying 50~150 mm coarse sand screeds, trial excavation pit bottom to the base elevation. Hydraulic jack is manual loading, beam erection heap load platform, sand bags stacked to provide a reaction force. The reference beam is used for the installation on the basis of separate piles, the center distance between the center of reference pile and test pile is 4.0 m, and the distance between the reference pile center and pier platform edge is 2.0 m.

2.2.2 *Pile capacity*

The 23 piles were tested in the bearing capacity of single pile. A typical test results for Q-s curve are shown in Figure 2. It shows gentle Q-s curves and no steep drop segment. Corresponding s-lgt curve is gentle regular arrangement. Therefore, all single pile bearing capacity of CFG pile detected value to meet the design requirements.

Figure 2. Single pile bearing capacity Q-s curve.

Figure 3 is a pile capacity eigenvalues corresponding settlement distribution. Table 1 shows the statistical results of pile capacity eigenvalues corresponding settlement. Visible, pile bearing capacity values correspond to the distribution of the main settlement in the range of 4~6 mm, this part of the settlement process will be the majority complication of the embankment, therefore, far less than the required value of the settlement after non-ballasted track structure subgrade work be equal or less than 3 cm.

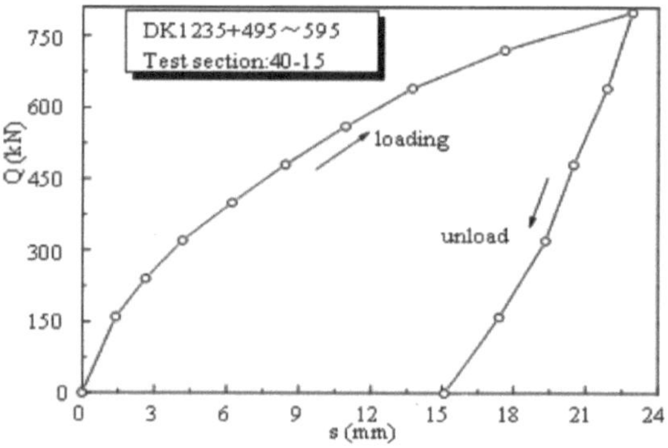

Figure 2. Single pile bearing capacity Q-s curve.

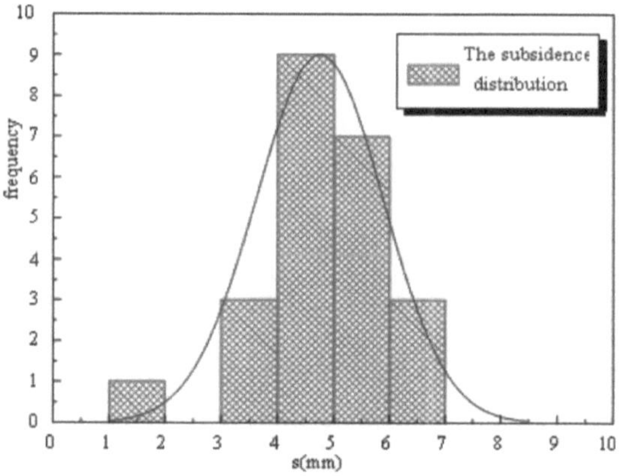

Figure 3. The settlement of single pile bearing capacity characteristic value of the corresponding.

Table 1. Single pile bearing capacity characteristic value (400 kN) corresponding to the settlement.

Maximum	Minimum	Mean	Standard deviation	Coefficient of variation
6.42 mm	1.73 mm	4.74 mm	1.11 mm	0.23

2.2.3 *The bearing capacity of composite foundation*

A total of 71 composite foundation loading tests. Termination of the test loads was loaded to achieve the desired maximum load. A typical test results are shown.

In Figure 4, the result shows gentle P-s curves and no steep drop segment. Detection results indicate that all composite foundation bearing capacity values are detected to meet the design requirements.

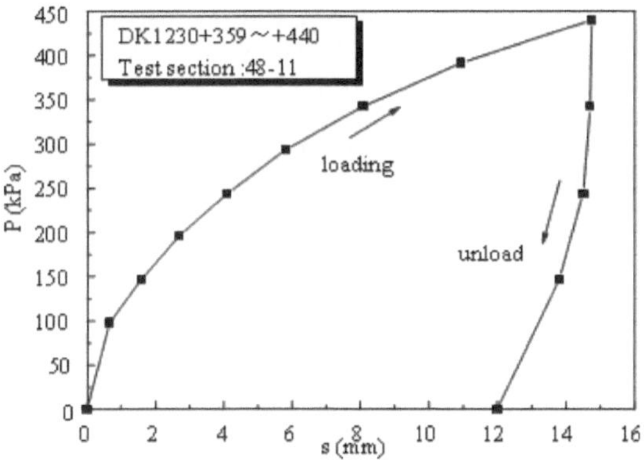

Figure 4. The bearing capacity of composite foundation P-s curve.

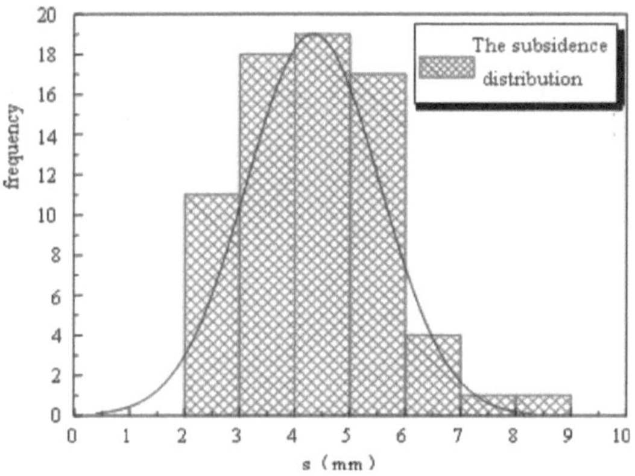

Figure 5. The settlement of composite foundation bearing capacity corresponding.

Table 2. Single pile bearing capacity characteristic value (220 kN) corresponding to the settlement.

Maximum	Minimum	Mean	Standard deviation	Coefficient of variation
8.26 mm	2.35 mm	4.33 mm	1.20 mm	0.28

Figure 5 is a composite foundation bearing capacity values corresponding settlement distribution. Table 2 shows the values of the corresponding bearing capacity of composite foundation settlement statistics. As it can be observed, the bearing capacity of composite foundation settlement eigenvalues distribution is slightly larger than the corresponding pile bearing capacity distribution of eigenvalues. Bearing capacity of composite foundation settlement value corresponds to the main distribution area of 2~6 mm. Settlement of composite foundation embankment process will

Table 3. Low strain test results.

Total	Testing	I type	II type	III type	IV
26286	2693	2215	473	5	0
Proportion	10.2%	82.3%	17.5%	0.2%	0

be completed in the majority, and theoretically, it would be less than the settlement after settlement under loading test conditions. Therefore, after the settlement of composite, the foundation will work after less than non-ballasted track roadbed settlement control values.

2.3 *Low strain reflected wave method*

Acceptance by the passenger line with regulatory requirements, low strain detection by 10% CFG total number of piles. Total construction of the Wuhan–Guangzhou project completed CFG pile test section is 26286, testing the number 2693, the ratio was 10.2%, with a pass rate of 99.8%. Low strain test results are shown in Table 3. It can be observed from the measured waveform low strain detection, II type pile on top of the pile 92% less 1.0 m~2.0 m occur within the scope of a slight constriction, where the soil is soft plastic or plastic-like flow of clay, soil lateral binding small, CFG pile construction sequence is one of the factors to produce necking. To avoid a new pile to twist and extrusion the existing piles, a construction technology can be used to make the piles, where a new pile is created in alternate lines and rows. while after construction, the new piling must become a pile body that strong enough with certain strength. To drill too fast will cause an insufficient release of the air, and may also cause the constriction. Therefore, when a pile of concrete output should be controlled to match the speed and promotion, always ensure that the height of the concrete pipe to avoid necking generation.

2.4 *Core-drilling experiments*

Wuhan–Guangzhou passenger line CFG pile test section core-drilling sample design strength be equal or greater than 15 MPa, in seven piles of sampling 12-bit, the measured result is 17.7 MPa~36.2 MPa, the results show that the compressive strength of core-drilling samples meet the design requirements, the maximum intensity of the pile more than two times the design requirements.

2.5 *Bits (vertical and horizontal) and the effective diameter of the pile detection*

The main detection of each pile's test is pile group distance, and the subject of the design or test the tolerance range, carried out mainly with a ruler to measure for the detection on the spot; pile mainly on the effective diameter of the pile manually selected excavation to the specified depth, diameter of the pile with a ruler be measured, to detect whether an effective diameter of the pile design pile diameter.

Wuhan engineering test section in addition to individual pile CFG pile test section had shifted to high office, basically reached the required standard deviation range test, the results were within the range of 5 cm. Deviation of excess pile, field-increasing loading test, designed checking, and an increase in the upper pile cap sizes all can meet the design requirements.

3 CONCLUSION

Through the construction and subsequent detection of a large construction site detection process test segment, based on a comprehensive analysis of the detection results, after which experts discuss on CFG pile testing procedures, test results of the three main aspects of treatment and detection strict control draw a practical conclusion.

Detection Program: In addition to strengthening the construction process, as well as the detection of the raw material mixture stirred for quality control, there are 28 days after the first test block on the pile strength test. After the strength meets the requirements, there will be low strain integrity testing and loading testing; strength in the case does not meet the strength test block, conducted on behalf of the pile to pile drill core-drilling samples to detect physical strength body again meets the design requirements, other subsequent projects above.

Test results processing: CFG pile for detecting the presence of class III low strain pile exceeds 10% or class IV pile appears when you want to pile all the batches for testing; proportion of post-test class III piles but not more than 10% before each construction unit 100; extend more than 5% to the detection range, around the pile the pile detecting class III; does not exceed 5% when the same is detected, but the end result in the same proportion to expand or verify eligibility pile must all testing.

Detection of the body: the first self-test detects the construction unit, reported a third-party testing unit supervision and self-qualified after re-examination, a process of construction in all three parties after passing before the next.

By following this detection scheme for quality control of CFG pile, effectively ensure the quality of the Wuhan–Guangzhou high-speed railway passenger dedicated pile foundation treatment, right across the board as a whole to ensure the overall quality control and sedimentation control embankment structures play a crucial role in safeguarding.

REFERENCES

Ding Xuan-ming, Chen Yu-min & Kong Gang-qiang. 2012. Formation mechanism of velocity waves in low strain integrity testing of PCC pile. *Rock and Soil Mechanics* 33(1):154–161.

Jun-Jie Zheng, Sari W. Abusharar & Xian-Zhi Wang. 2008. Three-dimensional nonlinear finite element modeling of composite foundation formed by CFG–lime piles. *Computers and Geotechnics* 35(4): 637–643.

Zheng Gang & Yu Zongfei. 2003. Standardization of bearing test and the determination of bearing capacity of composite foundation. *Journal of Building Structures* 24(1): 84–91.

Engineering Management and Industrial Engineering – Leung (Ed.)
© 2015 Taylor & Francis Group, London, ISBN 978-1-138-02773-2

Determination of functions for lightning impact point location on Transmission Lines

J.A. Morales Garcia
Carrera de Ingeniería Eléctrica de la Universidad Politécnica Salesiana, Cuenca, Ecuador

J.C. Montesdeoca Contreras
Carrera de Ingeniería Electrónica de la Universidad Politécnica Salesiana, Cuenca, Ecuador

ABSTRACT: Protection algorithms are related to what provides the highest accuracy in location of faults produced by lightning strokes on transmission lines. Thus, in this paper a novel methodology for lightning stroke location based on impedance estimation in new coordinate axes is presented. Principal Component Analysis (PCA) is used to project the voltage and current signals with data windows of ¼ cycles onto new coordinate axes, corresponding to the Principal Components (PCs) subspace. Then, these new voltage and current values are used to calculate the apparent impedance in the PCA subspace. This projection shows a relation between the new impedance value and the point of impact along transmission lines. Based on this, two functions that localize the flash point produced by lightning strokes are found. Results show that the proposed methodology for location of lightning stroke on transmission line presents satisfactory results.

1 INTRODUCTION

Transmission lines are the most important elements in Electric Power System (EPS), which are exposed to different conditions, where the possibility of faults produced by atmospheric discharges is higher than other phenomena [1]. Currently, events related to atmospheric discharges are considered as the main causes in producing Transmission Lines (TLs) outages [2]. Faults records on TLs are 50% higher than that produced by lightning strokes [2]. On this context, it is well accepted that the most important transmission lines outages are related to the lightning [3–5].

In the engineering environment, lightning strokes which hit directly on transmission lines or transmission towers are considered as serious inconveniences [6]. When a lightning produces faults, it is crucial to correctly determine the point of impact on TLs [7].

At the moment, algorithms based on the Fourier Transform are used for the localization of faults produced by atmospheric discharges [8]–[9].

As regard to the bibliographic review, different approaches using the Wavelet Transform and the travelling waves theory have been reported for the common faults localization, which are different from lightning strokes [10]–[14]. However, it is well known that travelling waves techniques are influenced by the multiple reflections, which cannot correctly detect the first wave front, with these protection devices being considerably influenced, especially for the faults near relays.

An algorithm based on the Fourier Transform (FT) and the k-nearest neighbors (K-NN) was proposed [11]. Still, this algorithm uses data windows corresponding to 1 cycle, thus their operation time is similar to the traditional protection algorithms.

A referent work for this issue is presented [12]. This work is based on the Multiresolution analysis (MRA), in which by using the mother wavelet daubechies 8, the signals are decomposed in levels and by using these levels, a function useful to the lightning localization is determined. Still, the MRA depends on the mother wavelet selected and their decomposition [13].

This consideration is crucial in making analysis, considering models as really as possible, such as transmission towers, ground-wires and live wires [14]. However, the work proposed in [15] analyzing transmission lines omits these elements, especially transmission towers which are crucial in lightning strokes studies due to that this element produces over-voltages. These over-voltages are produced by multiple reflections when a flash hits the transmission tower [15]–[16].

In section 2 the operation principle of protection relays based on the impedance estimation is presented. Section 3 presents the theoretical basis of Principal Component Analysis. Section 4 presents the electric power system simulation. The signal processing useful to determine the impact point is presented in section 5. Finally, section 6 presents the main conclusions of this work.

2 PRINCIPAL COMPONENT ANALYSIS

Principal Component Analysis (PCA) is an appropriate mathematical tool when a variables set represented by p has been obtained, and it is desirable to find a smaller variables number z called principal components [15]. Hence, PCA is an efficient tool to analyze a data set where m observations on p-dimensional space can be represented optimally on a new space smaller than the original space.

Assuming a correlated data set X, it is possible to represent them through a new space where new data bases z are uncorrelated. These new data called Principal Components (PCs) are represented by a linear combination, which are calculated as follows:

$$PC_{1,2,\ldots,p} = \left[eigv_1 \ eigv_2 \ldots eigv_p\right] \cdot \left[x - \bar{x}\right]^T \qquad (1)$$

where, $PC_{1,2,\ldots p}$ are projections corresponding to original variables $\boldsymbol{x}$, which are projected through using the eigenvectors ***eigv***. These eigenvectors are calculated through the variance-covariance matrix S as follows:

$$S = \frac{\sum_{i=1}^{m}\left(x_m - \bar{x}\right)\left(x_m - \bar{x}\right)}{(m-1)} \qquad (2)$$

The matrix form corresponding to principal components are calculated as follows:

$$[PC] = [U][X] \qquad (3)$$

where U represents the eigenvectors matrix, and X represents the original variables matrix.

3 ELECTRIC POWERSYSTEM SIMULATED

Alternative Transient Program ATP/EMTP [29] is used for the Electric Power System (EPS) simulation. This model corresponds to a 220 kV EPS.

The transmission line denoted as M-N is an element under study, which corresponds to three phases TLs with two ground wires.

As regard to the simulations, different elements are considered as follows:

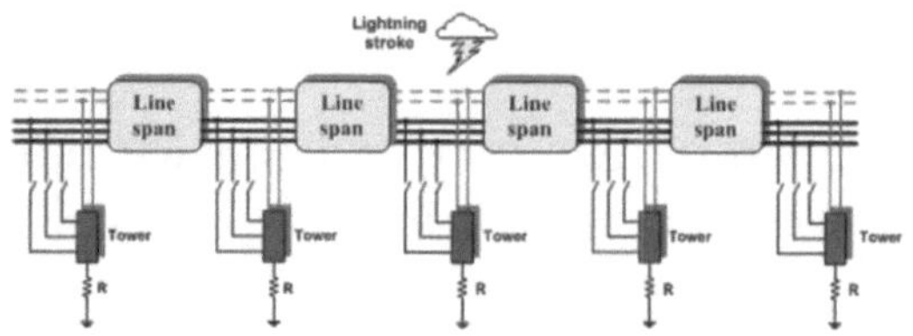

Figure 1. Overhead line model for lightning stroke over-voltages analysis.

Transmission line (TL): By using un-transposed distributed parameter line models J. Marti, shield wires and live wires of transmission lines are modeled by two or three spans at each side of the point of impact. See Fig. 1.

Transmission tower: In this work the tower is represented as a lossless distribute-parameter transmission line, characterized by its impedance and travel time [18].

Insulators: The insulator on TLs is simulated by a voltage-dependent flashover switch and a voltage-time characteristic, which is recommended and suggested by world organizations as EPRI and IEEE [6].

Lightning stroke: It has been accepted that when a lightning hits a transmission line, it injects current into the power system [6]. In this context, the Heidler model is used in this research.

3.1 *Lighting stroke signals data base*

In order to generate different signals produced by lightning on TLs, the standard lightning impulse 1.2/50 μs is used to generate direct faults (flashover) and indirect faults (back-flashover).

Lightning current of 10 kA and 150 kA is simulated to generate flashover y back-flashover, respectively. Furthermore, by using these lightning current values, different lighting strokes along the transmission line are simulated, where 26 transmission line stretches are considered. As regard to the lightning currents values, they are composed of the values range registered by different organizations [6].

4 PROPOSED METHODOLOGY

The methodology proposed in this paper has two analysis states. As regard to the first state, the voltage and current signals corresponding to normal state are processed thought PCA. Thus, by applying equations (1)–(3), these signals are projected onto a new space, and the impedance value is calculated. The voltage and current signals used in this section are discrete sinusoidal waveforms with a data window corresponding to 1 cycle. Still, in order to break down the variation of these signals, and extract the impedance by using a data window less than 1 cycle, different data windows corresponding to ¼ cycle are used. That is, the voltage and current signals corresponding to normal condition registered in 1 cycle are composed of data windows of ¼ of cycle.

In this context, a data matrix (1900×4000) registered by the relay R1 at bus M is analyzed through PCA. Thus, the impedance value is calculated onto a new space, using data windows corresponding to ¼ of cycle, decreasing the operation time of traditional algorithms.

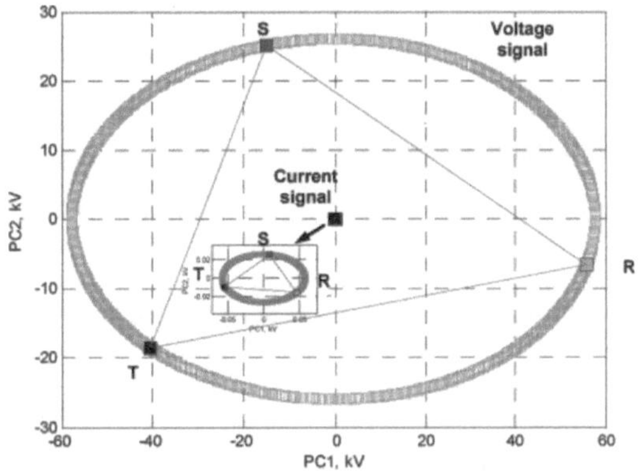

Figure 3. Voltage and current signals projected onto a 2D subspace of principal components.

By using equations 4 and 5, the real part and the angle are calculated, and thus the impedance value is determined. It is necessary to note that the previous procedure using normal state sinusoidal signals is developed in order to extract eigenvectors useful to make the projection of lightning stroke signals on a new space.

4.1 *Patterns analysis*

After the voltage and current signals corresponding to normal operation have been projected on the new coordinates, it is possible to observe that both the voltages and current signals describe well the defined ellipsoidal patterns as are presented in Figure 2 and Figure 3.

Later on the impedance values are calculated. However, in order to calculate this parameter, it is necessary to transform the values of principal components into their respective values in polar coordinates, and by using these values, the impedance can be calculated as follows:

$$real = \sqrt{PC_1^2 + PC_2^2} \quad (4)$$

$$\phi = \arctan\left(\frac{PC_2}{PC_1}\right) \quad (5)$$

where PC1 and PC2 represent the principal component values.

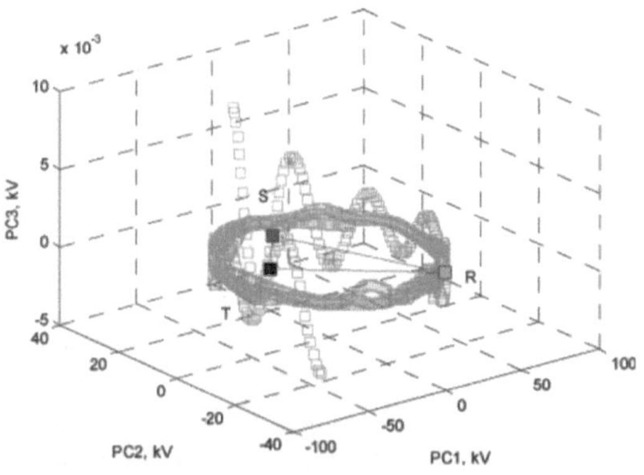

Figure 2. Voltage and current signals projected onto a 3D subspace of principal components.

4.2 *Lightning impact point determination*

After the eigenvectors are extracted, the voltage and current signals corresponding to new lightning stroke on transmission lines, which are registered by the protection relay as discrete signals f(v) and f(i) of 4000 points (row vectors 1×4000), are calculated and projected on the ellipsoidal patterns previously established. The previous projection is done by using the coefficients corresponding to the eigenvectors of the variance-covariance matrix; the projection of new lightning strokes is done as follows:

$$fv(pc) = \left(f(v) - \bar{x}\right) * U \quad (6)$$

$$fi(pc) = \left(f(i) - \bar{x}\right) * U \quad (7)$$

where U represents the projection matrix, represents the mean vector. Therefore, by using these eigenvectors, both direct faults (flashover) and indirect faults (back-flashover) produced by atmospheric discharges considering different impact points along the transmission line, are projected on a two dimensions space.

Later on, the impedance value onto the new space is calculated as follows:

$$Zpc = \frac{real\,V}{real\,I} \quad (8)$$

where *realV* and *realI* are the real part of the voltage and current signals produced by lightning strokes.

The relation or patron between the new impedance value and the lightning impact point along the transmission line is shown in Figure 4.

Thus, by using a regression procedure, it is possible to determine two functions useful to localize the flash impact point.

In order to extract the best patterns, different functions were analyzed on the two patterns as are presented in Figure 5.

However, after a detailed analysis it is possible to determine two functions corresponding to direct and indirect faults as are shown in Figure 5b. These functions are mathematically expressed as follows:

$$l_b = \left(10.60484 - 0.83363 * \sqrt{zpc}\right)^2 \tag{9}$$

$$l_f = \exp(-0.224504 + 0.006030 * zpc) \tag{10}$$

where *lb* and *lf* represent the lightning impact point corresponding to indirect and direct faults, respectively.

In addition, from Figure 4 it is also possible to observe that using the impedance value, it is possible to determine whether the lightning hits the tower or the phase conductor, respectively.

Finally, faults produced by lightning strokes are tested through the proposed methodology. Similar to the previous analysis, the lightning signals must be projected onto the principal components axes, which is developed through using equations (6) and (7). Later on, these signals projected are used to calculate the impedance value through equation (8). Finally, the impedance value is analyzed through the two functions of lightning localization equation (9) and equation (10).

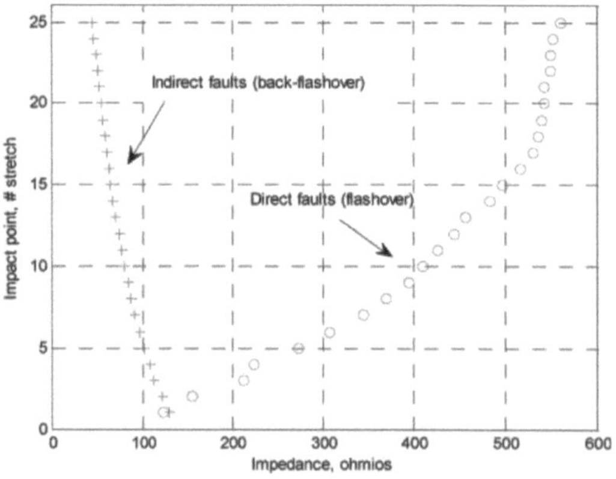

Figure 4. Relation between the impedance value and the lightning impact point.

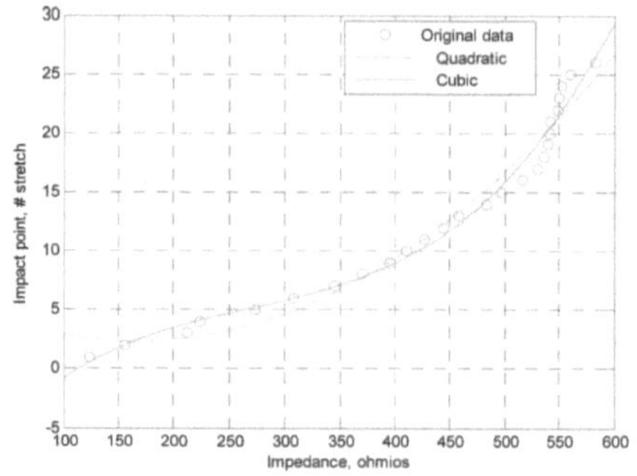

(a) Quadratic and cubic.

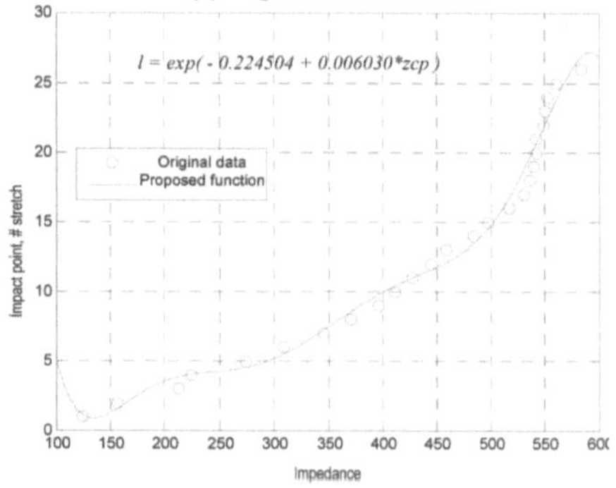

(b) Proposed function.

Figure 5. Function to direct faults (a) quadratic and cubic functions; (b) function proposed in this work.

5 CONCLUSIONS

In this work, a novel methodology useful to determine the lightning impact point on transmission lines based on principal components principal is presented. Thus, the impact point is determined through the relation between the voltage and currents signals with data windows of ¼ of cycle.

Results show that the methodology can estimate the impact point, considering both the most important elements of Electric Power Systems and crucial features of atmospheric discharges.

By using a regression process, two functions useful to localize the impact point of flashover and back-flashover are determined.

REFERENCES

Andersson, G., P. Donalek, R. Farmer, N. Hatziargyriou, I. Kamwa, P. Kundur, N. Martins, J. Paserbe, P. Pourbeik, J. Sanchez-Gasca, R. Schulz, A. Stankovic, C. Taylor, and V. Vittal, "Causes of the 2003 Mayor Grid Blackouts in North America and Europe, and Recommended Means to Improve System Dynamic Performance," IEEE Trans. Power Systems, vol. 20, no. 4, pp. 1922–1928, November. 2005.

CIGRE WG 33-02. "Guidelines for representation of network elements when calculating transients". 1990.
Costa, F.B., B.A. Souza, N.S.D. Brito, "Real-time detection of fault-induced transients in transmission lines," Electronics Letterns, vol. 46, pp. 753–755, May, 2010.
Diendorfer, G., W. Shulz, "Ground flash density and lightning exposure of power transmission lines," Power Tech Conference Proceedings, IEEE, Bologna, 2003.
EPRI AC Transmission Line Reference Book-200kV and Above, Third Edition, Electric Power Research Institute, December 2005
Farshad, M., J. Sadeh, "Accurate Single-Phase Fault-Location Method for Transmission Lines Based on K-Nearest Neighbor Algorithm Using One-End Voltage," IEEE Trans. Power Delivery, vol. 27, pp. 2360–2367, Oct., 2012
Johns, A.T., S.K. Salman, "Digital Protection for Power Systems," 1st ed. London: Peter Peregrinus, 1997.
Lee H. and A.M. Mausa, "GPS travelling wave fault locator schemes: Investigation into the anomalous measurements related to lightning strikes," IEEE Trans. Power Delivery, vol. 11, pp. 1214–1223, Jul., 1996
Lin, S., Z.Y. He, X.P. Li, Q.Q. Qian, "Accurate Travelling wave time-frequency characteristic-based fault location method for transmission lines," Generarion, Transmission & Distribution, IET, vol. 6, pp. 764–772, August, 2012
McLaren, P.G., M.A. Redfem, "Fourier-Series Techniques Applied to Distance Protection," Proceedings of the IEE, vol. 122. no.11, pp. 1301–1305, November. 1975
Mills, E. "Extreme Grid Disruptions and Extreme Weather, Lawrence Berkeley National Laboratory," U.S. Disaster Reanalysis Workshop, May 3, 2012
Morales, J.A., E.A. Orduña, "Patterns Extraction for Lightning Transmission Lines Protection Based on Principal Component Analysis," IEEE Latin America Transactions, vol. 11, no. 1, pp. 518–524, Feb. 2013.
Rebizant, W., J. Szafran and A. Wisniewski, "Digital Signal Processing in Power System Protection and Control," Springer-Verlag London, 2011
Sharafi, A., M. Sanaye-Pasand, P. Jafarian, "Ultra-high-speed protection of parallel transmission lines using current travelling waves," Generation, Transmission & Distribution, IET, vol. 5, pp. 656–666, June, 2011
The Working Group on Lightning Protection for Transmission System, "Lightning Observation on Japan Sea Coast in Winter, " CRIEPI Report No.T11 (1989) (in Japanese)
Zhicheng, W.U. "Analsis of transmission lines in special regions with high lightning strike rates and its countermeasures," 20th International Conference on Electricity Distribution, CIRED, Prague, 2099. pp. 8–11.

Engineering Management and Industrial Engineering – Leung (Ed.)
© 2015 Taylor & Francis Group, London, ISBN 978-1-138-02773-2

A mesh-adaptive data exchange method for solid body field coupling based on Radial Basis Function

L. Huo & T. Yang
College of Aerospace Science and Engineering, National University of Defense Technology, Hunan, Changsha, China

ABSTRACT: This paper focuses on the data interpolation between non-matching thermal-electric meshes. A mesh-adaptive interpolation algorithm based on radial basis function is proposed. The transfer matrix T is derived, and the calculation program used for data exchange between two non-matching meshes based on radial basis function is developed. Several groups of temperature interpolation examples between two non-matching meshes are tested, and the influence made by different parameters on precision is analyzed. The result shows that the algorithm in this paper is efficient and accurate for solid body field interpolation.

1 INTRODUCTION

As the rapid development of computer and coupling algorithm, the issue of multi-field coupling attracts increasing focus in engineering application field[1–4].

Thermoelectricity coupling is a hot topic in the design of reentry capsule. During the course of reentry, severe aerodynamic heating result in the large temperature gradient in the antenna window of the capsule. The temperature gradient can cause the gradient of material's dielectric coefficient of the antenna window, which can significantly influence the efficiency of the antenna.

The nature of thermoelectric coupling is solid body field coupling, and one of the problems in thermoelectric coupling calculation is the assignment of the dielectric coefficient according to the electric mesh nodes' temperature, while the temperature of electric mesh nodes is interpolated by the non-matching thermal mesh nodes' temperature. Therefore the efficient data exchange can be inverted into the issue of interpolation between two non-matching mesh in solid field. Coupling data exchange, which has significant influence efficiency, is a important step in multi-field coupling computation, and it is one of the major topics in the relevant field[5, 6]. The problem to solve in coupling data exchange is the two-way interpolation in two non-matching meshes generated by different calculation modules. Coupling data interpolation methods generally can be divided into two groups: local interpolation[7], such as Constant Volume Transfer (CVT) method and weighted residual method, and integral interpolation, such as spline function method and radial basis function method[8]. Local interpolation method is easy to understand and the method is relatively direct. However, some algorithm should be designed to get the mapping nodes, which will decrease efficiency when the non-matching meshes are complex. In integral interpolation method, all of the nodes of one mesh are utilized to interpolate one node of the other mesh. Integral interpolation method is relatively accurate but a huge linear system of equations needs to be solved and long computation time is cost. Therefore, a accurate as well as efficient data exchange method is required.

Compactly supported radial basis function method proposed by Wendland[9] can well solve the issues of above integral inerpolation method, while the compactly supported radius needs to be set artificially, which will lead to a abnormal result or computation termination when the dimension of the meshes or the density of grid is unknown. Therefore a mesh-adaptive interpolation method is needed.

In this paper, based on radial basis function method and the idea of mapping point selection in local interpolation method, a mesh-adaptive data exchange method based on radial basis function is proposed. Besides, the procedure is programed and 3-D solid body field data exchange examples are analyzed.

2 MESH-ADAPTIVE DATA EXCHANGE METHOD BASED ON RADIAL BASIS FUNCTION

2.1 *Radial basis function*

A group of points at different position in Euclidean space is given $X = \{x_1, x_2, \ldots, x_N\} \subseteq R^n$. Assuming that the functional value on these points is

$g_1, g_2, \ldots, g_N$ a continuous function passing through these points can be obtained. If the function is in the (1) form, it is called Radial Basis Function (RBF). RBF is a unary function, and the argument is the spatial distance. The expression is as follows:

$$f(x) = \sum a_j \phi\left(\left\|x - x_j\right\|\right) \tag{1}$$

where a_j is the coefficient to be solved, and x is the coordinate $X(x, y, z)$ of unknown node. x_j is the coordinate of the known node No. j and ϕ is the selected radial basis function. $\left\|x - x_j\right\|$ is the Euclidean distance between known node No. j and the unknown node. The expression is

$$\left\|x - x_i\right\| = \sqrt{(x - x_i)^2 + (y - y_i)^2 + (z - z_i)^2} \tag{2}$$

Major types of radial basis function are:

1. Gauss distribution function

$$\phi\left(\left\|x\right\|\right) = e^{-c^2\|x\|^2} \tag{3}$$

2. Multi-quadric function (β is a positive real number)

$$\phi\left(\left\|x\right\|\right) = \left(c^2 + \left\|x\right\|^2\right)^{\beta} \tag{4}$$

3. Inverse multi-quadric function (β is a positive real number)

$$\phi\left(\left\|x\right\|\right) = \left(c^2 + \left\|x\right\|^2\right)^{-\beta} \tag{5}$$

When β is defined, the shape of basis function is controlled by Multi-quadric function with c.

2.2 *Mesh-adaptive radial basis function method*

In this paper, multi-quadric function is selected as the radial basis function, and $\beta = 0.5$, $c = 10^{-3}$ is set.

It is assumed that the number of nodes in two non-matching meshes is M and N respectively, namely

$$\begin{aligned} X_1 &= \{x_1, x_2, \ldots, x_M\} \subseteq R^3 \\ X_2 &= \{x_1, x_2, \ldots, x_N\} \subseteq R^3 \end{aligned} \tag{6}$$

where X_1 is the source field, and X_2 is the interpolated field. The functional value on two meshes is as follows:

$$\begin{aligned} U_1 &= \{g_1, g_2, \ldots, g_M\} \\ U_2 &= \{g_1, g_2, \ldots, g_N\} \end{aligned} \tag{7}$$

For each $x_l \in X_2$, n nodes in X_1 that are most closed to x_l are selected and the radius of the corresponding region is set as the interpolation radius. Then a_j can be determined by the interpolation condition:

$$f(x_i) = \sum_{j=1}^{n} a_j \phi\left(\left\|x_i - x_j\right\|\right) = g_i, \quad i, j \in (1, M) \tag{8}$$

where g_l is the functional value at x_i. Then a simultaneous linear equations system is got:

$$\mathbf{AU} = \mathbf{G} \tag{9}$$

where

$$\mathbf{A} = [a_j]^T, \quad j = 1, 2, \ldots, n \tag{10}$$

$$\mathbf{U} = \phi\left(\left\|x_i - x_j\right\|\right), \quad i, j = 1, 2, \ldots, n \tag{11}$$

$$\mathbf{G} = [g_i]^T, \quad i = 1, 2, \ldots, n \tag{12}$$

By solving equation (12), we can get

$$\mathbf{A} = \mathbf{GU}^{-1} \tag{13}$$

Bring the solution of equation (13) into equation (14):

$$f(x_l) = \sum a_j \phi\left(\left\|x_l - x_j\right\|\right) \tag{14}$$

We can get the functional value at interpolated nodes x_l.

By the above method, we can get all the nodes' functional value in X_2, namely, the interpolation in two coupling non-matching meshes is completed.

3 ALGORITHM TEST AND ANALYSIS

It is assumed that the temperature distribution function is as follows:

$$\begin{aligned} T(x, y, z) = {} & \sin(x \times \pi/4) \cdot \sin(y \times \pi/4) \\ & \cdot \sin(z \times \pi/4) * 1000 + 1000 \end{aligned} \tag{15}$$

The first test example is a cubic computation domain, $0 \le x < 1, 0 \le y < 1, 0 \le z < 1$. The nodes in all the meshes are generated randomly. The node numbers of mesh1 to mesh4 are 500, 700, 900 and 1100 respectively.

First, the temperature distribution function is utilized to get the nodes' temperature of each mesh. Then the mesh-adaptive radial basis function method is employed to interpolate the temperature distribution (T_2^i) in one mesh by that of a other mesh, and T_2^i is contrasted to the temperature distribution solved before to get the error. Five groups are set. In Group1 (G1) mesh1 is used to interpolate mesh2. In G2 mesh2 is used to interpolate mesh3. In G3, mesh3 is used to interpolate mesh4. In G4 mesh1 is used to interpolate mesh3. In G5 mesh1 is used to interpolate mesh3.

It is assumed that the distance between the interpolated point p_i and the most closed node in temperature field is length_1, and the number of interpolation points is n, and the distance between the interpolated point p_i and the number n most closed node in temperature field is length_n. The parameter a is set as 0.001, length_1, $(\text{length}_1 + \text{length}_n)/2$ and length_n respectively. As the computation time is mainly related to the nodes number in the non-matching meshes, it is assumed that in each group the computation time is identical, which is equal to that when a is 0.001. The average error and maximum error in each group with different value of a are shown in Figure 1–Figure 8.

According to Figure 1–Figure 8, except for a = 0.001 in G4, the average error and maximum error in each groups are all decrease along with the increase of the interpolation nodes. In G5, the ratio of interpolation nodes number to interpolated nodes number is $\lambda = 500/1100 = 0.4545$, and it is the minimum in 5 groups, that is to say the known information (nodes number in mesh1) is the least compared with the unknown nodes' information (nodes number in mesh2). Therefore, the average error of G5 is the maximum in 5 groups. It is on the other way round that the ratio of G3 is $\lambda = 900/1100 = 0.8182$, which is the largest in 5 groups. And the average error of G3 is the minimum. When the value of a is different, the maximum error (about 7%) occurs when the inter-

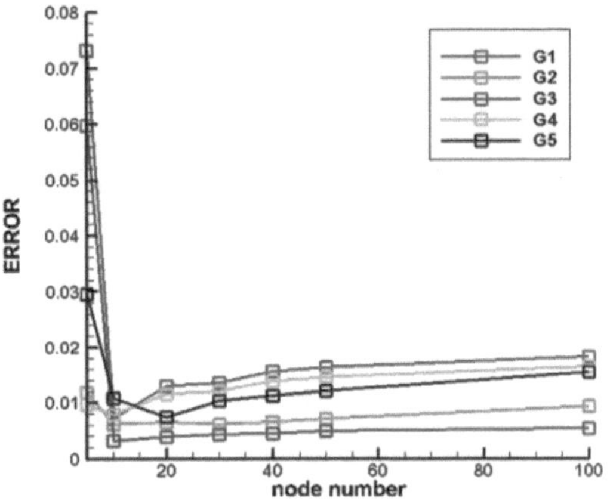

Figure 2. Relationship between maximum error and node number when a = 0.001.

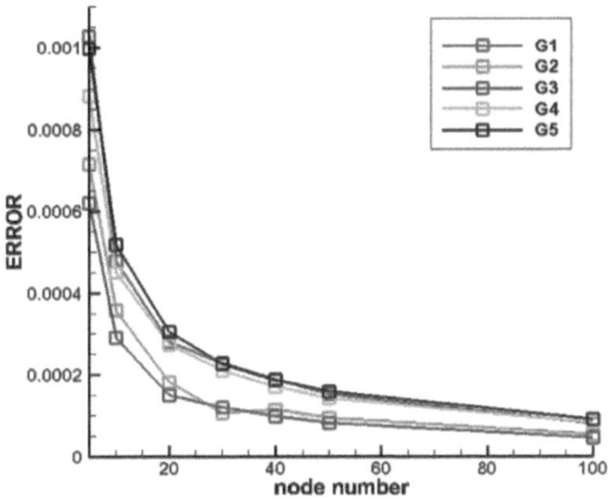

Figure 3. Relationship between average error and node number when a = length_1.

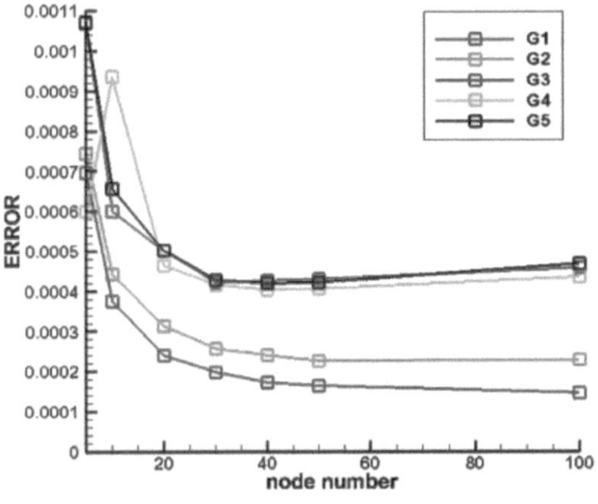

Figure 1. Relationship between average error and node number when a = 0.001.

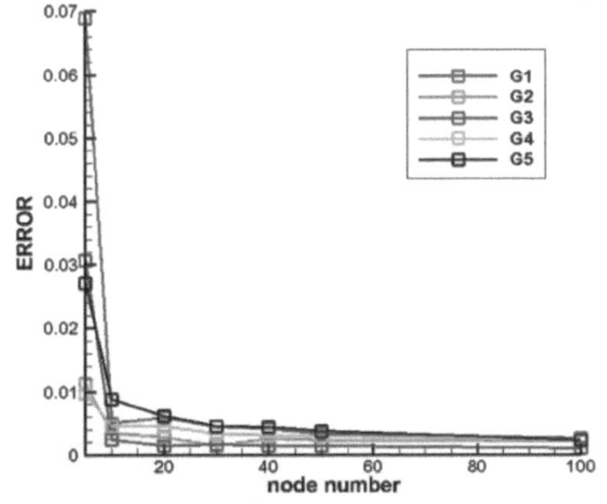

Figure 4. Relationship between maximum error and node number when a = length_1.

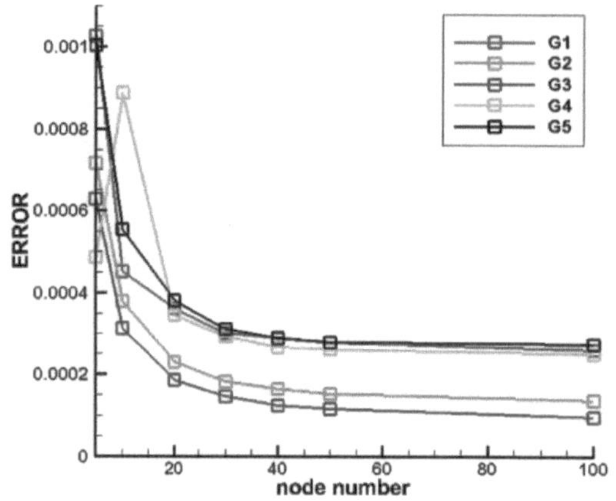

Figure 5. Relationship between average error and node number when a = ($length_1$ + $length_n$)/2.

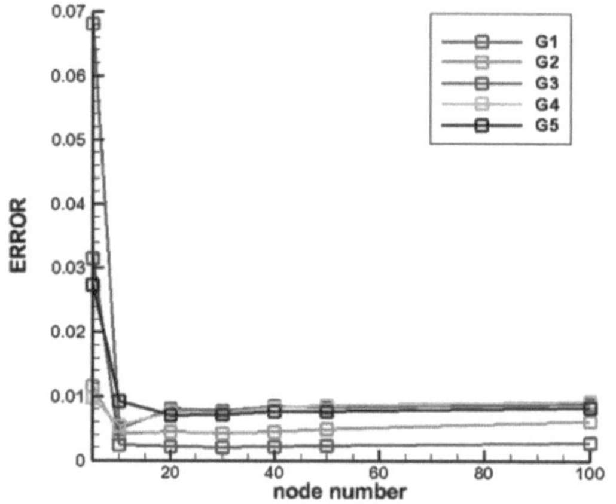

Figure 6. Relationship between maximum error and node number when a = ($length_1$ + $length_n$)/2.

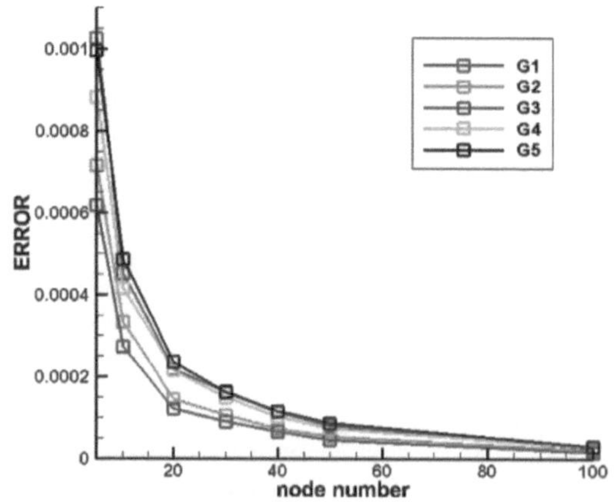

Figure 7. Relationship between average error and node number when a = $length_n$.

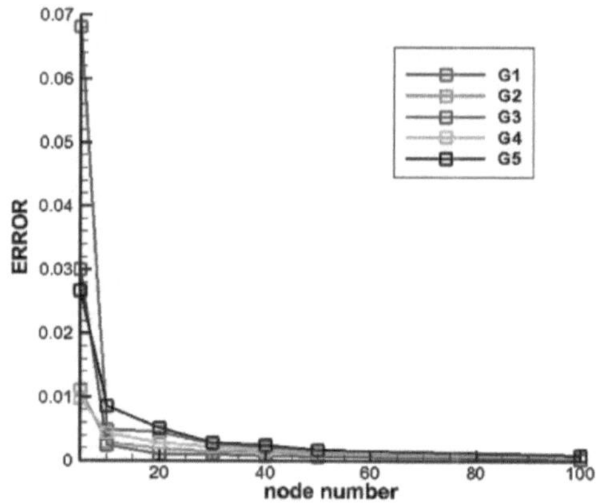

Figure 8. Relationship between maximum error and node number when a = $length_n$.

polation nodes number is the least in G1. This may caused by that the nodes numbers of the two meshes in G1 are the least and known information is insufficient. When the number of interpolation nodcs is larger than 20, the maximum error in each group is less than 2%.

Then the change of the average error and maximum error in each group along with a is analyzed. When a = 0.001, the average error and maximum error in each group reach their maximum. When a = $length_n$, the average error and maximum error in each group reach their minimum. While the average error and maximum error do not vary monotonically along with the increase of a: both of the average error and maximum error when a = $length_1$ and a = $length_n$ are less than that when a = ($length_1$ + $length_n$)/2. In all the groups, the average errors are less than 0.11%, and the maximum errors are less than 2% when the interpolation nodes number is large enough. Therefore the precision can fit engineering calculation.

To sum up, to solve the interpolation issue between two non-matching meshes in solid body field, a mesh-adaptive radial basis function method is adapted. Test examples are employed to validate the algorithm. The result shows that the computation error is related to the parameter a, the density of grid and the number of interpolation nodes. When an appropriate interpolation node number is set, the computation error is so small as to satisfy the engineering requirement.

A spheric domain test example is also employed to validate the algorithm. And the same conclusion can be brought.

The coordinate of the sphere is (0, 0, 0), and the radius is 1. The temperature distribution function is as equation (15) and the node numbers of mesh1 to mesh4 are 794, 484, 1562 and 1070 respectively. The interpolation groups are the same as the cubic domain example. The average error and maximum error in each group with different value of a is shown in Figure 9–Figure 16.

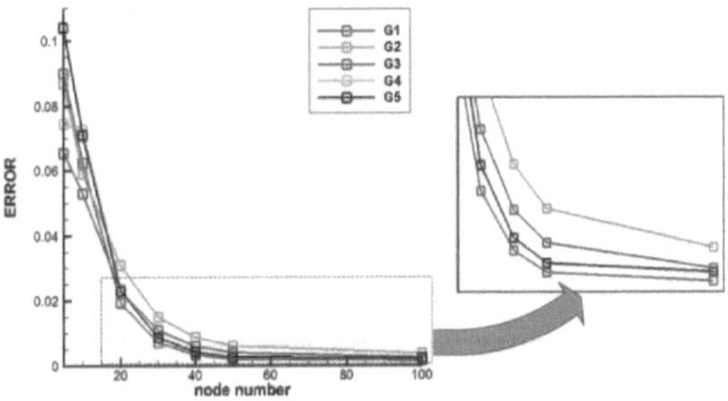

Figure 9. Relationship between average error and node number when a = 0.001.

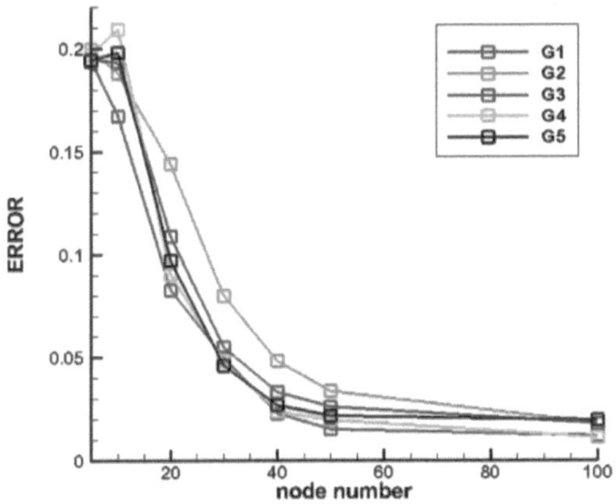

Figure 10. Relationship between maximum error and node number when a = 0.001.

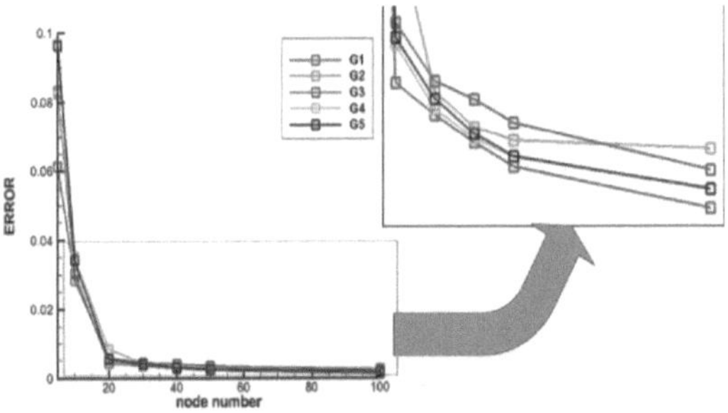

Figure 11. Relationship between average error and node number when a = $length_1$.

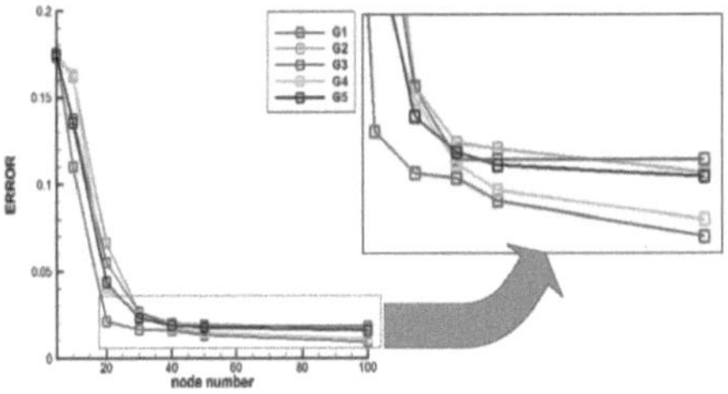

Figure 12. Relationship between maximum error and node number when a = $length_1$.

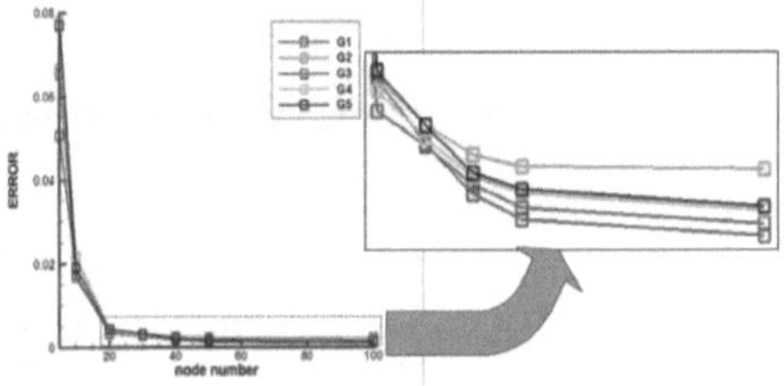

Figure 13. Relationship between average error and node number when a = $(length_1 + length_n)/2$.

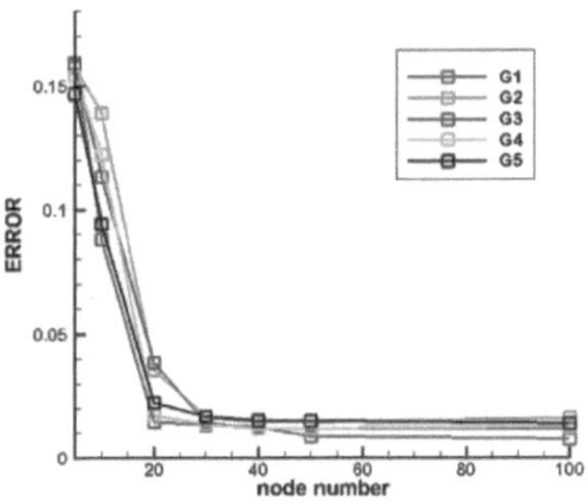

Figure 14. Relationship between maximum error and node number when a = $(length_1 + length_n)/2$.

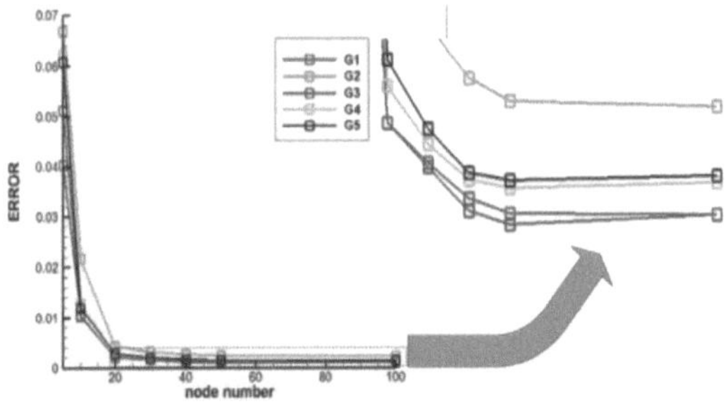

Figure 15. Relationship between average error and node number when a = $length_n$.

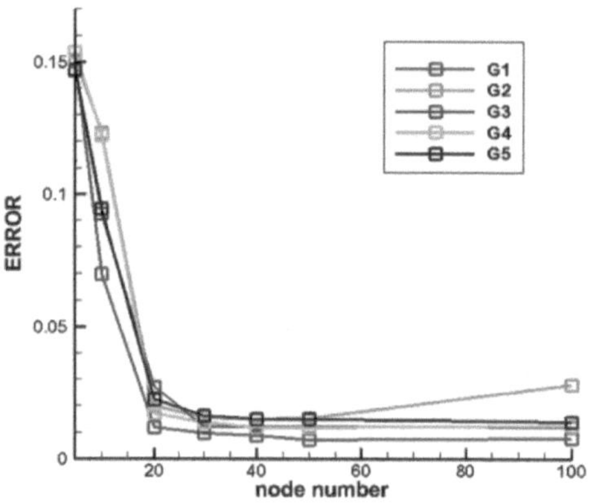

Figure 16. Relationship between maximum error and node number when a = $length_n$.

4 CONCLUSION

In this paper, a mesh-adaptive radial basis function method is adapted to solve the interpolation issue between two non-matching meshes in solid body field. The equations of mesh-adaptive radial basis function are deduced and the calculation program is developed. A cubic domain and a spheric domain test examples are employed to validate the algorithm. We can safely conclude from the result that the algorithm proposed in this paper shows good precision and can be suit for the typical solid body field. By set appropriate parameters, the computation efficiency and precision can be well satisfied in engineering application.

REFERENCES

[1] Gupta K.K. Aeroelastic Simulation of Hypersonic Flight Vehicles[J]. AIAA Journal, 2012, 50 (3): 717–723.

[2] Miller Brent A., Crowel Andrew R., Mcnamara Jack J. Loosely Coupled Time-Marching of Fluid-Thermal-Structural Interactions[C]. 54th AIAA/ASME/ASCE/AHS/ASC Structures, Structural Dynamics, and Materials Conference. Boston, Massachusetts, April 8–11, 2013, 2013.

[3] Ghoman Satyajit S., M. Salim Azzouz. Supersonic Aerothermoelastic Nonlinear Flutter Study of Curved Panels: Time Domain[J]. Journal of Aircraft, 2012, 49 (4): 1179–1183.

[4] Cheng Xing-hua. Coupled Aerothermoelastic Modeling and Analysis of Thermal Protection Panel for Hypersonic Vehicles[D]. NUDT, 2012.

[5] Han Xiang-ke, Qian Ruo-jun, Su Bo. Data Exchange Method for Fluid-structure Interaction Based on Interpolation Algorithm Adopting Compactly Supported Radial Based Function[J]. Journal of Tongji University (Natural Science), 2011, 39 (1): 48–51.

[6] Song Shao-yun, Li Shi-qi. Mixed Method of Node Load Interpolation in Collaborative Simulation of Coupled Problem[J]. Computer Simulation, 2006, 23 (8), 73–75.

[7] Wu Zong-min. Model, Algorithm and Theory for Scatter Data Fitting[M]. Beijing: Science Press, 2007.

[8] Wei Yi-kun, Yang Wei, Liu Jing. Interpolation Method of Radial Basis Function and Its Application[J]. Journal of Shen Yang University[J], 2008, 20 (1): 7–9.

[9] Wendland Holger. Piecewise polynomial, positive definite and compactly supported radial functions of minimal degree[J]. Advances in Computational Mathematics, 1995, 4: 389–396.

Engineering Management and Industrial Engineering – Leung (Ed.)
© 2015 Taylor & Francis Group, London, ISBN 978-1-138-02773-2

Data flow issues and BPMN mapping to Petri Net: Road map

Ahmed Elajeli Saleh Rgibi
Department of Data Analysis, Zawia University, Zawia, Libya

ABSTRACT: There are two important aspects of workflow to be considered: flow of control and flow of data. In terms of data in workflows, there is a little reported in literature. The problem of dataflow is becoming more and more prominent as the workflow system gets more complex. Verification of data flow has not been studied as extensively as verification of control flow. The dataflow verification in workflow processes is very important. Several possible errors in the dataflow are identified, such as the missing, losing, and the redundant data error. No means for checking these errors are provided. There are several challenging problems associated with dataflow issues such as: (1) how to discover errors relating to the dataflow; (2) how to analyze and verify the absence of dataflow errors in workflow system; (3) how to evaluate dataflow issues in workflow system. The objective of this paper is to provide a systematic technique of possible flows related to the flow of data in business process, check and analyze the dataflow issues. The research work of this paper illustrates dataflow issues approach procedure and mapping PBMN to Petri Net.

1 INTRODUCTION

Workflow is viewed as a set of tasks, which is systematized to achieve certain business goals by completing each task in a particular order under automatic control [1]. Resources are required for workflow implementation to support process execution. Resource allocation and resource constraint analysis are the popular workflow research topics.

The workflow concept has become the standard paradigm for process modeling in Business Process Management (BPM) systems. Workflows are typically looked from two important aspects of a workflow (also called a business process): flow of control and flow of data. The control flow describes the logical order of tasks; and the data flow describes the information exchange between tasks. There are a series of concepts that are applied to the representation, and utilization of data within workflow systems. These concepts not only define the manner in which data in its various forms can be employed within a business process and the range of informational concepts that a workflow engine is able to capture, but also characterize the interaction of data elements with other workflow and environmental constructs. A task makes use of current data to generate fresh data, which can be either a new data item or a new value of an existing data item. For example, an XOR split (decision) node needs data to enable it to choose one of the alternative outgoing paths. A task or decision node can execute successfully only if the required data is available. The inputs data requirements of individual decision nodes and the inputs/outputs data specifications of individual tasks are assumed to be known by the process designer. However, this does not ensure that every task and decision node in a workflow can get the data needed at the time of execution and that no data anomalies can occur. Data flow analysis is a testing method for detecting improper use of data in workflow system. The data flow is an essential part of workflow execution in conjunction with control flow, and it is important to make the system work. Activities in a workflow model provide necessary data to their underlying application components and human performers correctly to identify the context of the work they are supposed to carry out, and the dataflow-analysis-based workflow design starts with identifying the set of activities needed to be included in a workflow and their input and output data through analyzing existing business processes.

2 DATA FLOW ISSUES

Literatures report that very little work on the validation of data flow studied as extensively as verification of control flow. However, data flow validation has played an important role in boosting user confidence. Generally, there are six basic data validation issues which have been described in [2, 3, 4]. These issues are described as follows.

2.1 *Redundant data: (strongly and weakly)*

Strongly redundant data element is an element that is produced (written), but never read afterwards. Moreover, weakly redundant data strengthens/weakens the condition for the strongly redundant data error. The bad scenario is when some data are produced, but then they are not read in all possible continuations.

2.2 *Lost data*

Error occurs when the same data item *A* is assigned as output for two (or more) activities that may be executed in parallel as part of a fork structure in the process model. Thus, the activity responsible for the last update cannot be established, which may potentially cause a problem for activities that require the data item later in the process.

2.3 *Missing data*

The missing data error describes the situation where some data element needs to be used, that is read or destroyed, but it is not created or has been deleted without being created again. *A* data element *D* is thus missing if there is an execution path along which no writing to D happens before a reading or destroying of *D* takes place.

2.4 *Mismatched data*

It may arise when the structure of the data produced by the source is incompatible with the structure required by the activity, or more precisely by the application(s) invoked by the activity.

2.5 *Misdirected data*

It may arise when data flow direction conflicts with control flow in the same workflow model. The control flow determines the order and selection of activities that need to be performed to satisfy underlying business objectives of the workflow model. The data flow determines the flow of data from provider activities to receiver activities. Thus a provider activity must always precede a receiver activity in terms of control flow.

2.6 *Inconsistent data*

It arises when an activity retrieves a data item from an application or human participant and stores that item and its value in the process for data storage. This data item and its value may be utilized later in the workflow by another activity and be read from the storage. The problem is that if during that idle time (after the item has retrieved and before being read) the data item is updated externally to the workflow, it may cause an inconsistent view of the data. Data inconsistency also can be introduced when different activities deal with multiple sources for the same data item at different time in the process.

3 DATA FLOW ISSUES APPROACH

Figure 1 illustrates the dataflow issues approach procedure. It is divided into three steps: the translation into Petri Net, workflow net check, and using color Petri Net to data manipulation. A BPMN model should be well-formed. If it is a set of well-formed BPMN processes, the mapping of BPMN model must maintain the semantics of the original models. The workflow net is a very important sanity to check for workflow nets soundness [5]. For colored Petri Nets (CP-nets), it is possible to use data types and complex data manipulation. As business processes have become more complex, the probability of making errors in their design has increased. Therefore, the need for identifying and correcting the errors becomes critical. Business process analysis is often performed only through walkthroughs. Verification techniques have been developed for them to be used for the verification of business processes. A special class of Petri Nets has proven to be useful: the workflow nets [6] and most Petri Net tools can be used to analyze the translated BPMN models. In this paper, we will focus on mapping BPMN to Petri Net.

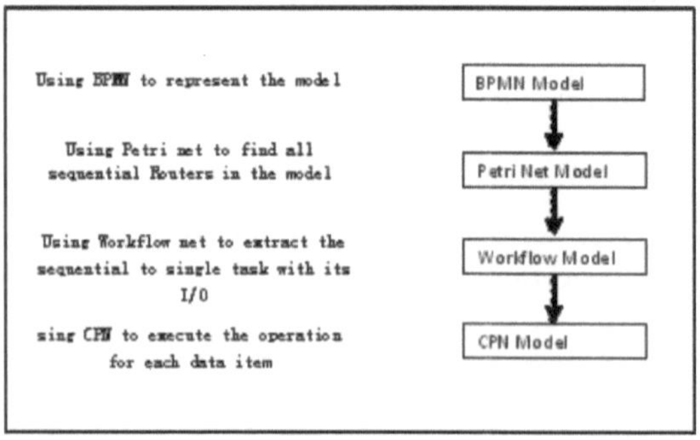

Figure 1. Overall procedure.

4 BPMN MODELING TECHNIQUE

BPMN provides a graphical notation for business process modeling, with an emphasis on control flow. It defines a Business Process Diagram (BPD), a kind of flowchart incorporating constructs tailored to business process modeling, such as AND-split, AND-join, XOR-split, XOR-join, and deferred (event based) choice. A BPD is made up of BPMN elements. We consider a core subset of BPMN elements shown in Figures 2, 3, 4 and 5. There are four types of diagrams: sequential flow diagram, parallel flow diagram, data objective diagram and BPMN iteration diagram.

Modeling with BPMN is essential for users to understand and communicate business processes across the enterprise. BPMN provides a powerful augmentation to other modeling techniques such as relational data modeling, and these diagrams are described as follows.

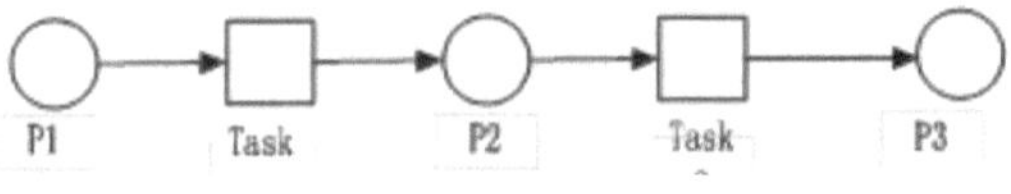

Figure 2. Sequence flow diagram.

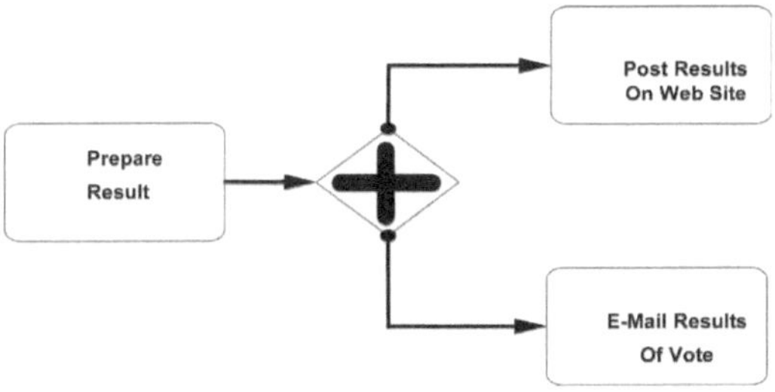

Figure 3. Parallel flow diagram.

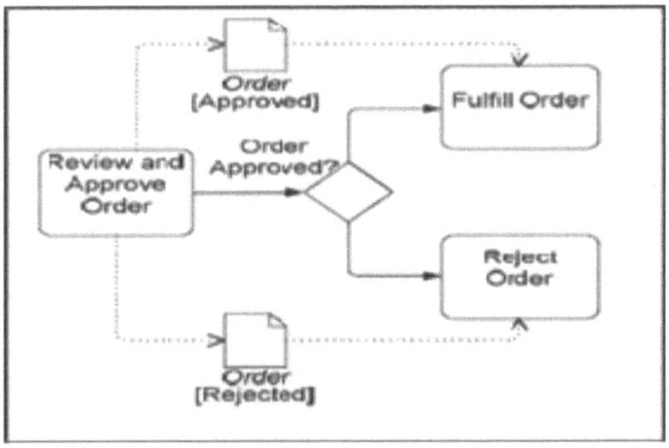

Figure 4. Data objects in BPM diagram.

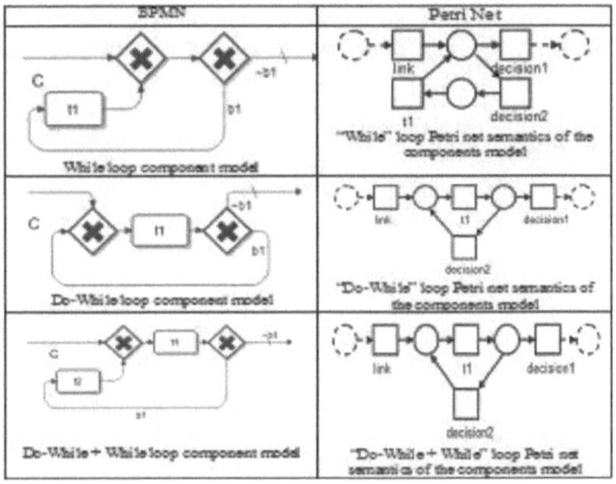

Figure 5. BPMN control statement mapping to Petri Net.

4.1 *Sequential flow diagram*

The simplest routing is the sequential execution of tasks. We refer to the sequential performance of tasks when these have to be carried out one after another. If two tasks need to be carried out sequentially, there should be a clear interdependence between them. Figure 2 shows an example of sequential routing.

4.2 *Parallel flow diagram*

If more than one task can be carried out at the same time or in any order, then we refer to parallel routing. If we confine ourselves to the situation with two tasks, $task_1$ and $task_2$, then there are three possibilities: both tasks can be performed simultaneously; $task_1$ can be carried out first, then $task_2$; or $task_2$ can be the first, followed by $task_1$. Figure 3 illustrates how we can model this situation. In order to enable the parallel execution of task post results on the web and task e-mail results of vote. We begin with a task prepare result, then we meet AND-split, which is added so as to allow more than one task to be managed at the same time, as shown in Figure 3.

4.3 *BPMN and data objective*

The BPMN specification considers data objects as a way to visualize how data is processed, as shown in Figure 4. The semantics of data objects remain unspecified even left to the interpretation of the modeler. However, the BPMN specification defines the notion of input sets for activities. Then each input set is a conjunction of data conditions that must be hold to perform the activity. If more than one input set are defined for an activity, one of them is sufficient to execute that activity. Similarly, output sets can be defined. Therefore, we introduce a particular semantics of BPMN data objects in this work, as a necessary precondition for formal verification that is inspired by the notion of input and output sets.

4.4 *BPMN iteration*

In programming language, loop is the first value expression evaluated. If the result is true, then the loop body is executed. Otherwise, the next statements after the loop body will be executed. Loop component model describes some kind of mechanism. In the middle of executing a task or event, first we evaluate the expression of truth. If it is true, perform the task; if it is false, the task is not performed, and the implementation of the follow-up activities or tasks may not execute even once. Figure 5 illustrates control statement mapping from BPMN to Petri Net.

4.5 *Mapping BPMN to Petri Net*

The BPMN specification considers data objects as a way to visualize how data are processed. The semantics of data objects remain unspecified and are even left to the interpretation of the modeler [7]. However, the BPMN specification defines the notion of input sets for activities. Then each input set is a conjunction of data conditions that must be hold to perform the activity. If more than one input set are defined for an activity, one of them is sufficient to be executed in that activity. Similarly, output sets can be defined. We introduce a particular semantics of BPMN data objects in this section. It is viewed as a necessary precondition for formal verification that is inspired by the notion of input and output sets. First, we assume a single copy of the data object that is handled within the process. The single copy will be assumed to exist from the moment of process instantiation. The mapping of BPM to Petri Net is introduced and covers BPMN's control flow [8]. We extend this mapping with data flow in the following way: first, we provide a separate data flow mapping for each data object. Each of these mappings represents preconditions and effects of tasks regarding the corresponding data object. In a second step, the control flow of Petri Net is obtained through merged with all data flow nets.

4.5.1 *The procedure of "Folding-Expand" model transformation*

Component based "Folding-Expanding" model transformation method refers to the work of Aalst et al, [9] where they transformed workflow into BPEL net. Aalst transformed a formal language (Petri Net) model to a non-formal language (BPEL), but this paper is going to transform a non-formal language (BPMN) model to a form language (Petri Net). The basic idea of component based "Folding-Expanding" model is as follows. First, it is necessary to replace a specific component of the well-formed BPMN model with a component containing the task nodes (called "folding"). Secondly, by repeating this step and ideally the final BPMN model will be "folded" into a most simple BPD. Third, it is easy to give the Petri Net semantic the simplest BPD, but you cannot simply treat it as a transition; you should replace it with components that contain Petri Net semantics. If the simplest task node in BPD is a component task node, it is a recursive process, which means that if the component contains at least one component task, it is also needed to give the component with its the Petri Net semantic (called "expanding"). Finally the transformed Petri Nets are shown. The flowchart of the component based "Folding-Expanding" algorithm is shown below in Figure 6.

4.5.2 *The formal description of the "folding" process*

You can observe from the preceding description, the important part in the component based "Folding-Expanding" algorithm is the "folding". Through the "folding", the connection relationship between the other nodes in the model components

Figure 6. The flow chart of the component based "Folding-Expanding".

remains unchanged, but the structure of the model is changed. There may be a new component, so that the "folding" process repeats until it gets the simplest BPD. Following is the formula description of the folding process.

$BPD=(O,F,Cond)$ is a well-formed core business process diagram,
$C=(O_C,F_C,Cond_C)$ is a component in BPD. Function *Fold* will replace the component C in BPD with a component task $t_C \notin O$, which means $Fold\,(BPD,C,t_C)=(O',F',Cond'$

where:
- O' = (O\Oc) U{tc}
- Tc is the task Set in C, Tc = Oc ∩T
- T' = (T\Tc) U{tc} is the task Fold (BPD, C, tc)

  $T^{R'}=(T^R\setminus T_C)$ is the receiving task set $Fold\,(BPD,C,t_C$, which means t_C would not accept task,

$$F'=(F\cap(O\setminus O_C\times O\setminus O_C))\cup\{(entry(C),t_C),(t_C,exit(C))\}$$

Cond =

$$\begin{cases}\text{Cond}[F'], \text{entry}(C)\notin G^D\\ \text{Cond}[F']\cup\{((\text{entry}(C),t_C),\text{Cond}(\text{entry}(C),t_C))\}.\end{cases}$$

4.5.3 *Reasonableness maintenance*

In the existing algorithm of transforming BPMN models to Petri Nets, most of them are based on "one-to-one replacement" which means that each BPMN graphical element is replaced by the corresponding Petri Net semantic. A major drawback of this method is that the meaning of BPMN graphical elements combination in Petri Nets may produce other semantic, which cannot maintain the reasonableness when transforming from BPMN source model to Petri Net target model.

Figure 7 shows an example of the theorem above. Figure 7 (a) meet the reasonable requirements of the workflow net, among which the transition t1 corresponds to a Petri Net "t(c)", and this Petri Net who is in accordance with the condition of components boundary *n* the literature[10] can

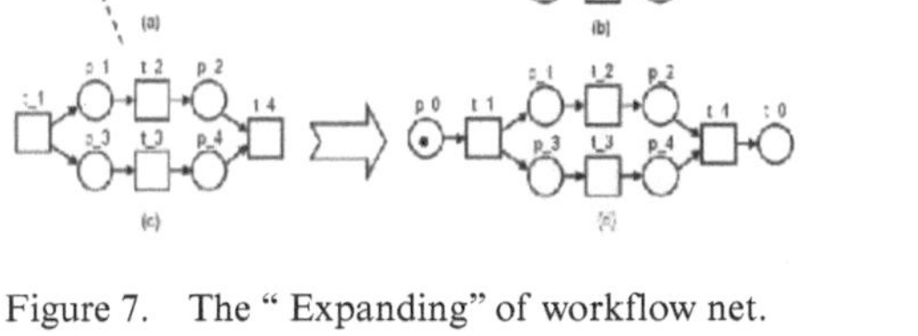

Figure 7. The " Expanding" of workflow net.

be rewritten to workflow nets "(d)" that satisfies the property of reasonableness. The transition "t1" in workflow net "(a)" could be replaced with "(c)", to get workflow net "(b)". According to the theorem, workflow net "(b)" also satisfies the property of reasonableness. The step "Folding" is completely based on BPMN model. BPMN is a non-formal graphical symbolic language, and a model containing no formal semantic. Reasonableness is defined from the viewpoint of model execution semantics, therefore it is impossible to give a rigorous formal proof of reasonableness before and after "Folding". The mapping of the simplest BPD or BPMN component to Petri Net involves BPMN and Petri Nets. The Petri Net is formalized, but BPMN is not, therefore it is impossible to give a strict proof of the mapping step. The step "Expanding" is based on Petri Nets. It could maintain reasonableness and consistence before and after "Expanding" under certain condition. The condition is: the Petri Net semantic of the component task node corresponding. According to the literature [10] workflow net rewrite rules, it can be rewritten as workflow nets. In the "Folding" process, the components task node is created, and if it is considered only as ordinary tasks node, it could be modeled by using a Petri Net transition. In the "Expanding" process, it is equivalent to use the Petri Net semantics of the components task nodes to represent transition, according to the theorem, and if the original Petri Net is a workflow net and that satisfies the property of reasonableness, the Petri Net semantics of the components task node can be rewritten to that which satisfies the property of reasonableness. The workflow after "expand" is also satisfying.

5 CONCLUSION

We illustrated dataflow issues approach procedure; it is divided into three steps: the translation into Petri Net, workflow net check, and using color Petri Net to data manipulation. A BPMN model should be well-formed. In this paper, the mapping of BPMN to Petri Net is introduced and covers BPMN's control flow, and we extend this mapping with data flow. The step "Folding" is completely based on BPMN model. BPMN is a non-formal graphical symbolic language, and is a model containing no formal semantic. Reasonableness is defined from the viewpoint of model execution semantics, therefore it is impossible to give a rigorous formal proof of reasonableness before and after "Folding". The mapping of the simplest BPD or BPMN component to Petri Net involves BPMN and Petri Nets. The Petri Net is formalized, but BPMN is non-formal, therefore it is impossible to give a strict proof of the mapping step. The step "Expanding" is based on Petri Nets. It could maintain reasonableness and consistence before and after "Expanding" under certain condition.

REFERENCES

Ahmed Elajeli Rgibi, Yao Shuzhen, detecting Data flow Issue using BPMN modeling and Petri net to analysis WF net by implementing RWD Boolean table. 2011 2nd International Conference on Electronics and Information Engineering (ICEIE2011).

Basu, A., R.W. Blanning. 1994b. Model integration using metagraphs. Inform. Systems Res. 5(3) 195–218.

Basu, A., R.W. Blanning. 2000. A formal approach to workflow analysis, Inform. Systems Res. 11(1) 17–36.

BPMN vs. UML, By Ismaël Ghalimi, Chief Strategy Officer, Intalio (www.intalio.com)—September 2002.

Van der Aalst, W.M.P.: Verification of Workflow Nets. In P. Azéma and G. Balbo, editors, Application and Theory of Petri Nets 1997, volume 1248 of Lecture Notes in Computer Science, pg. 407–426-Springer. 1997.

Van der Aalst, W.M.P., ter Hofstede, A.H.M.: Verification of workflow task structures: a Petri-net-based approach. Inf. Syst. 25(1), 43–69 (2000).

Sadiq, W. and Orlowska, M.E.: Analyzing Process Models using Graph Reduction Techniques. Information Systems 25(2):117–134. 2000b.

W.M.P. van der Aalst. A class of Petri nets for modeling and analyzing business processes. Technical Report 26, Eindhoven University of Technology, 1995. (Cited on pages 4 and 28.).

Engineering Management and Industrial Engineering – Leung (Ed.)
© 2015 Taylor & Francis Group, London, ISBN 978-1-138-02773-2

Genetic algorithm with local optimization of sub-contractor selection in multiple subprojects

Q.R. Qiu & Y.M. Di
School of Mathematical and Physical Science, North China Electric Power University, Beijing, China

ABSTRACT: Based on the analysis of main evaluation of bid evaluation factors, this paper established a quantitative evaluation index system and an integer programming subcontractor selection optimizing model with the serial number of subcontractor as the decision variable and relative benefits as the objective function. In accordance with the characteristics of subcontractor selection problem for multiple subprojects, a genetic algorithm with local optimization was proposed. At first, the subcontractors were sorted by their benefits to the enterprise and on this basis the single (double) project local tuning strategy for feasible solution and the single department tuning strategy for infeasible solution were given. Elitist strategy and tournament selection strategy were used to update the population. Numerical example shows that the algorithm is effective.

1 INTRODUCTION

With the economic development and social progress, large and medium-sized projects become more and more. If adopting self-supporting construction, the bid-winning enterprises will spend a lot of equipment and staff, which makes the cost to increase and construction cost difficult to control. To improve efficiency, the bid-winning enterprises can subcontract some of their projects explicitly before construction based on the characteristic of each project. Thus it not only reduces the enterprise investment and ensures the interests of the enterprise, but also meets the project quality and time limit in the main contract. Therefore, how to choose sub-contractor is a field that is worth studying extensively.

In paper [1], the author has established the 0–1 programming model, and used Excel "planning" function to solve the model. The model of paper [1] can also use the methods of 0–1 programming such as implicit enumeration, branch and bound method to solve [2]. But in this model, the number of variables is $\sum_{i=1}^{n} m_i$ and using the methods of 0–1 programming such as implicit enumeration, branch and bound method are very inefficient when there are too many qualified contractor units (m_i is too big) to select.

Nowadays, intelligent algorithms represented by GA, PSO and so on are increasingly used in a variety of combination engineering optimization problems and a lot of achievements have been achieved [3~7]. If adopt the 0–1 model in paper [1] for sub-contractor selection problem with many children projects, Binary encoding and GA can be used. But when the bid units of each sub-project are large, this encoding method will lead to long population individual species and have a bad effect on searching efficiency of the algorithm.

Based on the above issues, this paper uses the subcontractor's serial number as variables to build the model for multiple subprojects subcontractor selection problem. The target function and constraint conditions of this new model are simple and easy to calculate. But the variables are shown in the subscript, so it cannot be solved by the classic optimization method. This paper adopts the genetic algorithm to solve this problem. To improve the solving efficiency, the paper has adopted the local tuning strategy for the individual tuning according to certain probability. To reduce the size of searching scope, the contractors of each component are ranked according to their benefit from big to small.

2 MATHEMATICAL MODEL OF THE PROBLEM

2.1 *Index system of bidding evaluation and evaluation function for subproject*

Contractors are chosen through engineering bidding, and the standard is to select the best subcontractors on the principle of getting the maximum relative benefit for the tenders.

In tender evaluation, besides the tender offers, the completion date, quality, construction plan, material consumption, construction experience and enterprise reputation are also needed to be evaluated comprehensively[8]. For quantitative analysis of the evaluation process, each evaluation factor is needed to be quantified using certain target.

Suppose there are n projects for bidding, the subproject i has m_i units to participate in the bid. The related symbols of subproject i are shown in Table 1.

If the i-th project is contracted by the x_i-th unit, the relative benefit of tender is as follows:

$$c_{ix_i} = \sum_{k=1}^{6} p_{ki}(s_{kix_i} - \tilde{s}_{ki}) \tag{1}$$

2.2 *Mathematical model of the problem*

This paper uses the subcontractor's number as decision variable to build the integer programming model for multiple subprojects subcontractor selection problem.

If the i-th project is contracted by the x_i-th unit, the objective function of the problem is as follows:

$$\max g(x) = \sum_{i=1}^{n}\sum_{k=1}^{6} p_{ki}(s_{kix_i} - \tilde{s}_{ki}) \tag{2}$$

Variables are limited as follows:

x_i is an integer in $[1, m_i]$ (3)

The constraint conditions of each index are:

Table 1. The related symbols of parameters.

Quantative indicators	The bidding data of the j-th tendering units	Base bid price	Unit	The returns conversion
The sub-project price	s_{1ij}	$\tilde{s}_{1i}$	yuan	p_{1i}
Project period	s_{2ij}	$\tilde{s}_{2i}$	month	p_{2i}
Excellent project rate	s_{3ij}	$\tilde{s}_{3i}$	%	p_{3i}
The material consumption	s_{4ij}	$\tilde{s}_{4i}$	yuan	p_{4i}
The experience rate of similar projects	s_{5ij}	$\tilde{s}_{5i}$	%	p_{5i}
Contract completion rate	s_{6ij}	$\tilde{s}_{5i}$	%	p_{6i}

1. Total price limit:

$$M_{11} \le \sum_{i=1}^{n} s_{1ix_i} \le M_{12} \tag{4}$$

2. Total time limit:

$$(1-a)M_2 \le \sum_{i=1}^{n} s_{2ix_i} \le (1+a)M_2 \tag{5}$$

3. The limit of total rate of good engineering:

$$\sum_{i=1}^{n} \omega_{i3} s_{3ix_i} \ge M_3 \tag{6}$$

4. The limit of relative usage of main material:

$$\sum_{i=1}^{n} \omega_{i4} \frac{s_{4ix_i}}{\tilde{s}_{4i}} \ge M_4 \tag{7}$$

5. The limit of experience rate of contractor for similar projects:

$$\sum_{i=1}^{n} \omega_{i5} s_{5ix_i} \ge M_5 \tag{8}$$

6. The limit of contract completion rate of contractor:

$$\sum_{i=1}^{n} \omega_{i6} s_{6ix_i} \ge M_6 \tag{9}$$

Among them, M_{11}, M_{12}, M_{21}, M_{22}, M_3, M_4, M_5, M_6 are fixed constants.

Compared with 0–1 programming model, the new model has the following advantages:

1. The number of variables is n, far less than $\sum_{i=1}^{n} m_i$;
2. The objective function and constraint conditions are more concise, and also easy to calculate;
3. Resource-constrained testing is easier.

But in the constraint condition (4)~(9), the decision variables appear as the left subscript, so the problem cannot be solved by using classic optimization algorithm.

For the constrained optimization problem mentioned above, penalty function can be used to translate it into the following non-constrained optimization problem besides constraints (3).

3 PRETREATMENT AND LOCAL OPTIMIZATION METHOD FOR THE PROBLEM

Definition: Denote: $x = (x_1, x_2, \ldots, x_n)^T$ $y = (y_1, y_2, \ldots, y_n)^T$, x is in the back of y (denoted $x \succ y$), if

1. $x_i \geq y_i, i = 1, 2, \ldots, n$;
2. There is $i_0 \in \{1, 2, \ldots, n\}$, such as $x_{i_0} > y_{i_0}$.

According to the relative benefit obtained by those bidders, the bidders of each subproject can be arranged in the order of the largest for the purpose of improving the efficiency. After this treatment, we can obtain the conclusions as follows:

Theorem: $x^* = (x_1^*, x_2^*, \ldots, x_n^*)^T$ is the feasible solution to the problem; if $x = (x_1, x_2, \ldots, x_n)^T \succ x^*$, x must not be the optimal solution to problem.

In this paper, we use the local optimization of double project for feasible solution to get the local optimization solution according to certain probability. Specifically, we randomly generate two integers among $\{1, 2, \ldots, n\}$, recorded as p_1, p_2, fix the other components of $x = (x_1, x_2, \ldots, x_n)$ except x_{p_1}, x_{p_2}, to seek the local optimal solution.

According to the relative benefit obtained by those bidders, the bidders of each subproject are arranged in the order of the largest. Based on the above theorem, local optimal solution can be obtained by less schemes in the process of local optimization for the feasible solution. We don't need to use 0–1 programming, classic methods of optimization problems and modern optimization methods.

In this paper, we use the local optimization of single project for non-feasible solution $\tilde{x} = (\tilde{x}_1, \tilde{x}_2, \ldots, \tilde{x}_n)^T$ according to certain probability. We randomly generate a integer among $\{1, 2, \ldots, n\}$, recorded as p_3, fix the other components of $\tilde{x} = (\tilde{x}_1, \tilde{x}_2, \ldots, \tilde{x}_n)^T$ except p_3, and then seek the local optimal solution.

4 GENETIC ALGORITHM WITHLOCAL OPTIMIZATION

For genetic algorithm of this problem, this paper uses natural number encoding. Encoded $x = (x_1, x_2, \ldots, x_n)$ as a chromosome, in which x_i is the natural number between 1 to m_i and it represents that subproject i chooses the x_i-th contractor. The calculation of the fitness is according to the objective function of unconstrained optimization problem.

Algorithm flow:

Step1: Selection of parameters. Set $t = 1$. Generate the initial population A_t which are satisfied (3).

Step2: Calculate the fitness value.

Step3: According to the roulette, randomly select parent from the current parent population A_t. Do crossover operation and perform local tuning to get the population B_t according to the probability p_l.

Step4: Perform mutation on the individuals of A_t. Then perform local tuning to get the population C_t according to the probability p_l.

Step5: From $A_t \cup B_t \cup C_t$, select the best n_{best} unrepeatable into A_{t+1}; randomly select $NP - n_{best}$ individuals in the rest of the individuals into A_{t+1}.

Step6: Judge whether termination condition is satisfied. If satisfied, the algorithm is ended; otherwise, $t \leftarrow t + 1$, turn back to Step3.

5 NUMERICAL EXPERIMENT AND ANALYSIS

Suppose that a large-scale project will take bids for n subprojects. According to engineering planning, the base bid price of each index of subprojects will be estimated, whereas the benefit conversion rate of each subproject is getting from history data of the enterprises'. To test the effectiveness of algorithm, the j-th indicator of subproject i is generated by equation (10).

$$s_{j,i} = \tilde{s}_{j,i} \times [1 + (rand(1) - 0.5) \times a_{j,i}], \quad j = 1, 2, 3, 4, 5, 6 \tag{10}$$

The revenue conversion rate for the j-th index of subproject i is created by equation as follows:

$$p_{j,i} = -q_{j,i,1} \times [1 + (rand(1) - 0.5) \times q_{j,i,2}], \quad j = 2, 3, 5, 6 \tag{11}$$

$\tilde{s}_{j,i}, a_{j,i}$ $j = 1, 2, \ldots, n$ is base bid price, error bounds for bid value and base bid price respectively of the j-th index of subproject i. Through bid prequalification, the number of eligible subcontractors is $N_i = 10 + fix(rand(1) * 5)$ for project i. And $p_{i,1} = -1, p_{i,4} = 1$.

When selecting the subcontractors, the relative total benefit provided by candidate subcontractors must be non-negative. So the data will be regenerated until this condition is satisfied.

Appendix is given a data set. If the corresponding problem can use mathematical model in document [1], then it can also use the Matlab software to solve[9]. In the Matlab2013, a software platform, we can get the following solution according to the internal parameters: $\tilde{x} = (1, 5, 10, 6, 9, 9, 16, 1, 1, 7, 10, 12)^T$, the relative efficiency value obtained by renderer is 209.3563. Using the method proposed in this paper, we can get another solution, $x^* = (1, 5, 10, 6, 9, 9, 16, 2, 1, 7, 1, 12)^T$, and the relative efficiency value is 209.5101. Results show that the method in this paper obtains a better solution.

6 SUMMARY

In this paper, sub-contractor selection in multiple subprojects is discussed. This paper uses the subcontractor's serial number as the variables to build correlation model and proposes a genetic algorithm with local optimization. According to the relative benefit obtained by those bidders, the bidders of each subproject are arranged in the order of the largest. After this treatment, local optimal solution can be obtained by less schemes. We don't need to use 0–1 programming, classic methods of optimization problems and modern optimization methods. Elitist strategy and tournament selection strategy are used to update the population. The using of elitist strategy makes outstanding individuals of new species not worse than the outstanding individuals in original population. This guarantees that the excellent individuals obtained in the process of genetic cannot be destroyed by crossover and mutation operation so that local optimal individual is not easy to be eliminated. Random selection strategy is used for the other individuals, which can keep certain diversity for the next generation.

Numerical experiment shows that:

1. The algorithm proposed in this paper is effective in solving the problem.
2. Local optimization can improve individual fitness after basic crossover, mutation, and improve the algorithm's efficiency.
3. Calculation results show that there are many local optimal solutions, which can be obtained quickly after 20 iterations by this algorithm. Whereas subsequent objective function value is increasing slowly.

To get the optimal solution, the number of iterations is further needed. How to improve the computational efficiency still needs to be studied further.

REFERENCES

[1] T.J. Lai, 2013. The Transformation of Multi-dimensional Resource Allocation Problems and the Method of Solving Them by Excel Software, Value Engineering, 96–97.

[2] Jorge Nocedal, Stephen J. Wright, Numerical optimization, January, 2006. Beijing: Science Press.

[3] D.W. Wang, W.X. Ye, Q.L. Rong, 2002. Partner selection model and soft computing method mo for dynamic alliance of enterprises [J], Science in China, Series E, 32(6): 824–830.

[4] W. Zhang, J.H. Zhou, N. Wang, 2012.Integer-coded genetic algorithm with novel repairing mechanism for large scale unit-commitment problem [J]. CIESC Journal, 63(9): 2972–2978.

[5] Hansen N, Niederberger A.S.P, Guzzella L. Koumoutsakos, P. 2009. A Method for Handling Uncertainty in Evolutionary Optimization with an Application to Feedback Control of Combustion. IEEE Transactions on Evolutionary Computation, 13(1): 180–197.

[6] Adleman L.M, 1994. Molecular computation of solution to combinatorialproblems [J]. Science, 266(5187): 1021–1024.

[7] Sharapov R.R, Lapshin A V. 2006.Convergence of genetic algorithms [J]. Pattern recognition and image analysis, 6(3): 392–397.

[8] State Development Planning Commission, Drafting team of Public Bidding of the People's Republic of China.The Operation manual ofthe Law on Public Bidding [M] Beijing: Law Press.

[9] Y. Zheng, H.L. Zhao, Y. Xu, 2008. MATLAB-based Dynamic Programming. Journal of Chongqing Institute of Technology (Natural Science), 22(5): 152–155.

[10] E.J. Xia, G.L. Su. Study on Optimization Model of Bidding Assessment on Engineering Project, China Soft Science, 2003(10), P. 133–138.

Engineering Management and Industrial Engineering – Leung (Ed.)
© 2015 Taylor & Francis Group, London, ISBN 978-1-138-02773-2

Research and application of multi-dimensional partial differential model in environmental improvements

Y.L. Miao, X. Ji, X. Qiang, J. Zhang & J.T. Meng
Xi'an University of Technology, Xi'an, Shanxi, China

ABSTRACT: With the rapid development of society, the impact of the increasing population on the environmental quality of the city increased prominently. In order to eliminate the pollution fundamentally, the task that looks for pollutant source is very heavy, urgent also. According to the environmentally geological thinking in this paper, with the factors of heavy metal pollutant movement in various environment systems, this paper establishes the models. One dimensional partial differential model is applied to river water quality calculation and prediction, two dimensional partial differential model is applied to large estuaries, bays, shallow lake water quality calculation and prediction, and three dimensional partial differential model is applied to determine the heavy metal pollutant source in the city. The partial differential model is extended to N-dimensional in order to deal with complex and volatile environment system, which provides a practical way to protect human existence environment.

Keywords: Environment system, Multi-dimensional partial differential model, Heavy metal, Pollutant source

1 INTRODUCTION

Propagation characteristics of pollutants are often used to analyse pollution degree and center offset. We find out that pollutants spread from center to surrounding, and with the increase of the distance, the degree of pollution decreased. Given the present conditions, heavy metal pollutant move through the soil moisture migration, and it is also influenced by the soil matrix, physical and chemical properties of soil solute, and other chemicals in the soil. If only consider some of these factors, can it calculate the concentration of each direction by establishing low dimensional partial differential model [1].

However, with the dramatic changes of global environment, the pollutant movement is influenced by the following factors like soil water content, saturation, PH, soil water potential, pressure potential, matric potential, solute potential, wind speed, air temperature etc. The low dimensional partial differential model can't satisfy demand any more [2]. In response to this problem, a multi-dimensional partial differential model is established to determine the location of the pollutant source.

2 MATERIALS AND METHODS

Heavy metal pollutant move through the soil moisture migration [3], and it is also influenced by the soil matrix, physical and chemical properties of soil solute, and other chemicals in the soil. Physical mechanism of soil solute transport mainly includes convection, diffusion and mechanical dispersion. Due to the limited conditions, we choose the diffusion and gravitational potential as its propagation characteristics, and establish a partial differential model after analyzing the propagation characteristics.

2.1 *Analysis of propagation characteristics*

The phenomenon resulted from the mass migration of thermal motion is called diffusion, which shift from high concentration to regions of low concentration to achieve the characteristics of uniform concentration. In liberal solution, the diffusion of solute follows Fick's first law, the mathematical expression is:

$$J_d = -D_0 \frac{dc}{dx} \tag{1}$$

where J_d = molecular diffusion flux of solute molecules in liberal solution; D_0 = diffusion coefficient of solute molecules in liberal solution; c = volume concentration of diffusion substances; and D_c/D_x = solute concentration gradient [5]; '–' represents the solute molecules is diffused along the direction of concentration reduced.

In the soil solution, the diffusion of solute molecules also follow the Fick's first law, the mathematical expression is:

$$J_{ds} = -\theta D_s \frac{dc}{dz} \tag{2}$$

where J_{ds} = diffusion flux of solute molecules in the soil; D_s = diffusion coefficient of solute molecules in the soil. In neither saturated soil nor unsaturated soil, D_s is less than D_0. Generally, solute diffusion coefficient in the soil is only expressed as a function of water content, which has nothing to do with solute concentration, the empirical formula is:

$$D_s = D_0 \tau \tag{3}$$

where τ = bending factor, which is a dimensionless physical quantity [6]. According to Wagenet, for the majority of the soil, the range of variation τ is typically 0.3–0.7.

The gravitational potential of heavy metal pollutant of soil due to its gravity, and it equals to the value of the gravity acting which moved the unit weight heavy metal pollutant from the reference point coordinates to the research height. Gravitational potential will be influenced by the position of coordinate points and axis direction. In the normal course of the study, we often choose the water of groundwater or the Earth's surface as the origin, the axis direction may be upward or downward based on the calculated need [7]. If the study coordinates' height is z, the gravity potential of unit mass of heavy metal pollutant in soil can be expressed by units of length as:

$$\psi_g = \pm z, \tag{4}$$

where ψ_g = gravity potential of heavy metal pollutant.

2.2 *Establishment of the partial differential model*

2.2.1 *Derivation of the fundamental equations*

Assuming a point $P(x, y, z)$ in the environmental medium (like soil), put this point as an infinitesimal [8], its side-length is $2dx$, $2dy$, $2dz$, velocity $v(x, y, z)$ of point P in three directions components is u_x, u_y, u_z, pollutant fluxes in the three directions are as follows:

$$\begin{aligned}\Delta m_x &= \Delta m_{1x} + \Delta m_{2x} + \Delta m_{3x} + \Delta m_{4x} \\ \Delta m_y &= \Delta m_{1y} + \Delta m_{2y} + \Delta m_{3y} + \Delta m_{4y} \\ \Delta m_z &= \Delta m_{1z} + \Delta m_{2z} + \Delta m_{3z} + \Delta m_{4z}\end{aligned} \tag{5}$$

Now merge the molecular diffusion coefficient D_m, turbulent diffusion coefficient D_{1x} and dispersion effect term D_{2x} into a dispersion term [9], namely:

$$\begin{aligned}E_x &= D_m + D_{1x} + D_{2x} \\ E_y &= D_m + D_{1y} + D_{2y} \\ E_z &= D_m + D_{1z} + D_{2z}\end{aligned} \tag{6}$$

where E_x, E_y and E_z are respectively for the dispersion coefficients in the X-axis, Y-axis, Z-axis direction.

Mass flux of dispersion effect in the X-axis direction:

$$E_{2x} = -E_x \frac{\partial c}{\partial x} \tag{7}$$

Mass flux of direction Y-axis and Z-axis:

$$E_{2y} = -E_y \frac{\partial c}{\partial y}, \; E_{2z} = -E_z \frac{\partial c}{\partial z} \tag{8}$$

Total mass flux of direction X-axis, Y-axis and Z-axis:

$$F_x = E_{2x}, \; F_y = E_{2y}, \; F_z = E_{2z} \tag{9}$$

Mass flux into and out of the element in the X-axis direction:

$$4dydz\left(F_x - \frac{\partial Fx}{\partial x} dx\right) \tag{10}$$

Flux difference between end-faces:

$$-8dxdydz \frac{\partial F_x}{\partial x} \tag{11}$$

Similarly, flux difference in Y-axis and Z-axis direction [10]:

$$-8dxdydz \frac{\partial F_y}{\partial y}, \; -8dxdydz \frac{\partial F_z}{\partial z} \tag{12}$$

In this infinitesimal, the quality variation of pollutants by an infinitesimal period:

$$8dxdydz \frac{\partial C}{\partial t} \tag{13}$$

According to the law of conservation of mass:

$$\frac{\partial C}{\partial t} + \frac{\partial F_x}{\partial x} + \frac{\partial F_y}{\partial y} + \frac{\partial F_z}{\partial z} = 0 \tag{14}$$

Put the formulas (7) to (9) into (14), result in:

$$\frac{\partial c}{\partial t}+u_x\frac{\partial c}{\partial x}+u_y\frac{\partial c}{\partial y}+u_z\frac{\partial c}{\partial z}= \frac{\partial}{\partial x}\left(E_x\frac{\partial c}{\partial x}\right)+\frac{\partial}{\partial y}\left(E_y\frac{\partial c}{\partial y}\right)+\frac{\partial}{\partial x}\left(E_z\frac{\partial c}{\partial z}\right) \tag{15}$$

Namely:

$$\frac{\partial c}{\partial t}+\nabla(uc)=E\nabla^2 c \tag{16}$$

Taking into account the effects of physical, chemical and biological to the pollutant in the environmental medium, such as inflow and outflow of river, the degradation,sedimentation, raise and adsorption of pollutants [11].we can combine the above-mentioned as an additional item S(x,y,z,c,t), the above formula is:

$$\frac{\partial c}{\partial t}+u_x\frac{\partial c}{\partial x}+u_y\frac{\partial c}{\partial y}+u_z\frac{\partial c}{\partial z} =\frac{\partial}{\partial x}\left(E_x\frac{\partial c}{\partial x}\right)+\frac{\partial}{\partial y}\left(E_y\frac{\partial c}{\partial y}\right)+\frac{\partial}{\partial x}\left(E_z\frac{\partial c}{\partial z}\right) +S(x,y,z,c,t) \tag{17}$$

2.2.2 *Simplified of the basic equations*

Formula (17) is a basic equation about the pollutants movement and transformation, it's used to combine with the equation of environmental fluid medium, then we can simulate pollutants movement and transformation process in various environmental media. Although formula (17) is complete in theory, it is necessary to master some materials which contain many parameters of locations and model coefficients in any time, and these materials also include the extremely complex relationship between space and time. So it is very difficult to solve this equation. In daily work, we often use the predicted target, hydrology, and medium conditions or pollutant characteristics to simplify it.

1. Zero-dimensional partial differential model: Under some certain conditions, we can take the study of environmental medium as a complete reactor. In the space of environment medium, pollutants distribution is uniform, that is to say, changes in the concentration do not exist in any directions, the pollutants disperse to all parts instantly. Small lakes and box-type atmospheric model belong to zero models [12].
2. One-dimensional partial differential model: If the pollutant is only varying along the X-axis direction, Y-axis and Z-axis is uniform, it can be simplified to one-dimensional partial differential model. In a uniform flow field, u_x and E_x can be used as a constant, the above formula:

$$\frac{\partial c}{\partial t}=\frac{\partial}{\partial x}\left(E_x\frac{\partial c}{\partial x}\right)-u_x\frac{\partial c}{\partial x}+S(x,y,z,c,t) \tag{18}$$

One-dimensional model can be used for river water quality simulation and prediction.

3. Two-dimensional partial differential model: When the concentration of pollutants in X-axis and Y-axis directions are changed, the model can be simplified to two-dimensional partial differentiation,

$$\frac{\partial c}{\partial t}=\frac{\partial}{\partial x}\left(E_x\frac{\partial c}{\partial x}\right)+\frac{\partial}{\partial y}\left(E_y\frac{\partial c}{\partial y}\right) -u_x\frac{\partial c}{\partial x}-u_y\frac{\partial c}{\partial y}+S(x,y,z,c,t) \tag{19}$$

Two-dimensional model is commonly used for river, estuaries, bays, shallow lake water quality simulation and prediction.

4. Three-dimensional partial differential model: When the X-axis, Y-axis and Z-axis directions pollutant concentrations change, we can get the location of pollution source by three dimensional partial differential model. And we find out that this problem is belonging to three dimensional partial differential model through research soil pollution data of heavy metal elements in urban areas [13]. Due to the limit of conditions and data, assuming the pollution sources is the location of the largest concentration, without considering the enrichment of elements, calculating the maximum point of heavy metal concentrations. The formula is:

$$\frac{\partial c}{\partial t}=\frac{\partial}{\partial x}\left(E_x\frac{\partial c}{\partial x}\right)+\frac{\partial}{\partial y}\left(E_y\frac{\partial c}{\partial y}\right)+\frac{\partial}{\partial z}\left(E_z\frac{\partial c}{\partial z}\right) -u_x\frac{\partial c}{\partial x}-u_y\frac{\partial c}{\partial y}-u_z\frac{\partial c}{\partial z}+S(x,y,z,c,t) \tag{20}$$

where $\partial_c/\partial_t = 0$ when the point source under steady-state conditions, we can ignore the effects of diffusion and longitudinal flow in X-axis direction and Y-axis direction, namely $E_x = 0$, $u_y = 0$, $u_z = 0$, the formula can be simplified as:

$$u_x\frac{\partial C}{\partial x}=E_y\frac{\partial^2 C}{\partial y^2}+E_z\frac{\partial^2 C}{\partial z^2}+S(x,y,z,c,t) \tag{21}$$

The analytical solution of this formula is:

$$C(x,y,z)=\frac{M}{4\pi x\sqrt{E_yE_z}} \exp\left[-\frac{u_x}{4x}\left(\frac{y^2}{E_y}+\frac{z^2}{E_z}\right)\right]\exp\left(-\frac{Kx}{u_x}\right) \tag{22}$$

Taking 319 sampling points (see Table 1 and Table 2) into the formula (21), calculating the

Table 1. Part 1 of the original data.

x(m)	y(m)	Altitude z(m)	Function area	As (μg/g)	Cd (ng/g)	Cr (μg/g)	Cu (μg/g)
74	781	5	4	7.84	153.80	44.31	20.56
1373	731	11	4	5.93	146.20	45.05	22.51
1321	1791	28	4	4.90	439.20	29.07	64.56
0	1787	4	2	6.56	223.90	40.08	25.17
1049	2127	12	4	6.35	525.20	59.35	117.53
1647	2728	6	2	14.08	1092.9	67.96	308.61
2883	3617	15	4	8.94	269.80	95.83	44.81
2383	3692	7	2	9.62	1066.2	285.58	2528.4
2708	2295	22	4	7.41	1123.9	88.17	151.64
2933	1767	7	4	8.72	267.10	65.56	29.65
4233	895	6	5	5.93	201.40	45.19	24.90
…	…	…	…	…	…	…	…
5101	4080	13	1	8.23	756.40	42.73	87.52
5438	3994	10	2	9.35	407.50	55.54	61.83
5382	3012	50	1	8.90	307.30	54.39	57.21
5314	2060	40	4	3.77	242.10	30.93	32.13
5503	1127	6	1	5.41	178.90	29.54	23.73
5636	133	17	1	7.78	315.50	49.76	28.03
6605	374	6	1	5.62	134.60	25.33	19.10
7093	1381	45	4	5.41	235.60	36.88	48.80
7100	2449	89	4	4.58	203.80	39.03	24.18
6837	3490	28	4	6.91	568.50	54.59	113.46
7906	3978	22	4	5.00	506.50	59.45	70.71
8045	3052	39	4	5.62	880.00	78.29	121.12
…	…	…	…	…	…	…	…
14318	13569	30	5	8.23	121.30	43.29	31.63
10352	17133	31	5	10.74	479.20	96.28	29.23
9095	16414	29	5	11.68	870.50	70.84	35.17
10510	15314	19	5	7.34	279.00	51.25	27.95
13954	5615	61	5	6.05	162.00	36.22	17.91
10142	1662	8	5	5.41	907.00	43.08	36.48
17765	3561	8	5	6.26	132.90	42.59	16.58
6924	5696	7	5	6.47	197.00	38.18	21.09
4678	3765	40	5	6.47	100.70	36.19	13.31
6182	2005	25	5	4.79	119.10	35.76	19.71
5985	2567	44	4	7.56	63.50	33.65	21.90
7653	1952	48	5	9.35	156.00	57.36	31.06

Table 2. Part 2 of the original data.

x(m)	y(m)	Altitude z(m)	Function area	Hg (ng/g)	Ni (μg/g)	Pb (μg/g)	Zn (μg/g)
74	781	5	4	266.00	18.20	35.38	72.35
1373	731	11	4	86.00	17.20	36.18	94.59
1321	1791	28	4	109.00	10.60	74.32	218.37
0	1787	4	2	950.00	15.40	32.28	117.35
1049	2127	12	4	800.00	20.20	169.96	726.02
1647	2728	6	2	1040.0	28.20	434.80	966.73
2883	3617	15	4	121.00	17.80	62.91	166.73
2383	3692	7	2	13500	41.70	381.64	1417.8
2708	2295	22	4	16000	25.80	172.36	926.84
2933	1767	7	4	63.00	21.70	36.94	100.41
4233	895	6	5	259.00	14.60	35.88	102.65
…	…	…	…	…	…	…	…
5101	4080	13	1	63.00	19.26	88.74	184.69
5438	3994	10	2	112.00	24.05	66.82	208.27
5382	3012	50	1	326.00	25.72	131.93	256.94
5314	2060	40	4	28.00	11.56	50.60	144.69
5503	1127	6	1	52.00	9.89	49.84	118.88
5636	133	17	1	550.00	18.95	45.73	109.29
6605	374	6	1	45.00	11.66	40.50	87.14
7093	1381	45	4	43.00	14.06	53.61	213.47
7100	2449	89	4	87.00	16.66	53.09	138.88
6837	3490	28	4	264.00	23.22	82.40	399.90
7906	3978	22	4	202.00	26.13	78.01	334.39
8045	3052	39	4	293.00	25.61	171.14	540.00
…	…	…	…	…	…	…	…
14318	13569	30	5	86.00	11.40	33.21	46.86
10352	17133	31	5	98.00	25.30	80.36	112.35
9095	16414	29	5	302.00	29.10	78.15	435.44
10510	15314	19	5	44.00	22.50	51.20	117.66
13954	5615	61	5	35.00	14.20	36.41	61.02
10142	1662	8	5	10.00	14.50	41.02	121.20
17765	3561	8	5	27.00	16.20	35.52	63.31
6924	5696	7	5	64.00	18.60	40.18	168.05
4678	3765	40	5	42.00	11.50	34.34	56.23
6182	2005	25	5	44.00	9.90	39.66	67.06
5985	2567	44	4	60.00	12.50	41.29	60.50
7653	1952	48	5	59.00	25.80	51.03	95.90

maximum concentrations, the results are *As*(4400, 7400, 8), *Cd*(4500, 1110, 6), *Cr*(3400, 5910, 4), *Cu*(2400, 3600, 20), *Hg*(2600, 2350, 22), *Ni*(3200, 5900, 4), *Pb*(2000, 3300, 7), *Zn*(13600, 9600, 18), the unit is meter.

5. Multi-dimensional partial differential equations: The movement of heavy metal pollutant in the soil is through many factors, and here only two of them are selected, it may cause some bias. In order to have a better study in the evolution mode of geological environment, we read the geological environment-related information, then we find out Geological environment contains the lithosphere, atmosphere and hydrosphere, the geological environment is based on the evolution of these three circles. With multiple factors, we utilize partial differential equations to get each of them, we can get the propagation coefficient of concentration, so that we can analyze the evolution model.

$$\frac{\partial c}{\partial t}+u_x\frac{\partial c}{\partial x}+u_y\frac{\partial c}{\partial y}+u_z\frac{\partial c}{\partial z}= \frac{\partial c}{\partial x}\left(E_x\frac{\partial c}{\partial x}\right)+\frac{\partial}{\partial y}\left(E_y\frac{\partial c}{\partial y}\right)+\frac{\partial}{\partial x}\left(E_z\frac{\partial c}{\partial z}\right) + S(x, y, z, c, t) \tag{23}$$

It can be extended to N-dimensional equation as:

$$\begin{aligned}\frac{\partial c}{\partial t}+u_x\frac{\partial c}{\partial x}+u_y\frac{\partial c}{\partial y}+u_z\frac{\partial c}{\partial z}\cdots u_n\frac{\partial c}{\partial n} &= \frac{\partial}{\partial x}\left(E_x\frac{\partial c}{\partial x}\right)+\frac{\partial}{\partial y}\left(E_y\frac{\partial c}{\partial y}\right)\\ &+\frac{\partial}{\partial z}\left(E_z\frac{\partial c}{\partial z}\right)+\frac{\partial}{\partial n}\left(E_n\frac{\partial c}{\partial n}\right)\\ &+S(x,y,z\cdots n,c,t)\end{aligned} \tag{24}$$

3 CONCLUSIONS

For the purpose of solving environmental pollution in this paper, we analyzed the evolution model of the geological environment, collected the isotopic concentration distribution index of heavy metals in different parts at various time periods, and then proposed gravitational potential migration characteristics of the material and the diffusion coefficient. In the end, we established a three-dimensional model of partial differential equations. We can obtain specific location sources by solving the model. In order to respond to future environmental problem and climate, soil moisture, soil saturation, soil PH value, soil water potential, pressure potential, matric potential, solute type, atmospheric wind speed, air temperature and land humidity etc. It can be used as material changes in the characteristics of migration. It can build a more comprehensive multi-dimensional partial differential equation models to find the source of contamination with these characteristics factors.

The multi-dimensional partial differential model can be extended to determine the noise source, seismic source, the spread of radiation sources etc. The diffusion characteristics approximately meet the feature which is diffused from the center to weeks, that is to say, the closer sources, the more serious pollution, on the contrary which is also true. Finally, we hope the theoretical innovation in this article can improve the environmental pollution.

ACKNOWLEDGEMENTS

We would like to thank the anonymous reviews for their valuable comments. This program is supported by Shaanxi Provincial Natural Science Basic Research Program (2009JQ1010), Shaanxi Provincial Department of Education Science Research Program (2010JK702), and Xi'an University of Technology PhD project started (104-210901).

REFERENCES

[1] Lv, J., Bian, C. & Li, Y. etc. Analysis of the heavy metal pollution in urban surface soil. Science and technology innovation herald, 2012, 26(2): 147–149.

[2] Yang, J., Sun, L. & Yang, X. etc. Assessment on heavy metal pollution in the farmland soil along Xi River in Shenyang. Agricultural journal of environmental science, 2007, 26(5): 1933–1936.

[3] Xu, S., Guo, S. & Hu, X. Analysis of heavy metal pollution evaluation and speciation of cadmium in Shenyang Zhang Shi irrigation. Journal of applied ecology, 2007, 19(9): 2144–2148.

[4] Ma, L., Yang, X.B. & Tong, C.Z. etc. Element geochemical characteristics of heavy metals in sediments Hunhe basin in Liaoning province. Journal of testing, 2008, 27(3): 184–188.

[5] Liu, T. & Guo, S.M. Soil heavy metal content of sewage irrigation for farmland in Shenyang. The influence of the environment, 2003, 29(117): 51–52.

[6] Wang, G.X. & Sun, L.B. Survey and evaluation of soil heavy metal pollution in east Gao sewage irrigation area of Inner Mongolia. Journal of Shanxi science and technology, 2007, 22(2): 115–116.

[7] Yue, X. & Wang, N. Analysis of urban topsoil heavy metal pollution. Technology Innovation and Application, 2012, (33): 116–117.

[8] Li, Y. Analysis of the heavy metal pollution in city surface soil. Science, technology and engineering, 2012, 12(11): 2772–2774.

[9] Zhu, Z., Fan, X. & Ling, H. etc. Analysis of the heavy metal pollution in city surface soil. Quality and Goods, 2012, (02): 271–272.

[10] Li, Z.Z. The improved analytic hierarchy process and the ideal solution sorting method combined with the application in the evaluation of heavy metal pollution. China health statistics, 2010, 27(2): 166–16.

[11] China Geological Survey. Technical Specification for Sample Analysis in Eco-geochemistry Assessment (DD2005-3). Beijing: China Geological Survey, 2005, 9–10.

[12] Lin Y.P. Rock evaluation forms and pollution of soil heavy metals. Mineral in Quanzhou, 2011, 31(2): 115–120.

[13] Zhou, M., Chen, C.H. & Wang, H.L. etc. Autumn in Shanghai typical atmospheric particulate matter in the process of polluting chemical composition characteristics. Journal of environmental science, 2012, 32(1): 81–92.

Engineering Management and Industrial Engineering – Leung (Ed.)
© 2015 Taylor & Francis Group, London, ISBN 978-1-138-02773-2

Bike-and-Rail promoting in Xi'an city

L. Yang, Y.Q. Wang, C. Li & Y.X. Li
School of Highway, Chang'an University, Xi'an, China

ABSTRACT: This paper proposed the Bike and Rail choice model composed of two layers of the logistic models to simulate the travellers' mode choice behaviors. The model had a good test result and the estimation results were analyzed as the basis of the policy making to encourage the Bike and Rail mode. The suggestions of the Bike and Rail promotion policies in Xi'an include: simplify the transfer process, shorten the transfer distance, promote smart card that can be used in the public transit and public bikes, and set bike lanes. This can attract more users of the Bike and Rail, and even more people with high income for their conveniences and safeties. The findings in this paper can also be the references for other cities expecting to promote the cycling.

1 INTRODUCTION

Cycling is widely recognized as an environmentally friendly and healthy mode of transport. During the past decades, cycling has once witnessed a long term decline in most countries for years (Mark Wardman, et al, 2007). However, at present, cycling infrastructure has grown dramatically in many cities, and more programs are now in place to encourage cycling (Elliot Fishman, et al, 2012), and Xi'an is not an exception. The local government introduced public bike service firstly inside the Economic and Technological Development Zone in the northern part of the city since the year of 2009. Meanwhile, similar to other fast developing metropolitan cities, Xi'an has introduced metro system to ease serious traffic congestions and has begun operation with one line in the year of 2011, with two lines till to the year of 2013. Since then, Bike and Rail (B+R) promoting has become the object of the urban transport development of the local government. Now there are 34 spots and 680 bicycles with the max 1597 person trips and 2.34 times per bicycle per weekday in this district. Compared with other cities (20.08 in Taiyuan, 12.48 in Kunshan, 6.22 in Zhongshan, 5.6 in Wenzhou and FoshanSanshui District, 5 in Suzhou, 3.72 in Hangzhou, and 3.1 in Guangzhou) in China, the public bike program has more spaces for its improvement and needs to be further promoted in Xi'an in order to largely reduce the motorized transport. Thus, this paper works on analyzing the factors that impact the use of B+R through the mode choice models established by using the data collected in the Economic and Technological Development Zone. Besides, the suggestions based on the model result analysis are proposed in this paper as well. The findings in this paper can be of reference values in the policy makings to encourage more use of B+R in Xi'an and also to encourage other cities' public bike program and cycling developments in China.

2 RELATED WORK

The previous B+R mode choice models varied among the researchers. Lin Junhong applied the regression method to establish the B+R model. The independent variables include: the number of the boarding in the metro station in the peak hours, number of the transit lines around the metro station, working people density within 500 m around the metro station, the percentage of the slops and the number of the working people and the students above 16 ages between 500 m and 1700 m to the metro stations (Lin Junhong, 1991). Zhang Jie suggested that the transfers to the bicycles equaled to the number of the working people multiply by the bicycle mode share and the percentage of the long distance trips above 7 km within the area between 1 km and 5.5 km around the transfer stations (Zhang Jie, et al, 1995). Based on the random utility theory and the mode choice model, Li Tingting chose travel time, travel cost, age, income, car owning, trip distance and distance from the trip origins to the station as the independent variables to establish the public bike demand forecasting model (Li Tingting, 2010).

While in the meantime, the previous impact factors of the bicycle mode choice are various in kinds. Yang Chen divided bike driving factors into individual characteristics such as age, gender, job, income, education and family attributes, trip information such as purpose and covering distance, and subjective perception of mode service level such as time consumption, convenience, safety, comfort and economy (Yang Chen, et al, 2007). Shan Xiaofeng considered individual characteristics, such as gender, age, time urban living time, education, job, income, family structure and whether having children, and travel information including distance and time consumption will impact people's bicycle mode choice (Shan Xiaofeng, et al, 2006). Ma Pei took the factors of the bicycle facility into consideration of the bicycle mode choice (Ma Pei, et al, 2011). Although there are many factors that affect the use of the bicycle, they can be classified as follow: socio-economic characteristics of the individuals and households, trip characteristics and transport facility and its service.

3 BIKE AND RAIL MODE CHOICE MODEL

3.1 *Model structure*

The proposed B+R modeling method is based on the discrete choice theory put forward firstly in 1985 (Ben-Akiva, et al, 1985). Then the impact factors can be the independent variables in the model. Since there exists one time of transfer during the travelling, we need to simulate the two stage process of the mode choice of the travellers. Firstly, the travellers will choose whether or not use the metro mode, and secondly, after going out of the metro station, he or she will whether or not choose the bicycle mode for arriving at the destinations. Thus, two layers' discrete choice models are needed for the B+R modeling and the Nested Logistic Model is introduced in the modeling (Chen Tuansheng, 2007, Yang Liu, 2010). The upper layer model is used to analyze the mode choice behaviors between the metro, bus, car and bicycle, while the lower layer analyzes the mode choice behaviors between the bicycle, car, bus and walking during the transfers. The multiple logistic regressions are applied in each layer for the calculation of each mode's probability, and the upper layer model would restrain lower layer model through conditional probability while the natural logarithm value of the total utilities of the lower layer model participates in the upper model calibration as a variable.

3.2 *Variable choice*

The socio-economic characteristics of the individuals and households, trip characteristics and transport facility and its service are considered in the model.

3.3 *Data collection*

The survey was conducted in the Economic and Technological Development Zone in Xi'an, since it is the earliest district promoting the use of the public bikes from the year of 2009 and the Metro Line 2 passed through this district in 2011. Thus the metro station in this district are surrounded by the public bikes, and moreover, the B+R demands and users' characteristics have entered a mature stage and become more stable than other districts in Xi'an. Based on the above reasons, the travellers around the three metro stations (Shi TuShu Guan Station, Feng Cheng 5th Street Station, and Xing ZhengZhongXin Station) with large number of boarding and alighting in this district were chosen to be surveyed randomly. Besides, the neighborhoods within 500 meters near the public bike spots in Xi'an Economic and Technological Development Zone were also surveyed, and the sampling rate was determined according to the number of the households in the neighborhoods. The questionnaire included: socio-economic characteristics (gender, age, job, income and car owning), travelling characteristics (traffic mode, trip distance, travel time and trip purpose), the assessment and the subjective perceptions of the transport facility and its service. In total, 1310 effective questionnaires were surveyed.

3.4 *The lower layer model*

Multinomial logistic regression model is applied and if the fourth mode of walk and rail is taken as a base case, the other three modes' choice probabilities (Bike and Rail, Bus and Rail, Car and Rail) can be presented as the following.

$$\ln\left(\frac{P_i}{P_3}\right) = \alpha^i + \sum_{k=1}^{K} \beta^i_k X^i_k = V_i (i = 0, 1, 2) \tag{1}$$

Note: $i = 0, 1, 2$, which represents the travellers going out of the metro stations and transfers to the modes of bike, bus and car respectively, except walk. α^i is the constant of the mode i, β^i_k is the coefficient of X^i_k, X^i_k is the numbered k variable of the mode i, V_i represents the relative utility of walk and rail.

By using the SPSS19.0 multinomial logistic regression module, and eliminating the explanatory variable whose significance value is more than 0.1 in likelihood ratio test, the model estimations can be obtained.

$$\begin{cases} \ln\left(\frac{P_{Bike+Rail}}{P_{Walk+Rail}}\right) = -2.846 - 0.899purposeB + 1.162distanceB \\ \qquad + 1.034convA + 0.897safeA \\ \ln\left(\frac{P_{Bus+Rail}}{P_{Walk+Rail}}\right) = -3.175 + 0.846ageB + 1.716carA + 1.316carB \\ \qquad + 1.114bikeA + 0.997bikeB - 1.263purposeC \\ \qquad - 1.187distanceA + 0.793safeA \\ \ln\left(\frac{P_{Car+Rail}}{P_{Walk+Rail}}\right) = -18.069 + 16.898ageB + 1.157bikeA \\ \qquad - 0.997incomeA - 2.362distanceA \end{cases} \tag{2}$$

3.5 *The upper layer model*

The natural logarithm value of the total utilities of the lower layer model of each sample is calculated by using the estimation results of the lower layer model and the survey data. This value is taken as one of the variables of the upper layer model in the model estimation. The fourth mode of bike is considered as the base case. The other three modes' choice probabilities can be presented as the following:

$$\ln\left(\frac{P_i}{P_3}\right) = \alpha^i + \sum_{k=1}^{K} \beta_k^i X_k^i = V_i (i = 0, 1, 2) \tag{3}$$

Note: $i = 0, 1, 2$, which represents the modes of metro, bus and car respectively, except bike. α^i is the constant of the mode i, β_k^i is the coefficient of X_k^i, X_k^i is the numbered k variable of the mode i, V_i represents the relative utility of bike. Similarly as the lower layer model estimation, by using the SPSS19.0 multinomial logistic regression module, and eliminating the explanatory variable whose significance value is more than 0.1 in likelihood ratio test, the model estimations can be obtained.

$$\begin{cases} \ln\left(\frac{P_{Rail}}{P_{bike}}\right) = 15.212 + 6.429carA + 4.360carB \\ \qquad + 2.925distanceA + 4.938comfortA \\ \ln\left(\frac{P_{bus}}{P_{bike}}\right) = -0.664 + 18.360carA + 2.562bikeA + 1.259bikeB \\ \qquad - 1.337purposeA - 2.280purposeC \\ \ln\left(\frac{P_{car}}{P_{bike}}\right) = 21.940 + 1.705bikeA - 2.418incomeA \\ \qquad - 1.723purposeA - 2.582purposeC \\ \qquad - 1.680distanceA \end{cases} \tag{4}$$

4 RESULT ANALYSIS

From the estimation results in the upper layer model, car availability and comfort level of the traffic mode make much contribution to the choice of the rail mode: one who doesn't own cars is more likely to choose rail than the people who own cars; people who believe rail more comfortable are more likely to choose the rail mode. Number of cars and bikes impacts a lot in the choice of the bus mode: people who don't have cars and bikes will be more likely to choose bus. Income and travel distance play important roles in choosing the car mode: people with income less than 40,000 RMB per year and travel distance less than 3 km have little possibilities of choosing the car mode.

From the results of the lower layer model, trip purpose, trip distance, convenience and safety of the transport facility have significant impact on choosing the Bike and Rail mode: the Bike and Rail mode attracts people who are touring, whose trip distance is between 3 km and 5 km, and who believes Bike and Rail is more convenient and safe. As to the mode of Bus and Rail, age, number of the owned cars and bikes, and safety of the transport facility will have great impact: people who are between 18 and 35 years old, owning no cars or bikes and believing Bus and Rail much safer will be more likely to choose the Bus and Rail mode. And as to the mode of Car and Rail, income and number of the cars will have more influence: people with less than 40,000 RMB per year will be not likely to choose this mode, however, people who own cars will probably choose the Car and Rail mode.

Therefore, to encourage the B+R mode in Xi'an, not only do the district focus on official, business and commercial development need public bike facilities and services, but also the districts around the tourist attractions and scenic spots. In policy making of promoting the B+R mode, the convenience and safety of using this mode should be the top objectives. This means: simplify the transfer process, shorten the transfer distance, promote smart card that can be used in the public transit and public bikes, promote to set bike lanes to keep safety of cycling. All of the convenient and safe services will attract more people to use the B+R mode, even people with high income level.

5 CONCLUSION

The nested logistic model of B+R mode choice was established and the model estimation results were analyzed in this paper. The trip purpose, trip distance, convenience and safety of the transport facility have significant impact on choosing the B+R mode, and attract more people who are touring, whose trip distance is between 3 km and 5 km, and much requirement of convenience and safety of the transport facilities is important to the cycling encouragement. These findings suggest that the tourist attractions and scenic spots need more public bike services and the convenience and safety of using this mode is crucial to its promotions in Xi'an. Moreover, it can also be the basis of the policy making to encourage more use of this mode. Thus, simplifying the transfer process, shortening the transfer distance, promoting smart card that can be used in the public transit and public bikes, and setting bike lanes are effective measures in the improvement of the convenience and safety of cycling. This is not only useful in Xi'an's public bike program development, but can also be the references for other cities expecting to promote the cycling.

ACKNOWLEDGMENT

This research was funded by Scientific Research Project of the Education Department of Shaanxi Provincial Government (Project No.12 JK0799) and National Natural Foundation of China (Project No.51178055-E0807).

REFERENCES

Ben-Akiva, Moshe, E., and Lerman, S.R., (1985). Discrete choice analysis: theory and application to travel demand. Vol. 9. MIT press.

Broach, J., Dill, J., and Gliebe, J., (2012). Where do cyclists ride? A route choice model developed with revealed preference GPS data, Transportation Research Part A, 2012(46), 1730–1740.

Chen, T.S., (2007). Travel behavior characteristics analysis for commuter, Doctoral dissertation of Beijing Jiaotong University.

Fishman, E., Washington, S., and Haworth, N., (2012). Barriers and facilitators to public bicycle scheme use: A qualitative approach, Transportation Research Part F, 15, 686–698.

Martens, K., (2004). The bicycle as a feedering mode: experiences from three European countries, Transportation Research Part D, 2004(9):281–294.

Martens, K., (2007). Promoting bike-and-ride: The Dutch experience, Transportation Research Part A, 2007(41):326–338.

Ma, P. and Wu, H.Y., (2011). Research on the bicycle transfer rail traffic behavior mechanism and model, Journal of Beijing University of civil engineering and architecture, 27(2):36–40.

Li, T.T., (2010). Study on locating and planning of urban public bike rental station., Master thesis of Beijing Jiaotong University.

Lin, J.H., (1991). Study on the bicycles parking demand forecasting in the rail transit station, Master thesis of Taiwan University,

Shan, X.F., Wang, W., and Wan, Q., et al, (2006). Research on the mode choice of bicycle based on the behavior analysis, Journal of the modern urban study, 2006(11):81–88.

Wardman, M., Tight, M., and Page, M., (2007). Factors influencing the propensity to cycle to work, Transportation Research Part A 2007(41), 339–350.

Yang, L., (2010). Study on the travel behaviors and characteristics, Master thesis of Chang'an University.

Yang, C., Lu, J., Wang, W., et al, (2007). A study on the influencing factors of bicycle transportation based on individual mode choice, Journal of transportation systems engineering and information technology, 7(4):131–136.

Zhang, J., Ren, F.T., Liu, X.M., and Shi, Y.Z., (1995). Study on the transferring between bike and bus in Beijng, China journal of highway and transport, 8(1):153–157.

Engineering Management and Industrial Engineering – Leung (Ed.)

Research on music push model based on terminal situation model

H.Y. Han, S.H. Guo & Y. Han
School of Information, Central University of Finance and Economics, Beijing, China

ABSTRACT: The main current information push model is based on the users' interest, lacking theoretical support to terminal situation model. To improve the approach of pushing music, realizing mobile-termination-oriented pattern, this paper presents a music push model that focuses on terminal situation. Firstly, the paper defines the concept of terminal situation. Secondly, the paper puts forward the music push model based on Bayesian network and collaborative filtering algorithm. Next, the paper defines this model's establishment process. Finally, the paper validates the effects of terminal-situation-oriented model through living examples, indicating that this model is more applicable and effective.

Keywords: Terminal situation; Pushing music; Bayesian network; Collaborative filtering algorithm; Push model

1 INTRODUCTION

With the rapid development of mobile Internet technology and the increasing power of mobile intelligent terminal function, the wide use of mobile terminals can even be called a kind of dependence on a mobile terminal. Confusing music resources in Mobile music market has far exceeded a user's music demand and capacity for independent choice. In a general situation, users usually, especially in a demanding moment, search for online music based on keyword search in the music platform. By measuring musical played record, ranking or word of mouth, user comments, they select the final network music that they decide to listen to. It is difficult to fully combine with rating and visit terminals such as shrimp music when searching a song. About this issue, the current solutions are usually based on the personalized information of users, such as downloading history, recommending the music that may cause their interests. Then they improve the network music's "appearance" probability, which is regarded as one of the methods for precise delivery. However, one of the problems existing in the solution is the uncertainty of both the user's access to the music platform and access to music.

The observation of actual case shows that the users' use of mobile search and behavior of browsing pages are different from PC. Therefore, to know more accurately users' interest and push information, it is necessary to modify the above variables, such as adding, deleting new variables based on the characteristics of the mobile end's page views, and build a new model. People like Lv Yang[1], in their paper, through the novel analysis of the behavior of mobile Internet users, update the scope of data sources in the traditional sense, especially for "fragmentation" characteristics of the use of mobile Internet, and put forward the precision push strategy based on fragments' period.[2] The purposes of this study are, therefore, according to the characters of the behavior of the mobile terminal users and the use of habits, building both terminal situation model and a network of music push model, and designing algorithm and validation.

2 DEFINITION

As is mentioned above, the analysis of characteristics of mobile terminals, for the recommendation of songs to users, can no longer be simply measured by the dominant information such as the user ratings, the past log information, but songs' real listening time, repeat play count per unit time, and other factors such as the use condition of personalized service and the external information generated by the user in a particular situation should be considered to form a model of feedback and appraisal system in order to filter collaborative algorithms. [7] So, on the basis of predecessors' research, this article will improve the traditional two dimensional model based on the users—extension project to a three dimensional model of the context of the user—projects—situation information. Among the information, the situation information consists of the user situation, the task situation and physical situation. The user information includes the basic information such as age, gender, social status, history search and download, etc.; Task information is made of all influential factors during the period of listening to a song for a user, including bandwidth, network connectivity, available resources and equipment, etc. The physical environment includes time, location, weather and date. All these situations externally perform in the past listening to music of target users, and the personalized mobile terminal situation shown includes listening time, times, frequency, the timing of the preferences, etc. The external situation, which can be obtained by implicit and explicit measures, determines the music category under the user preferences according to the algorithm.

3 ALGORITHM DESIGN

3.1 *Algorithm models based on mobile characteristics*

3.1.1 *User interest matrix based on the user – item evaluation*

In research of user interest model, User – Item matrix (U – I) has been used widely. This method uses matrix Rmn to express User model, and uses User and Item to determine the specific User (U)'s Preference (P) for a particular Item (I). The model is shown in figure 1:

According to the analysis of the physical situation and user situation above, the User dimension is defined as: $User = <U_{age}, U_{sex}, U_{status}, U_{history}>$. and $U_{age}, U_{sex}, U_{status}, U_{history}$ are several properties of user's, $U_{age}, U_{sex}, U_{status}$ means user's age, gender and social status in User situation; and $U_{history}$ means user's history search and download in User situation. For each $u \in User$, each property has a series of values to match.

The definition for Item dimension is similar, and through the analysis of the task situation, properties for items are as follows:

$$Item = < I_{name}, I_{type}, I_{size}, I_{system} >$$

$I_{name}, I_{type}, I_{size}, I_{system}$ Means the song name, type, size and the mobile terminal system is adapted.

$$UI = \begin{bmatrix} u_1p_1 & u_1p_2 & u_3p_3 & \cdots & u_1p_{n-1} & u_1p_n \\ u_2p_1 & u_2p_2 & u_3p_3 & \cdots & u_2p_{n-1} & u_2p_n \\ u_3p_1 & u_3p_2 & u_3p_3 & \cdots & u_3p_{n-1} & u_3p_n \\ u_4p_1 & u_4p_2 & u_3p_3 & \cdots & u_4p_{n-1} & u_4p_n \\ u_5p_1 & u_5p_2 & u_3p_3 & \cdots & u_5p_{n-1} & u_5p_n \\ u_{...}p_1 & u_{...}p_2 & u_{...}p_3 & \cdots & u_{...}p_{n-1} & u_{...}p_n \end{bmatrix}$$

Figure 1. User interest model based on U – I model.

If the User and Item can be expressed as two Dimension for the model (D1, D2), so in the case of $u \in$ *User* and $i \in$ *Item*, U – I model utility score equation can be expressed as: $R_{(u,i)} = D_1 \times D_2$.

3.1.2 *Three-dimensional model based on mobile characteristic*

In the analysis of the physical situation above, we can figure out the behavior of mobile terminal users has obvious features such as mobility, use time fragmentation. And through the analysis of user's personalized situation, mobile music push users will have their own music preference according to user situation, task situation and physical situation, so we put the dimension3 into the traditional two-dimensional user—item matrix to consider the preference of music category. So the three-dimensional model is shown in Figure 2.

For each $c \in$ *Condition*, utility score function is expressed as $R(u,i,c) = D_1 \times D_2 \times D_3$, and as the supplement of user—item two-dimensional grading system, three-dimensional can be understood as: under the specific Situation, user's(u) preference for item I, and the specific measures of situation will be described in the followings.

3.2 *Collaborative filtering algorithm based on terminal user situation*

1. Category preferences of the user's characteristic resource
First of all, we need a lot of user situations based on different categories of music; these cases show that in a specific situation, the user prefers a certain type of music. Here are the definitions of the situation factors: V represents the data set of songs, and vi represents a song of a specific musical type, we express the location, network status, time and whether it is an important festival with B1, B2, B3, B4. After consulting

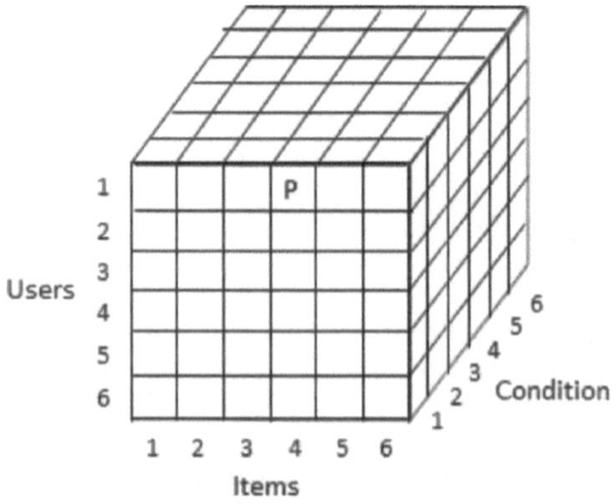

Figure 2. Three-dimensional user interest models expressed as U – I – C model.

a number of documents, we decide to use the Bayesian method to combine the characteristic information with traditional collaborative filtering recommendation algorithm. In further research, the characteristic properties should be dynamically updated with the number of songs and download changes to ensure the precision and real-time performance of description.

In addition to the definition above, the probability of choosing the type of music for a user with the properties of < b1, b2, b3, b4 > is expressed with Bayesian method as:

$$P\left(vi|b1,b2,b3,b4\right) = \frac{P(vi) * P(b1,b2,b3,b4|vi)}{\sum_{vi \in V} P(vi) * P(b1,b2,b3,b4|vi)} \tag{1}$$

$P(vi)$ And $P(b1, b2, b3, b4 | vi)$ all can be obtained by a lot of experimental data. $P(b1, b2, b3, b4 | vi)$ Probability needs to analyze the characteristics ($b1, b2, b3, b4$) of users, and in the case of a large number of information data, the probability can be considered accurate and effective, but due to the limited number of experimental data, we assume that $b1, b2, b3, b4$ represent user's location, condition of network, time, whether it is an important festivals is unrelated, so the formula can be expressed as follows:

$$P(b1,b2,b3,b4|vi) = \prod_{j=1}^{4} P(bj \mid vi) \tag{2}$$

Take formula (3) into (2):

$$P\left(vi|b1,b2,b3,b4\right) = \frac{P(vi) * \prod_{j=1}^{4} P(bj|vi)}{\sum_{vi \in V} P(vi) * \prod_{j=1}^{4} P(bj|vi)} \tag{3}$$

After consulting the references, we combine the users' category preferences based on their own situation with the traditional collaborative filtering algorithm; multiply formula (1) with formula (3):

$$Pai^* = P\left(vi|b1,b2,b3,b4\right) * Pai = \frac{P(vi) * \prod_{j=1}^{4} P(bj|vi)}{\sum_{vi \in V} P(vi) * \prod_{j=1}^{4} P(bj|vi)} * \left(\overline{ra} + \frac{\sum Sim(a,b) * (rb_i - \overline{rb})}{\sum |sim(a,b)|}\right) \tag{4}$$

Formula can be divided into two parts to understand, on the left side Pai^* means prediction score after considering the preference for category based on user behaviors characteristic, on the right side it means prediction score under traditional collaborative filtering algorithm and the preference of category vi with properties < $b1, b2, b3, b4$ >. For recommending the songs from same category, the appetite is the same. After correction, to recommend songs of former recommendation list based on the final score, it will improve the precision. The overall process steps are shown in Figure 3.

3.3 *Model based on naive Bayesian classification algorithm*

3.3.1 *Description of naive Bayesian classification algorithm*

Naive Bayesian classification algorithm usually has definitions formulas are shown in Table 1.

In the algorithm of the last section, the naive Bayesian classification model can be disassembled as follows under the condition that each property is unrelated:

$$P(yi|x) = \frac{P(x|yi) * P(yi)}{P(x)} \tag{5}$$

In order to get the highest classification probability for the category, we just need to calculate the maximum value for the right of equation. And for P(x), it is the same value for all categories in classification C, it is constant, so we only need to calculate the maximum molecules. Since the properties are unrelated, the formula can be transferred as follows:

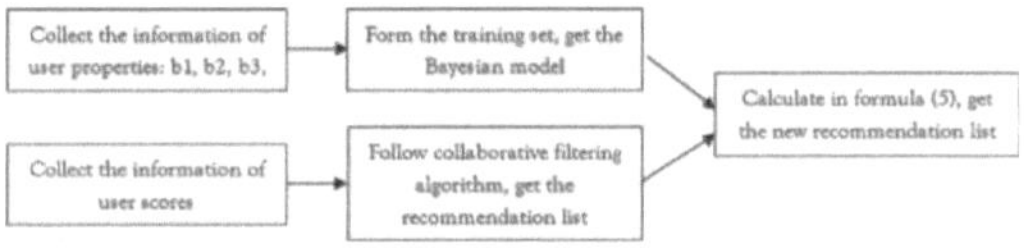

Figure 3. The steps of collaborative filtering algorithm based on terminal user situation.

Table 1. Naive Bayesian classification algorithm expression and definition.

Expression	Definition
$X = \{a1, a2, a3 ... am\}$	X means unclassified item, ai means the property for X
$C = \{y1, y2, y3 ... yn\}$	C means category set, yi means category.
$P(y1 \mid x)$ $P(y2 \mid x)$...	$P(y1 \mid x)$ $P(y2 \mid x)$.... Means under the condition X, the probability for category yi.
$P(a1 \mid y1)$ $P(am \mid yn)$	Probability of each property of each category.

$$P(x \mid yi) * P(yi) = P(yi) * \prod_{j=1}^{m} P(aj \mid yi) \quad (6)$$

So in order to get the final classification results, we only need to calculate each category of data $P(yi) * \prod_{j=1}^{m} P(aj|yi)$; the maximum value is the classification results.

Based on the theory above, we can summarize the process steps of naive Bayesian classification algorithm as shown in Figure 4.

3.3.2 *Structure of the algorithm based on naive Bayesian classification model*

Under the guidance of the steps above, we apply it to the algorithm application of "musical category preference for terminal users" in this paper. First of all, we need to define the expression that we need in the process is shown in Table 2.

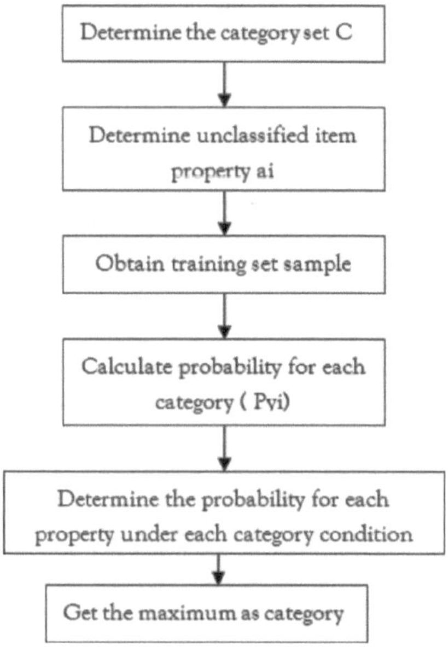

Figure 4. Naive Bayesian algorithm process steps.

Table 2. Expression and meaning of algorithm in this paper.

Expression	Definition
$X = \{b1, b2, b3, b4\}$	X means unclassified item, ai means the property for X
$V = \{v1, v2, v3, \ldots vn\}$	C means category set, yi means category of music.
$P(v1 \mid x)$ $P(v2 \mid x)$...	$P(y1 \mid x)$ $P(y2 \mid x)$... Means under the condition X, the probability for category yi.
$P(b1 \mid v1)$ $P(b3 \mid yn)$	Probability of each property of each category.

Firstly, we need to determine the category of music V.

First of all, we need to divide numbers of songs into several music category, we get the category of music V is shown in Table 3.

According to the data above, we can see that pop, rock, country, R&B, hip hop are several kinds of music taking up large percentage in the music market, so they are music categories that users always download. Therefore, category set Vis defined is shown in Table 4.

V = {pop, rock, R&B, hip-hop, country}

Secondly, we need to determine the property of terminal music situation for target user:

In actual situations, there are many factors to describe habits of listening to music. So we pick up several special properties to collect data and calculate.

Thirdly, we need to choose the training sample:

According to user's existing data, we obtain the values for properties of different vi, and here is the data set obtained from experiment:

$$P(bi \mid vj)(1 \leq i \leq 3\ ;\ 1 \leq j \leq 7)$$

Table 3. The proportion of each music category in music library.

Category	Proportion
Pop	65.04%
Rock	11.88%
Country	2.54%
R&B	6.16%
Hip-hop	5.44%
Electronic	4.51%
Jazz	1.23%
Classical	0.84%
Folk	2.36%

Table 4. Characteristic properties and the result set.

Symbol	Meaning	Result set
b1	Location	{traffic, relaxation, home, work, bar}
b2	Network	{terrible, bad, normal, preferable, excellent}
b3	Time	{morning rush, lunch time, tea time, rush hour, after work}
b4	Whether it is an important festival	{yes, no}

$P(v1) = 25181/38717 = 0.6502$
$P(v2) = 4599/38717 = 0.1188$
$P(v3) = 2385/38717 = 0.0616$
$P(v4) = 2105/38717 = 0.0544$
$P(v5) = 1748/38717 = 0.0451$

Fourthly, calculate the conditional probability of the characteristic properties in each category according to the sample of training set.

4 EXPERIMENTAL VERIFICATION

4.1 *Generation of the training set*

In this part, each tester will be offered 30 unscored songs to score (from different categories), and the system will ensure the music list of each tester to be repeated at least 50%. In order to get data of traditional collaborative filtering algorithm, we require the tester to evaluate each song in a 10-point scale. In order to guarantee the authenticity of the data, we require that the amount of the same score of 30 songs shall not exceed six of each tester. If the tester hasn't listened to the target song, he can choose not to evaluate the song. However, to avoid the sparsity of real data, the experiment requires each tester to evaluate 50% of the songs at least, and delete the non-standard data until the qualified data reach 30, then stop collecting data. To collect the situation of tester, we use the speed test and GPS positioning to capture the message of mobile in the background while listening to music, including the location, the condition of network, time, and whether it is an important festival, etc.

4.2 *Evaluations of the recommended songs through algorithm*

In this part, firstly, we summarize the training set obtained from the former part, and transfer the data into category selection probabilities form described in the Bayesian formula. Then, we let the testers provide their own characteristic information of the mobile <b1, b2, b3, b4> when they are listening to music. Then, according to the training, we set to determine their preferences of category, and according to his music score we choose similar neighbors for them, and use the formula (5) to calculate the score, finally, we set the highest score of TOP—10 songs as the recommended preference songs of the new algorithm. By rating the songs of the recommended list, we can get the information of whether the preferences are accurate. Through the study of reference, most of the collaborative filtering recommendation algorithms are measured through calculating accuracy or recalling rate of recommended effect, so we choose accuracy for effect assessment here.

4.3 *Data process*

For the convenience of decrypting the statistics, we define the classification of b1, b2, b3, b4 as shown in Table 5.

4.4 *Computing collaborative filtering algorithm*

In all the collected data of 30 testers, we choose 5 testers to recommend list, and the parameters are listed in Table 6.

Table 5. Character parameters.

Character parameter	a	b	c	d	e
b1	Traffic	Relaxation	Home	Work	Bar
b2	Terrible	Bad	Normal	Preferable	Excellent
b3	Morning rush	Lunch time	Tea time	Rush hour	After work
b4	Yes	No			

Table 6. Situation of 5 testers.

Tester / parameter	Average score	b1	b2	b3	b4
1	7.00	d	d	b	a
2	7.33	b	c	c	b
3	7.39	a	b	d	a
4	6.84	c	c	d	a
5	6.07	e	b	e	b

Table 7. Average score of each category tester 1.

Tester1	b1 = b	b2 = d	b3 = c	b4 = a
Pop	7.83	7.83	7.83	7.83
Rock	5.68	5.68	5.68	5.68
R&B	6.32	6.32	6.32	6.32
Hip-hop	6.76	6.76	6.76	6.76
Country	7.47	7.47	7.47	7.47

Table 8. Probability of different characters of different preference in each music category.

Pop	p(b1|v)	p(b2|v)	p(b3|v)	p(b4|v)
bi = a	28.32%	17.59%	24.50%	68.00%
bi = b	25.50%	10.33%	16.33%	32.00%
bi = c	23.18%	25.00%	13.00%	
bi = d	10.00%	32.56%	24.50%	
bi = e	13.00%	14.52%	21.67%	

Calculating the preference of music and proportion of characters under Bayesian algorithm, we use the average score of all songs as the tester's score for the category. Considering the score of all categories, we choose 7 points as the preference horizontal, which means if a piece of song is scored more than 7 points, the category of music will be regarded as the tester's preference.

According to the test template, we get the information of tester's score as shown in Table 7.

The data shows that the testers in a (b, d, c, a) mobile application characters have preference for pop and country music (score > 7). We do the same process to get data of other 29 testers, and get the probability of different characters of different preference in each music category, which means $P(b_i \mid v_i)$, $vi \in V, bi \in B$.

4.5 *Experimental results*

According to the method above, we get the recommended results as shown in Tables 9 and 10.

4.6 *Compare of accuracy*

Recommending the top-7 of two lists to the testers, and the testers will score them in 1–7 preference ranking. Then we compare the accuracy of

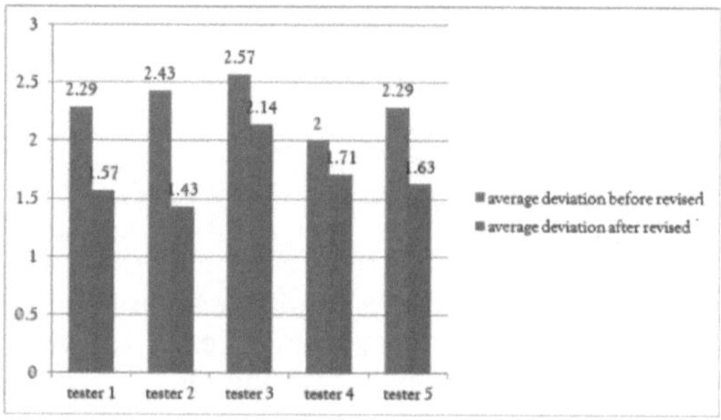

Figure 5. Compare the accuracy of two lists.

two lists by comparing the ranking list of average deviation method. And the following tree diagram is the data of 5 testers; the sixth is the average of the former 5.

4.7 *Experimental summary*

Through the experiment, we can see that the revised ranking improves the accuracy of recommending music, and after considering the terminal situation, the ranking is much closer to the real preference of tester.

Table 9. Corrected factors of Bayesian data set (take tester1 for example).

Tester1	P(vi)	P(b1 = b|vi)	P(b2 = d|vi)	P(b3 = c|vi)	P(b4 = a|vi)	(*10000)	Log
Pop	0.6502	0.2550	0.3256	0.1300	0.6800	47.7226	1.6787
Rock	0.1188	0.3786	0.3200	0.0658	0.5300	5.0194	0.7006
R&B	0.0616	0.2598	0.2900	0.2437	0.5800	6.5600	0.8169
Hip-hop	0.0544	0.1889	0.3678	0.3258	0.6200	7.6346	0.8828
Country	0.0451	0.2445	0.1659	0.5656	0.4700	4.8631	0.6869

Table 10. The recommended list of the traditional collaborative filtering algorithm (take tester1 for example).

Music	Type	Predicted score	Corrected score	Final score	Ranking
Rolling in the deep	R&B	7.95	0.8168	6.494	3
Yesterday	Rock	7.56	0.7006	5.297	7
Anesthesia	Pop	7.82	1.6787	13.127	1
End of May	Pop	7.23	1.6787	12.137	2
Satisfaction	Rock	7.18	0.7006	5.030	8
Love story	Country	7.13	0.6869	4.898	9
Mine	Country	7.08	0.6869	4.863	10
Magic	R&B	7.05	0.8169	5.759	6
Not afraid	Hip-hop	6.98	0.8828	6.162	4
Love the way you lie	Hip-hop	6.59	0.8828	5.818	5

5 CONCLUSION

To improve the accuracy of music recommendation, this paper presents a music push model based on the musical behaviors of terminal user. In the traditional collaborative filtering algorithms, two dimensional Users – Items model (U – I model) is a common algorithm model. By looking for similar neighbors' item evaluation we determine the position of target user in the U – I model, and recommend applications. Now we add Situation (Situation) dimension to the model, users need to look for the similar Situation first, and then they determine the value of dimension of Situation, finally, they look for similar neighbors in the Situation. The third dimension situation model is composed of four factors that represent the musical situation of user: location, the condition of network users, time and whether it is an important festival. By polymerizing these four elements through the Bayesian algorithm, we can get the user's preference for a specific music type. Through describing the situation more specifically, we can get a much more accurate music push model.

REFERENCES

[1] Dongmei Mu. The research based on context-awareness recommender system for mobile terminals. [J] Beijing University of Posts and Telecommunications. 2012.

[2] Xinwen Gong. The design and implementation of mobile learning platform based on context-aware. [J] Tientsin University, 2012.

[3] Xiaodong Li, Hong Zhang. Research on English M-Learning Software based on context-awareness [J], modern educational technology, 2013(5).

[4] Pengcheng Feng. Research on context aware personalized recommendation algorithm [J]. Donghua University, 2014.

[5] Ziming Zeng, Xin Li. Context-aware personalized information recommendation in mobile environment [J] Journal of Intelligence, 2012(8).

[6] Muhai Hu. A study on dynamic context-oriented recommendation methods and system. [J] Huazhong University of Science and Technology, 2011.

[7] Yuanyuan Chen, Zhengjie Liu. Research on mobile context awareness applications and interactions. Application research of computers, 2011(12).

Engineering Management and Industrial Engineering – Leung (Ed.)
© 2015 Taylor & Francis Group, London, ISBN 978-1-138-02773-2

Design and implementation of dance model based on 3D motion capture system

J.Y. Fu, X.Z. He, F.C. Wan, L.L. Duan & Y.M. Shan
National Language Information Technology, Northwest University for Nationalities, Lanzhou, Gansu, China

ABSTRACT: Thanks to the unique characteristics of Chinese national culture, a unified multi-ethnic country came into being. It is characterized by the old and traditional culture. As time goes by, the culture is gradually fading, which makes people worried. When talking about national intangible cultural heritage protection, deputy director of expert committee said that: "the broad and profound Chinese intangible cultural heritage is facing a huge threat due to improper development." This shows that the Chinese consciousness of protecting traditional ethnic culture gradually strengthened. At the same time, more and more scholars have focused on the research of intangible cultural heritage protection. Thus it is significant to protect the culture of folk dance.

1 INTRODUCTION

3D dynamic capture technology in the academia's research is mainly distributed in human motion analysis, character animation generation, biomechanics, sports study, robot simulation and other sports level. For example, the "warlords moon" project of Shanghai company, "double down" opera of the Institute of Chinese opera, and the realization of these cases is based on the analysis of the optical principle and structure of human body and the camera coordinate data collection in the space and reconstruction of 3D model. Since the 1970s, the social people from all walks of life began to focus on the application of the motion capture technology. However, the study about the protection of the traditional folk dance art has also gradually valued. This research hopes to save each national dance art essence.

In addition, 3D modeling is applied to implement digital protection of cultural heritage research. It began in the 1970s. Typical examples include "digital" the imperial palace of Beijing University, "digital Dunhuang art" of Zhejiang University and "three gorges cultural heritage digital" of Nanjing University research projects, and so on. Xin Li, a professor of the Zhejiang Normal University, mentioned in a book that Freeform Modeling 3D touch design system and BS Contact 3D virtual browser technology are analyzed. The book is called "the new road to digital protection of intangible cultural heritage". But these studies are conducted for small modeling.

2 THE EXPERIMENT DESIGN SCHEME

The value and significance of dance will be studied in this paper. By combining the Vicon infrared 3D motion capture system to create dance sport collection database adhering to the principle of human physiological mechanics to establish the database, four marked points of the head, five markers of the torso, 14 markers of the upper limbs, 12 markers of the lower limbs, four markers of the pelvic, total 39 mark point, and the space coordinates of the tags are collected, as the research object. Set up the database and use 3D Max and Maya virtual reality environment and role model, the model of bone and binding of the collected data and complete each moving point trajectory to copy and transfer. The next is the model skin and modeling design. Finally, experimental results show the platform. Specific experimental design scheme is described as follows.

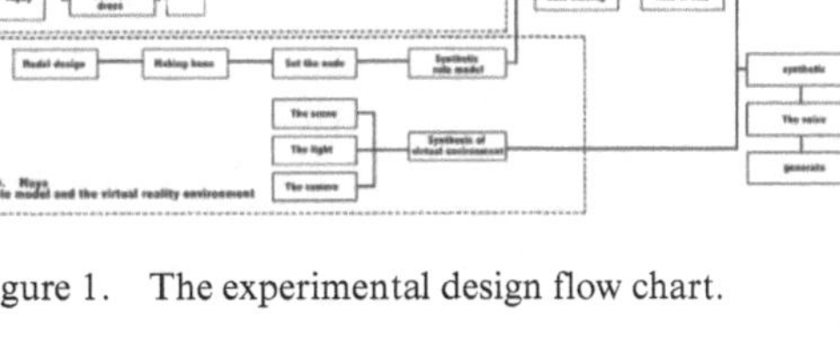

Figure 1. The experimental design flow chart.

2.1 *The dancers selected standards and choreography*

National dance is used as experimental object. Both the factor of regional cultural differences and the idea of performer about folk dance culture historical significance are considered when a dancer is chosen. The dance movements arrangement is used as representative. Because the space to collect accurate data is limited, and the relevant requirements are very demanding.

2.2 *Vicon infrared environment of 3D motion capture system*

2.2.1 *The light*

Generally, the indoor lighting to collect action is not like in the film and television production strict, but also need to make the light weak, and the color moderate. Choose indoor fluorescent lamp or incandescent lamp can conform to the requirements.

2.2.2 *The camera*

Under the condition of good indoor lamplight, 12 cameras can be divided into two layers, six direction, in a 1.5 m × 1.5 m × 2 m of the edges of the cuboid space. The first layer of camera and the second camera stagger the 30 degrees placed respectively. The purpose is to take care of the body to any location of the movement of each landmark data, and try to avoid blind angle and blind spots.

2.2.3 *The space*

Make sure the size of the data acquisition room is enough and equipment placement space reasonable as far as possible. Capacious and comfortable work environment are sought. It won't produce too much redundant space.

2.2.4 *The other*

Environmental temperature and humidity should be as far as possible in order to ensure the normal operation of equipment for the standard. It guarantee the power supply, and ensure that the work environment quiet. And there are no other interference factors.

2.3 *Dance moves to collect*

2.3.1 *Acquisition device connected*

It will be 12 groups of Vicon MX camera mounted to Yuntai. We have to fix it on a tripod and adjust its position. The camera by MX cable connected to MX Giganet, one end of the next RJ45 ethernet cable insert MX Giganet network connection interfaces, the other end is connected to the PC, complete collection devices connected.

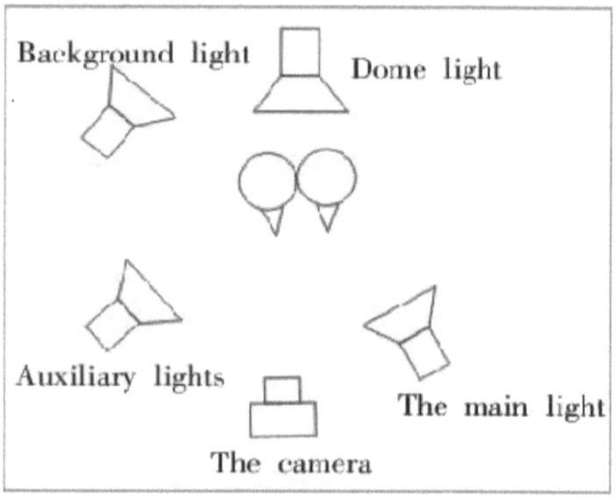

Figure 2. The indoor lighting scheme.

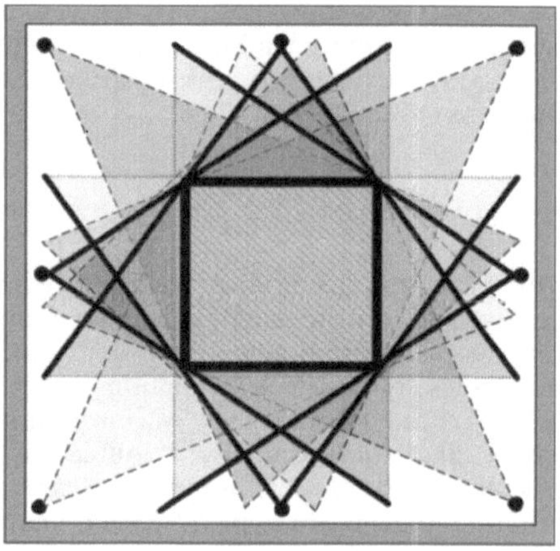

Figure 3. The camera imaging range diagram.

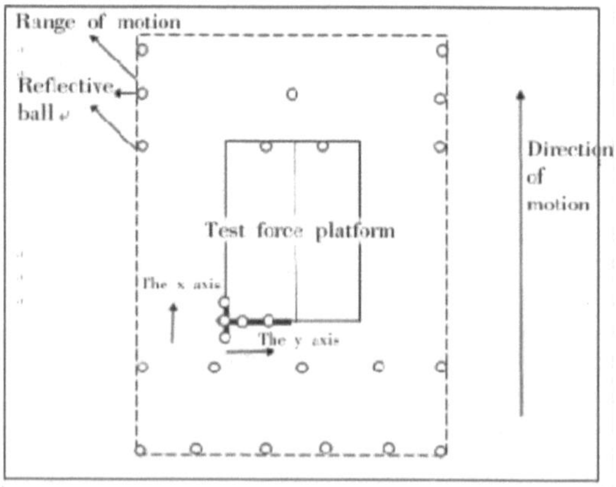

Figure 4. The actor movement space diagram.

Open the power supply, open the MX Giganet switch, make sure all the camera lights flashing blue light, and show the camera system boot up normally.

Then open the PC, install the software as required and open the Nexus program software, in view of the software interface to select the right window, enter the real-time mode. View all the camera connection state; to make sure all cameras are in good connection state after first to adjust the camera settings, is often done in view of the aperture and focusing, make sure the markers in the image is clear, and light saturation; next to remove redundant glitter and calibration system, after everything is ready, we are going to begin data collection.

2.3.2 *Performers prepare*

Performers must first put on special movement data collection tights, and then choose proper specifications of the markers, affixed to the set position of performer's body. Determine the mark point paste, and in the Nexus of software, this can be observed by different cameras, 39 marked points for acquisition, now the work can begin.

2.3.3 *Start collecting*

After everything is ready, establish the data management of libraries in the software, and then click the start button in the software online mode to start gathering.

2.4 *Collect the data extraction*

After completion of data collection, click the end button. We can observe in the software of data management window acquisition of data, and can choose the data types, which is suitable for this study from different data types, finally export data backup.

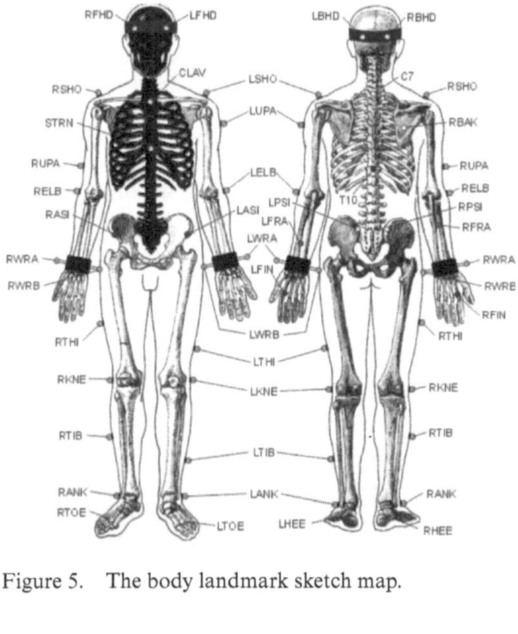

Figure 5. The body landmark sketch map.

2.5 *Role model creation and virtual environment design*

2.5.1 *Role model*

According to the principle of biology, and the human body proportion, Maya software is used to create a skeleton model with mobility. The model is selected according to the actual situation of dance movements.

2.5.2 *Virtual environment light setting*

The environment lighting and the characters in 3D Max and Maya software and, mainly adopt the method of combination of the main and auxiliary light decorate. The main lights are placed in the upper left in 45 degrees. Downward projection and two auxiliary lighting set the left and right of the characters, the activity level direction. The left side of the auxiliary light is higher than the height of the characters' shoulder. On the right side of the

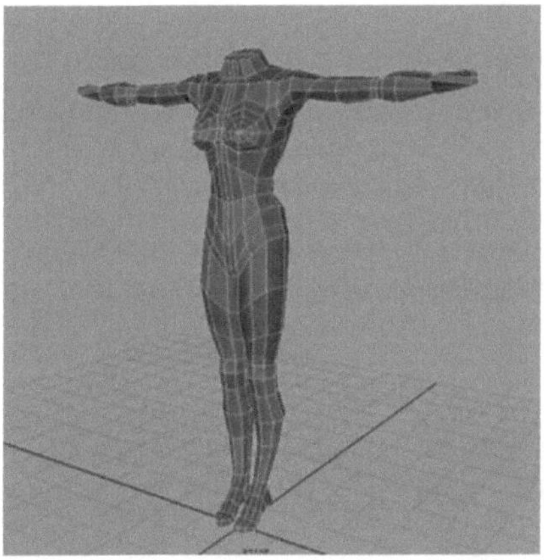

Figure 6. The 3D models.

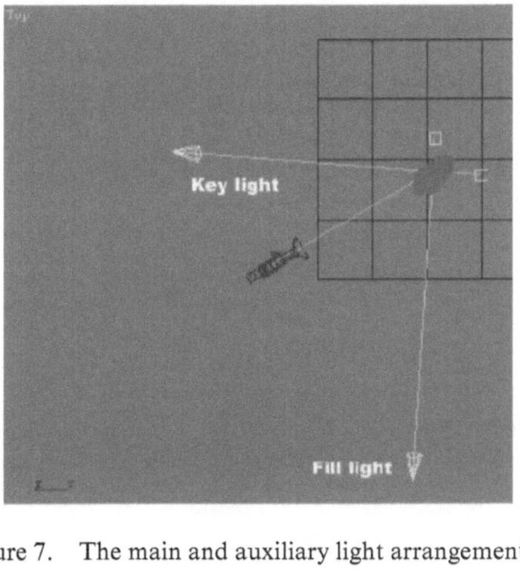

Figure 7. The main and auxiliary light arrangement.

height of the auxiliary light is below the waist of the characters. And auxiliary brightness is lower than the main light brightness. Then light intensity was set to 1 to 1.5.

2.5.3 *Binding bone trajectory*

According to the principle of human anatomy, from more than 200 movable joints in the human body skeleton for choice, the most representative national dance movements of key points are chosen as the research object.

Nexus motion capture software after the data is stored in the format of the CSM. This document is used to record all the ball Marker marks on space at different times in the space coordinates of ASCII files. Then the CSM file was loaded in the Maya software to generate intermediate data. You need to determine its degree of verisimilitude, and optimize the key frame interpolation technology, through repeated changes, the resulting new CSM file, complete the experiment.

2.6 *The role of skin and post processing*

Take Maya software as basis, the skeleton mode for the operation of the skin is established. The adjustment of Skin skeletal hierarchy, the adjustment of the Binding way and the maximum number of influence and other parameters. Skin are made to look natural gentle and has the very good smooth degree.

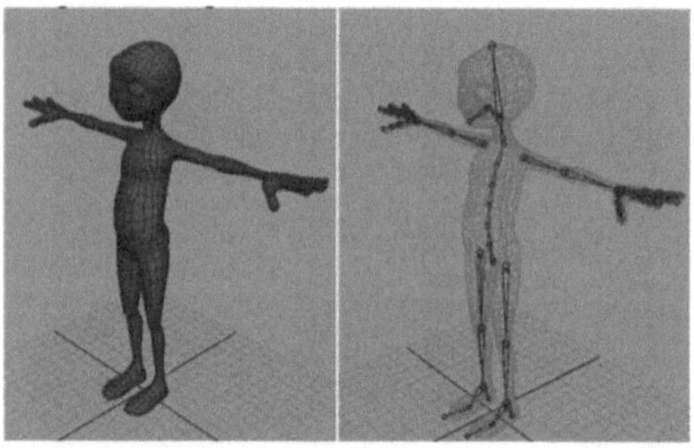

Figure 8. The model bone.

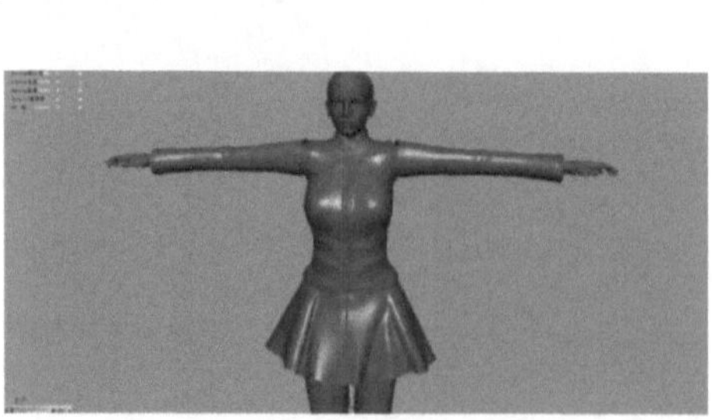

Figure 9. The model skin.

2.7 *Background music*

This part of the background music will be divided into two levels. The first level is the music of dance. The music must conform to the traditional culture, and make the music combine the dance moves which has choreographed. The second level is the monologue in video files, whose content is about national culture related introduction. The purpose is to let the viewer see not only the vivid folk dancing, but also make the national characteristics and national culture essence and the historical significance combine with each other echo.

3 THE EFFECT OF DISPLAY AND ANALYSIS

After the realization of finished products demo and operation on different equipment, error correction and optimization need to be repeated. Finally, it can achieve good effect through continuous improvement in digital protection, education and cultural propaganda, and so on. We expected that it will be successful in online games, film and television media, virtual reality, and other areas of the industry.

4 CONCLUSION

This research is a combination of theory and practice, proactive and scalable innovative research. This is a study of intangible cultural heritage protection and cultural industry. New concepts can be promoted and new theories are put forward. It is of great practical significance. And there is still a big challenge.

This paper is expected to achieve the constructtion of the virtual reality environment, combined with 3D model to create national dance moves, permoment, authenticity of reduction, and record the path of motion capture and track data. According to the data body mechanics analysis, three dimensional motion records marked points coordinates, tried to save, and set up perfect representative of the national dance movement data of database, through the film and television media platform and digital reduction technology at a higher level of science and technology research in the field of application. After thousands of year's people still can feel the multi-ethnic culture lasting appeal.

It is hoped that the Vicon infrared 3D motion capture technology can be used to create model experimental study of the dance and make due contribution for our country education, culture, life, other industries to provide concrete practices, and the minority cultural heritage.

ACKNOWLEDGEMENTS

This research is supported by Key Laboratory of National language intelligent processing, Gansu Province (Northwest University for Nationalities).

REFERENCES

Aggarwal J.K., and Cai Q. Human Motion Analysis: A Review. Computer Vision and Image Understanding, 73(3), 1999, 441–454.

A Survey of Computer Vision-Based Human Motion Capture. Computer Vision and Image Understanding 81, 2001, 231–268.

A survey of advances in vision-based human motion capture and analysis. Computer Vision and Image Understanding 104, 2006, 90–126.

Nakaoka S., Nakazawa A., Yokoi K., and Ikeuchi K. Leg Motion Primitives for a Dancing Humanoid Robot [C]. In Proc of IEEE International Conference on Robotics and Automation, 2003, 610–615.

Nagata N., Okumoto K., Iwai D., Toro F., and Inokuchi S. Analysis and synthesis of Latin dance using motion capture data [C]. Lect Notes Comput Sci, 3333, 2004, 39–44.

Engineering Management and Industrial Engineering – Leung (Ed.)
© 2015 Taylor & Francis Group, London, ISBN 978-1-138-02773-2

The musical instrument classification algorithm based on phase space reconstruction

X.M. Wang, D.F. Zhuo, Y.N. Guo, Q.J. Zhang & X.L. Guo
Taiyuan University of Science and Technology, Taiyuan, Shanxi, China

ABSTRACT: In this paper, audio time series produced by a variety of instruments are processed based on the phase space reconstruction theory and principal component analysis theory. The differences of each instrument in phase space are described by using the probability density function. In addition, a method of musical instrument classification is developed by using flexible neural trees to combine the parameters of probability density function with other features. The FNT model can solve highly structured dependent problem of the artificial neural network and also has a higher recognition rate. We compared the proposed method with the BP neural network and the support vector machine. The experiment indicates that the categorizer of the proposed method has higher average accuracy rate and lower RMSE value.

1 INTRODUCTION

In recent years, data analysis and classification with music have become a hot topic at home and abroad. In the past decade, a variety of different analytical methods have been mentioned and applied in the music signal. The MFCC features, characteristics and time-frequency spectral characteristics of the instrument in the performance categories were evaluated by Eronen (Eronen 2001). The result showed that MFCC got good results in the classification of the instrument family. MPEG-7 is essential in the music signal feature extraction as the standardization frame for the purpose of audio feature extraction and description (Lindsay et al. 2001). It is especially better for similar music signal. However, existing research results show that some classification accuracy rate of the audio signal is generally low. It shows that the differences between signals are not well depicted using the above audio features.

A wide range of classifiers in the audio category has also been researched in depth. Achieving two or more than two instrument separation by using Gauss modeling single channel speech separation method was realized by Pablo et al. (Sprechmann et al. 2012). Agostini and others (Agostini et al. 2003), who extracted the spectral characteristics of the music, combined the characteristics of the music on the sound SVM classification. Athanasia and Petros (Zlatintsi et al. 2012) used AM and FM modulation characteristics to analyze and recognize musical signals and the accuracy rate can reach 70%. Brown (Brown et al. 2001) and others used MFCC, constant q transform and autocorrelation function to extract features. The recognition accuracy rate can reach 79%–84% when it combines with Bayesian decision classification.

An audio signal has a time series of typical nonlinear characteristics. System dynamics feature of robustness can be extracted from the input of the time series by nonlinear theory. Phase Space Reconstruction (Yang et al. 2011) (PSR) is a common method for nonlinear signal analysis. A new classification has been proposed based on instruments in this paper. The characteristics of the various instruments are analyzed with phase space reconstruction method and the differences between the different instruments are characterized with the probability density function. The probability density function of the parameter values is combined with other timbre characteristic quantities. Flexible neural tree classifier is adopted as the classification. Experimental results showed that the algorithm complexity proposed in this paper is reduced and the accuracy of the classification of different instruments is enhanced to some extent.

2 THEORY

Recognition model of musical instruments is shown in Figure 1. The sample signal delays time and embedding dimension should be determined by the phase space reconstruction part. And mutual information method and false nearest point method should be used respectively. Then, the Principal Component Analysis (PCA) mainly removed the redundant information and dimension reduction instrument on musical signal was reconstructed,

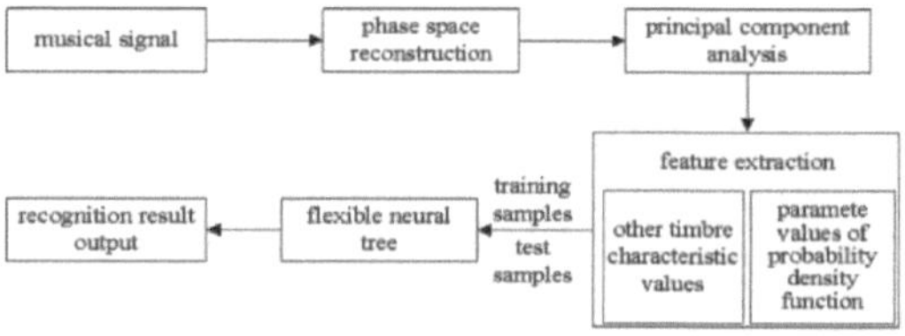

Figure 1. The identify model of the instrument.

which is displayed in three-dimensional space. The parameter value of the Probability Density Function (PDF) is mainly extracted in the feature extraction section. In order to achieve a better recognition result, the time-domain features and MPEG-7 features, which constitute a collection of features, are extracted as timbre characteristics. The classifier that has a high recognition rate uses the Flexible Neural Tree (FNT).

2.1 *Phase space reconstruction*

The phase space of a system with a driving force could be reconstructed from the one-dimensional time series in the topological structure. Where $x = (x_1, x_2, \ldots, x_k)^T$ (Rui et al. 2012) is audio dimension of time-sequence and τ is delay time. There is a m-dimensional embedded space Y. Map x to the embedded space. The reconstruction of the phase space vector is as follows:

$$Y = (y_1, y_2, \ldots, y_N) = \begin{pmatrix} x(1) & x(2) & \cdots & x(N) \\ x(1+\tau) & x(2+\tau) & \cdots & x(N+\tau) \\ \vdots & \vdots & & \vdots \\ x(1+(m-1)\tau) & x(2+(m-1)\tau) & \cdots & x(N+(m-1)\tau) \end{pmatrix} \tag{1}$$

where K is the length of audio time sequence x; the phase space vector number is $N = K - (m - 1)\tau$. It shows that the reconstruction of selection and determination of embedding dimension and delay time are key factors in phase space.

2.1.1 *The determination of delay time* τ

It plays a vital role in performing data model and selecting the appropriate τ to make the coordinates neither linear nor completely independent and make to some extent. The optimal delay time τ is determined by using the average mutual information method, for example equation (2).

$$I(\tau) = \sum_{n=1}^{N} P(x_n, x_{n+\tau}) \log_2 \frac{P(x_n, x_{n+\tau})}{P(x_n)P(x_{n+\tau})} \tag{2}$$

where $P(x_n)$ and $P(x_n, x_{n+\tau})$ are the probabilities. For the audio time series $P(x_n)$, it can choose the τ as the best delay time which makes $I(\tau)$ to achieve the local minimum first.

2.1.2 *The determination of embedding dimension m*

The embedding dimension m could be solved by selecting False Nearest Neighbor method (FNN). The starting point of FNN is: with the increase of embedding dimension m, the movement trajectories would be gradually opened and false nearest neighbor would be gradually removed in the procedure of reconstructing phase space so that the trajectory of the movement could be restored.

Whether each vector neighbors in m-dimensional space are false or not to calculate false nearest neighbor point ratio is to be judged. When the false nearest neighbor point is less than the ratio of 5%, it can be considered that trajectories in phase space are completely open. At the same time m is the best embedding dimension.

2.2 *Principal component analysis*

Principal component analysis is a method of mathematical statistics (Huang et al. 2013), which is based on K-L transform, and is mainly used to reduce the correlation between the various elements of random variables and highlight the differences. After K-L transform, musical signal can be reconstructed phase space by turning a high-dimensional space transformed into a low-dimensional space representation to preserve the original data information and achieve effective classification.

After PCA treatment, this paper selects the first three principal components to portray each instrument samples on the distribution of three-dimensional plane. The phase space structure is different after it is reconstructed for different kinds of musical signals.

2.3 *Feature extraction*

To describe the differences between various types of instrument family, the paper extracts the PDF and other features as feature musical tone signal collection. PDF portrayed the intensity of the phase space trajectories in the vicinity of each of the regional distribution.

2.4 *Flexible neural tree*

The FNT can solve the problem of the traditional high dependence on artificial neural network structure (Guo et al. 2012). The process model is generated as follows: first, it randomly generated a flexible neural tree and a corresponding set of parameters. After the optimum tree structure is found, its parameters and structure are optimized. Repeat cycles until a satisfactory solution is found (Chen et al. 2006).

The paper uses probability of evolutionary algorithms enhanced program to optimize the neural tree structure and uses the simulated annealing algorithm to optimize neural tree variable parameter. The fitness function makes the flexible neural tree map into real scalar fitness value. MSE or RMSE is selected as the fitness function usually.

3 SIMULATION AND ANALYSIS

Four common musical signals were selected in the experiment: French horn, piano, trumpet and flute. The sample is taken from a CD of piano solo sequence from Western and the rest of the experimental data are taken from UIOWA University's database. All the samples in the same condition are recorded. The sample data length is 2000 for each instrument signals, which selected 10 groups measured sample data. The first 8 groups were the training samples and the other two groups were the test samples. A vague classification model is trained according to the training sample eigenvalues and the initial structure in FNT and then the test sample signal is input into the trained FNT in performance validation. Recognition rate of the i class of the instrument is as follows:

$$R_i = \frac{n}{N} \times 100\% \tag{3}$$

where i is the sample class. n is the identified number of musical instruments samples to this class. N is the total number of the i-type instruments samples.

GUI is designed by using Matlab software, which is shown in Figure 2. Click "Load data" to import a mono instrument data, then data PSR will be reconstructed. Click the "Plot data" to output data. Next, click on the figure "Reduce dim", PCA will perform a data operation and the data will be put

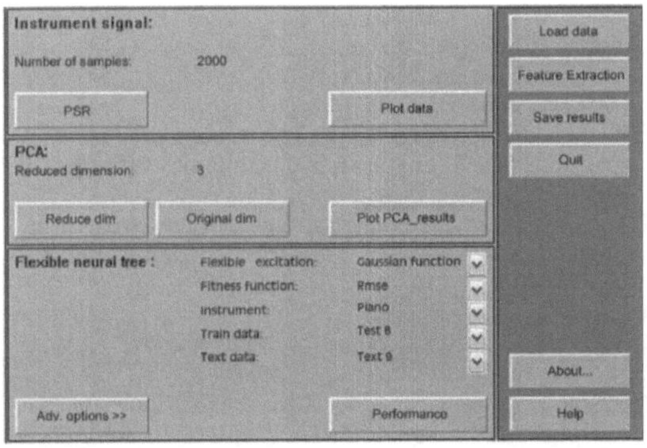

Figure 2. Matlab GUI achieve.

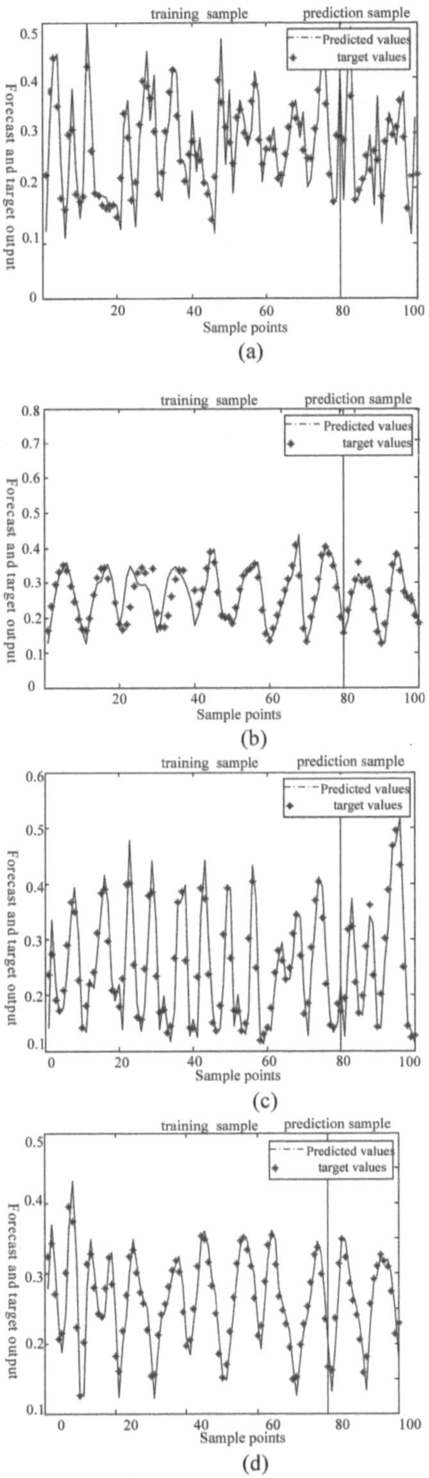

Figure 3. (a) Horn, (b) piano (c) trumpet and (d) flute the predicted output and the target output of training and testing samples.

Table 1. Instrument identification.

	Recognition rate			RMSE		
	BPNN	SVM	FNT	BPNN	SVM	FNT
Horn	64.0%	82.4%	84.9%	0.3158	0.0951	0.0887
Piano	77.3%	100%	98.7%	0.2746	0.0347	0.0456
Trumpet	66.7%	85.1%	84.4%	0.4173	0.0705	0.0720
Flute	57.3%	78.7%	88.9%	0.5421	0.1052	0.0530
Average value	66.3%	86.6%	89.2%	0.3875	0.0764	0.0648

down to three dimensions and be displayed in a three-dimensional figure. Then click on the figure "Feature Extraction" to extract the signal processing. The samples—the instrument 8 are trained by using the calculated characteristics, and the remaining two samples are tested one by one. Finally, click "Performance" to output results (Figure 3).

Figure 3. The predictive output and the target output of training samples and testing samples among (a) horn (b) piano (c) trumpet (d) flute.

RMSE value is set as a threshold value to determine whether the classification result is correct. When the RMSE is greater than the set value of the output threshold value, it is regarded as the identification of other instruments. On the contrary, it is considered correctly identified. FNT choose interval in which population size is 100 and the neural tree structure optimization iterations is 50. Select RMSE = 0.1 as the threshold and predict output in Figure 3 of the four instruments. You will find it is consistent with the target output test sample. Though there is some deviation, it is very close in general.

Using single classification model is not a good way to illustrate the performance of this algorithm. For that reason we choose BP neural network and supported vector machine classifier to compare with it. Table 1 shows the experimental results of several classification models. According to the results, BP neural network recognition rate is generally lower while RMSE values are higher, which is influenced by BP related to the neural network. SVM recognition rate on the piano, though up to 100%, has a slightly lower RMSE values, but the proposed algorithm has a higher average classification accuracy rate. RMSE values in the paper are lower.

4 CONCLUSION

This paper is based on the phase space reconstruction theory and principal component. By using the probability density function to describe the difference between the different instruments in the phase space and using the flexible neural tree as a classifier, it solves the problem of high degree of dependence on artificial neural network structure effectively. Data samples collected by the operation results shows that the algorithm is compared with the BP neural network and support vector machine classifier method has a lower RMSE values and higher average classification accuracy rate. Studies show that the instrument signal classification algorithm is very effective. Finally, how to improve the recognition rate of different instruments profitably is the focus of the future research.

ACKNOWLEDGEMENTS

This work was supported by the 2013 National Science Foundation for Young Scientists of China [number 61301250], the 2012 International science and technology cooperation project of ShanXi Science and Technology Department [number 2012081036] and 2011 research project of Department of Human Resources and Social Security of Shanxi Province [number 20121030].

REFERENCES

Agostini, G. & Longari, M. 2003. Musical instrument timbres classification with spectral features. *EURASIP Journal on Applied Signal Processing*: 5–14.

Brown, J.C. & Houix, O. 2001. Feature dependence in the automatic identification of musical woodwind instruments. *The Journal of the Acoustical Society of America* 109: 1064.

Chen, Y. & Abraham, A. 2006. Feature selection and classification using flexible neural tree. *Neurocomputing* 70(1): 305–313.

Eronen, A. 2001. Comparison of features for musical instrument recognition. *Applications of Signal Processing to Audio and Acoustics, IEEE Workshop on the IEEE*: 19–22.

Guo, Y.N. & Wang, Q.H. 2012. The gesture recognition model based on flexible neural tree and surface EMG signals. *Computer Applications and Software* 29(4): 170–173.

Huang, S.H. & Zhuo, D.F. 2013. An adaptive single input multiple output blind source separation methods. *Computer Applications and Software* 30(8): 170–173.

Lindsay, A.T. & Herre, J. 2001. MPEG-7 and MPEG-7 Audio' An Overview. *Journal of the Audio Engineering Society* 49(7/8): 589–594.

Rui, R. & Bao, C.C. 2012. Instrument classification method based on nonlinear dynamics. Journal of Electronics 4(7): 1481–1488.

Takens, F. 1981. Detecting strange attractors in turbulence. *Lecture Notes in Mathematics, Springer, Berlin* 898: 366–381.

Yang, D. & Ren, W.X. 2011. Non stationary evaluation of vibration signal based on the recursive analysis. *Vibration and Shock* 30(12): 39–43.

Zlatintsi, A. & Maragos, P. 2012. AM-FM modulation features for music instrument signal analysis and recognition. *Signal Processing Conference (EUSIPCO), 2012 Proceedings of the 20th European. IEEE*: 2035–2039.

Engineering Management and Industrial Engineering – Leung (Ed.)
© 2015 Taylor & Francis Group, London, ISBN 978-1-138-02773-2

Analyzing statistical properties of enterprise information systems users' access behavior

J.J. Ren, N.X. Wang & S.L. Ge
Jiangsu University of Science and Technology, Zhenjiang, Jiangsu, China

ABSTRACT: In order to explore the time statistical rule of user's access behavior in information systems, this paper investigates interval time distribution, activity, burstiness and memory of user's access to information systems, and human dynamics was used to analyze 226795 access records of 30 users in a ship building manufacturing enterprise. The results show that interval time distribution of individual user's access to information systems has serious fat tail and follows a power-law distribution. The results also indicate that the behaviors of these users' access to information systems show strong burstiness and weak memory. And there is a monotonous relation between the activity of individuals and power law exponent of interval time distribution.

1 INTRODUCTION

Research and analysis of user access behavior have become an essential problem which they must do everyday among the large portal websites, social networking and e-commerce websites. From the investigative report, they will know of the customers' area differentiation of the original regions, hobbies and behavior characteristics, and so on. Then they could improve the quality of service and the operation maintenance of websites to cater to the demand of customers and attract more customers. But there is no research on user's behavior in the field of management information systems of enterprises. Meanwhile, new information technologies are in a constant appearance, especially the emergence of cloud computing. The information system on cloud is becoming inevitable, (Petter et al. 2012). The lack of understanding the characteristics of user behavior in the information systems seriously hinders the development of management information system.

Through a quantitative statistical analysis of a great many human behavior events, digging out the statistical law and characteristics of human behavior has become a hot research issue now. Scholars used the method to make a lot of achievements through empirical research in many areas, such as mail communication (Barabási 2005, Oliveira & Barabási 2005), e-mail patterns (Johansen 2004), web browsing (Vázquez et al 2006), rating of movies (Zhou 2008 et al), mobile communication (Candia et al 2008, Jiang et al 2013), financial activity (Scalas 2006), blog forum (Guo 2010), face-to-face interaction networks (Starnini et al 2013). To do a human behavior research involving the work pattern is relatively rare in the present. Only paper (Scalas 2006) discussed the trade in security market and paper (Gao 2010) does a statistical analysis on the purchase orders of a world top 500 enterprise. However, there is no study on human behavior in enterprise information systems.

Therefore, this paper analyzes user's access log of an enterprise management information system and gives a quantitative analysis on user's access behavior in information systems by exploring time interval, activity, burstiness and memory of user's access distribution. By studying the statistical properties of information system user's access behavior, on one hand it can effectively predict the information system user's access behavior and then it will provide the theoretical basis for the optimization and upgrading of their information systems, and it can help to expand the research situation of user's access behavior on the other.

2 RESEARCH METHODS

2.1 *Human dynamics*

Scholars have been studying and trying to explore how to understand complex human behavior profoundly. However, due to the backward means of data records and lack of modern statistical tools and methods, scholars often assume that human behavior is based on Poisson process, which is a stationary random process to research the characteristic of human behavior. This model predicts that the interval time between two consecutive actions follows an exponential distribution:

$$P(\tau) \sim \lambda e^{-\lambda\tau} \tag{1}$$

Poisson processes can be observed as an exponential distribution. That is to say, human activity patterns are random and smooth, and the interval time between two consecutive actions is a stationary random process and a long time interval is very rare.

With the development of information technology and the progress of modern statistical tools, the means of data recording and analysis for human behavior are becoming more and more efficient and convenient. Then it is becoming possible to take advantage of big data to analyze human behavior. In 2005, Barabási (Barabási 2005) published a paper titled: "The origin of bursts and heavy tails in human dynamics" in Nature. This signifies the formal proposition of human dynamics. By analyzing the time interval of mail sending and replying process, the paper reveals that the human behavior deviates from the Poisson distribution, characterized by events occurring rapidly separated by long periods of inactivity. The interval time distribution between two consecutive behavior has serious fat tail, so power-law distribution function is better to be used:

$$P(\tau) \sim \tau^{-\alpha} \tag{2}$$

Human dynamics is a new crossing science, which digs up the statistical rules through the quantitative statistics of many human activities and then builds models to explore the production mechanism and possible dynamics impact on these rules. A large number of empirical studies reveal that human activities follow non-Poisson statistics, characterized by events occurring rapidly separated by long periods of inactivity.

2.2 *Characteristic indexes of time*

In human behavior dynamics, the characteristic indexes of time in human behavior include interval time, activity, burstiness, memory, and so on.

1. The interval time refers to time interval between two consecutive actions. For example, A accesses to an information system at '2012\01\01 08:08:08', and B accesses to the information system at '2012\01\01 08:08:09', so the interval time is 1 second. If the number of all users accessing to the information system is *K*, so the number of time intervals is *k*–1. The unit of time is second in this paper.
2. Burstiness is a physical quantity that is vaguely corresponding to significantly enhanced activity levels over short periods of time followed by long periods of inactivity. The formula from the Goh and Barabási (Goh & Barabási 2008) is used in this paper. The burstiness parameter is defined as follows:

$$B = \frac{\sigma_\tau - m_\tau}{\sigma_\tau + m_\tau} \tag{3}$$

where m_τ and σ_τ are the mean and the standard deviation of the time interval distribution. For the exponential distribution, the standard deviation and mean are equal, so the burstiness parameter is 0; For the serious fat tail distribution, the standard deviation is greater than mean, and the burstiness parameter is close to 1.

3. Memory is defined as the correlation coefficient of all consecutive interevent time values. When a long (short) interevent time tends to be followed by a long (short) on, the time series of human behaviors are regarded as memory. The interval time for all behaviors is arranged according to time sequence. The first $n_\sigma - 1$ time interval is as sequence 1, the after $n_\sigma - 1$ time interval constitute sequence 2, m_1 and m_2 are the mean of sequence 1 and sequence 2, σ_1 and σ_2 are the standard deviation of sequence 1 and sequence 2, respectively. This paper uses the formula from the Goh and Barabasi (Goh & Barabási 2008) which is defined as follows:

$$M = \frac{1}{n_\tau - 1} \sum_{i=1}^{n_\tau - 1} \frac{(\tau_i - m_1)(\tau_{i+1} - m_2)}{\sigma_1 \sigma_2} \tag{4}$$

4. Activity is the frequency of events of an individual. The article (Zhou et al 2008) puts forward the conception of activity and defines activity as the frequency between the first and the last event of an individual. N_i is the total number of records of *i*, and $P(\tau) \sim \tau^{-\alpha}$ is the time between the first and the last event of *i*. We measure the activity as follows:

$$f(x) = ae^{\alpha x}\left[M(x+b)^{\beta} + N(x+c)^{\gamma}\right] \tag{5}$$

3 DATA COLLECTION AND ANALYSIS

3.1 *The choice of sample companies*

In this paper we selected Company X, the domestic famous ship building enterprise with more than 10000 employees and workers, as the research object for revealing the time characteristics of user's access to the enterprise information system. Company X has always attached great importance to the construction of the enterprise information system and invested heavily to acquire the ERP system of the Oracle in 1999. Due to the influence of the global economic crisis, the shipping market was in downturn, then Company X implemented the cost management system in September 2011 in order to enhance its enterprise competitiveness. The cost of management system included the quotation cost, target cost, cost accounting, cost analysis and financial management, and so on. There are 314 registered users, daily 3 accesses per person and daily 3 function functions accessed per person in this system. The number and volume of the visitors represent the information-based application situation of the Company X.

3.2 *Data acquisition*

From the database log files of the cost management system in Company X, we extract the operations of user access to the cost management system from September 14, 2011 to June 14, 2013 and delete the records without login names and the wrong ones. In total we acquire 348,122 records from 314 system users. Then, we pick out 30 active users who have the most access volumes to research the statistical characteristics of the individual user access behaviors. Each access record contains nine elements (ID, UserID, LoginID, LoginName, LoginTime, LogoutTime, MachineID, MachineName, and ModuleName), and we also regard them as serial number, user number, login number, login name, login time, logout time, machine number, machine name, and function module.

3.3 *Data analysis*

We mainly use the SQL server, Matlab and Excel are used to process and analyze the sample data in the process of data acquisition and analysis in this paper. First, based on the database log files of Company X, we obtained the login time of all users access behavior and select the access volumes, activity, burstiness and memory from the 30 active users which are shown in Table 1. Then we present the

Table 1. 30 active users.

Rank	Access amount	Activity	Power index	Burstiness	Memory
1	39943	71.59	1.3730	0.7407	2.0030e-021
2	12210	18.73	0.9622	0.7440	1.3543e-022
3	11270	20.72	0.9395	0.6464	1.6426e-021
4	10894	19.63	0.9246	0.6687	2.2910e-022
5	10818	19.71	0.8644	0.6994	3.5174e-022
6	9516	17.40	0.7798	0.6588	2.7248e-022
7	9211	16.02	0.7653	0.6372	3.7562e-022
8	8563	15.46	0.8334	0.6036	1.3773e-021
9	7650	13.82	1.0428	0.6084	1.0539e-021
10	7471	13.39	0.8558	0.5954	7.0301e-022
11	7329	13.66	0.6557	0.5744	7.6276e-022
12	6267	10.86	0.6522	0.5882	3.9077e-022
13	5953	12.22	0.6401	0.8452	1.4575e-022
14	5863	10.53	0.6373	0.5885	3.1696e-022
15	5662	10.35	0.6067	0.5214	8.3925e-022
16	5578	10.12	0.6517	0.5929	6.2538e-022
17	5505	10.12	0.5801	0.5604	7.1235e-022
18	5416	9.83	0.7980	0.8999	2.4627e-021
19	4995	8.98	0.6422	0.5598	7.3829e-022
20	4779	8.67	0.5316	0.4705	3.4293e-022
21	4772	8.58	0.6787	0.5578	9.6334e-023
22	4573	8.19	0.6233	0.5969	7.0246e-023
23	4411	13.57	0.9022	0.5833	1.9624e-022
24	4407	7.91	0.5326	0.4228	3.5458e-022
25	4326	10.40	0.7032	0.5992	8.8835e-022
26	4176	7.51	0.5224	0.4566	8.1456e-022
27	4142	7.66	0.5136	0.5531	2.3038e-022
28	3859	7.13	0.6339	0.6617	2.5222e-022
29	3794	6.82	0.5101	0.5387	3.8674e-022
30	3478	6.12	0.7364	0.7809	1.3442e-022

interval time distribution in the double logarithmic coordinates by Matlab. At last, the main data are fitting to get the power exponents and the probability density functions by using the least square method to discuss the time statistical properties of the user access behaviors of information system.

4 ANALYSIS OF RESULTS

In order to explore the time statistical rule of user's access behavior in enterprise information system, this paper studies 30 users who ranked in the top in the database log and use Matlab software to get the interval time distribution of these users' access to the information system in log-log plot. Then the least squares curve fitting method is used to obtain the best fitting curve on the main data. The results show that the statistic of individual behavior is not randomly distributed in time, characterized by events occurring rapidly separated by long periods of inactivity. The results also indicate that interval time distribution of individual user's access to information systems has serious fat tail and follows a power-law distribution $P(\tau) \sim \tau^{-\alpha}$. The exponents of each individual user are different.

For example, the SA has 12210 visitors to the cost of management information system from September 14, 2011 to June 14, 2013. The number is at the second place of all users' visits. The interval time distribution between two consecutive rating of access visits is shown in Figure 1 (the blue points is the original data of interval time for individual user's visit to the cost management information system, the red line is the fitting curve, and the power exponent is 0.9622). Figure 2 and Figure 3 show

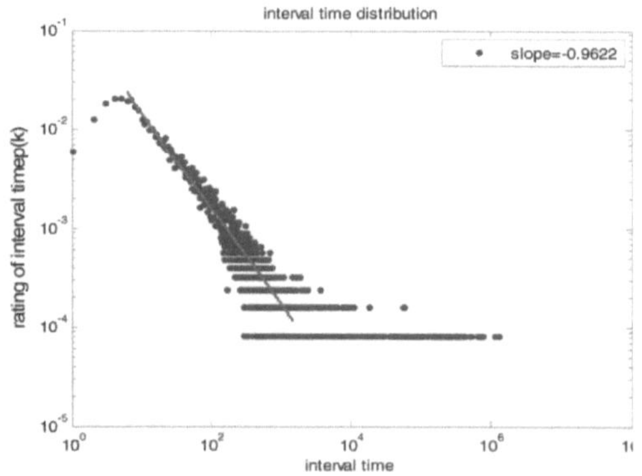

Power index α=0.9622

Fit Index: SSE=4.463e-005

R-square=0.9752

RMSE=0.0003024

Figure 1. Distribution of the access time for SA.

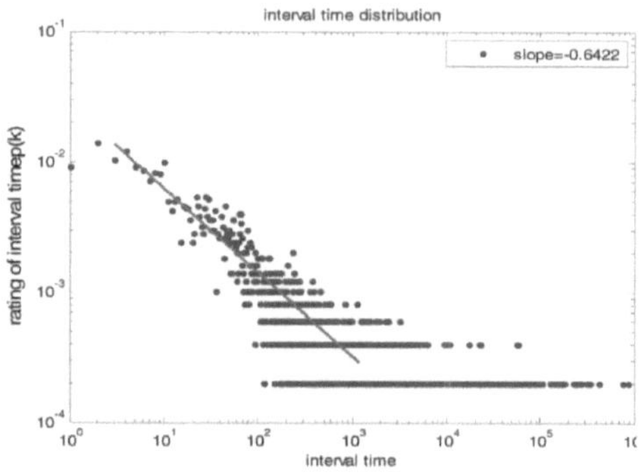

Power index α=0.6422

Fit Index: SSE=0.0001116

R-square= 0.8625

RMSE=0.0006172

Figure 2. Distribution of the access time for user 19.

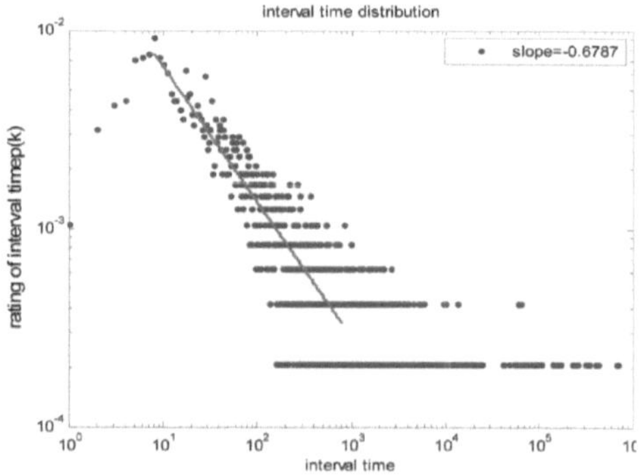

Power index α=0.6787

Fit Index: SSE=6.717e-005

R-square=0.8377

RMSE=0.0004821

Figure 3. Distribution of the access time for user 21.

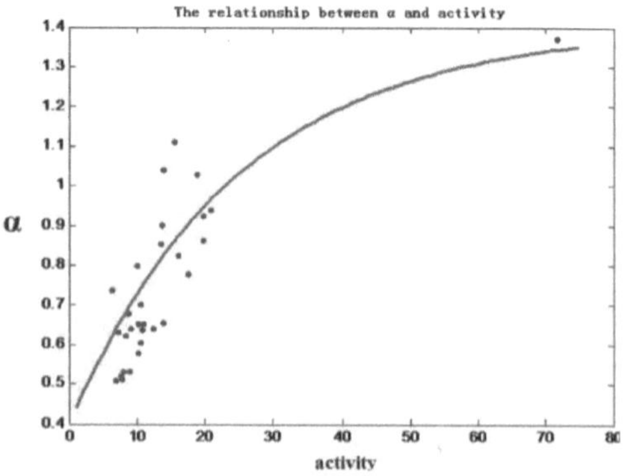

Figure 4. The relationship between α and activity.

the interval time distribution of randomly selected users who ranked 19 and 21 respectively.

At the same time, human behavior is highly complex. The characteristic of user's behavior is not only reflected in the distribution of the interval time. In order to excavate the time characteristics of users' access behavior in information system, in this paper, the burstiness, memory and activity of each user's interval time between two consecutive rating of access are counted. The study has found the following points:

1. The behaviors of these users' access to the cost of management information system show strong burstiness and weak memory. Compared with other studies of burstiness value in complex system, the burtiness value of an individual user's access to information system is far large. It shows that interval time distribution of individual user's access to information systems has more serious fat tail.
2. We can learn, from the experiences of daily life, a daily online shopper compared with an occasionally shopper, and the activity is more frequent. Therefore, there may be an influential relation between the activity of individuals and power law exponent of their interval time distribution. The result in this paper shows that the user with lower activity has smaller power law exponent, that is to say, there is giving a longer average interval time. The power law exponent will increase along with the increase of activity. But the power law exponent will not increase largely, and it will remain stable when it is up to 1.5 (see Fig. 4). For example, the number of the user's visit to information system is up to 72, but the power law exponent is 1.3.

5 CONCLUSION AND DISCUSSION

In order to explore the time statistical rule of user's access behavior to information system, empirical statistics and analysis at 30 active users show that in this paper:

1. Interval time distribution of individual user in information systems has serious fat tail and follows a power-law distribution. The power law exponents of interval time distribution of each user are different.
2. Behaviors of these users' access to information system show strong burstiness and weak memory.
3. There is a monotonous relation between the activity of individuals and power law exponent of interval time distribution.

The results of this paper demonstrated a highly complexity of human behaviors. There is a deeper meaning in research rule of user's behavior in enterprise information system. And it provides certain reference value for us to understand human dynamics more profoundly. At the same time, there are still some problems worthy for further research: the analysis of the frequency of user's access to information system; the difference of user's access to each function module. Also building a dynamics model to predict the user's behavior in information system is an urgent problem to be solved for us now.

ACKNOWLEDGEMENTS

Supported by Project of National Natural Science Foundation of China (Grant No. 71331003, 71271104, and 71101065) and Universities in Jiangsu Province plan to graduate research and innovation (1254051404).

REFRERENCES

Barabási A L. The origin of bursts and heavy tails in human dynamics [J]. Nature, 2005, 435: 207–211.

Candia J, Gonzalez M C, Wang P, et al. Uncovering individual and collective human dynamics from mobile phone records [J]. J Phys A: Math Theor, 2008, 41(22): 224015.

Gao Lei, Guo Jin Li, Fan Chao, et al. Individual and group dynamics in purchasing activity [DB/OL]. [2010–11–10]. http://arxiv.org/abs/1010.3815v2.

Goh K I, Barabási A L. Burstiness and memory in complex systems[J]. Europhys Lett, 2008, 81(4): 48002.

Guo Jin-li. Weblog patterns and modeling human dynamics with decaying interest. Computer Science. 2010, 1008: 0042v3.

Jiang Z-Q, Xie W-J, Li M-X, et al. Calling patterns in human communication dynamics[J]. Proceedings of the National Academy of Sciences, 2013, 110(5): 1600–1605.

Johansen A. Probing human response times[J]. Physica A, 2004, 338(1): 286–291.

Oliveira J G, Barabási A L, Human dynamics: Darwin and einstein correspondence patterns[J]. Nature, 2005, 437(27): 1251–1251.

Petter S, DeLone W, McLean E R. The Past, Present, and Future of "IS Success"[J]. Journal of the Association for Information System, 2012, 13(5): 341–362.

Scalas E, Kaizoji T, Kirchler M, et al. Waiting times between orders and trades in double-auction markets[J]. Physica A, 2006(366): 463–471.

Starnini M, Baronchelli A, Pastor-Satorras R. Modeling human dynamics of face-to-face interaction networks[J]. Physical review letters, 2013, 110(16): 168701.

Vázquez A, Oliveira J G, DezS Z, et al. Modeling bursts and heavy tails in human dynamics[J]. Phys Rev E, 2006, 73(3): 036127.

Zhou T, Kiet H A T, Kim B J, et al. Role of activity in human dynamics[J] Europhys Lett, 2008, 82(2): 28002.

Engineering Management and Industrial Engineering – Leung (Ed.)
 ISBN 978-1-138-02773-2

Integral estimation method of the accelerated performance degradation pseudo failure data

L. Ao
China Aero-Polytechnology Establishment, Beijing, China

ABSTRACT: The traditional reliability assessment methods are based on failure lifetime data. However, for high reliability and long lifetime products, it is very difficult to obtain their failure lifetime data through life test and accelerated life test. From this view, accelerated performance degradation test is used to verify the reliability of the high reliability and long lifetime products, the pseudo failure data of the products is first confirmed by the performance degradation data in this method, then the reliability of the products can be assessed by the obtained pseudo failure data. The two-step method is always used by the traditional reliability assessment based on accelerated performance degradation and pseudo failure lifetime data. What is the characteristic life of the products under different stress levels is confirmed by the pseudo failure data first, then the accelerated model parameters are confirmed by the regression method with the obtained characteristic life, at last, the reliability of the product under normal stress level can be confirmed by the obtained accelerated model. An integral estimation method for the accelerated performance degradation pseudo failure lifetime data is presented in this paper, the method can make the pseudo failure lifetime data from different stress levels an integral one for statistical analysis, and it can reach higher precision than the traditional two-step method do.

1 INTRODUCTION

With the continuous improvement of the level of design and manufacturing, the reliability of product is much higher than before, which leads to a problem for the reliability verification of the products, because the failure data are difficult to obtain during the reliability test, even in the accelerated life test[1].

Accelerated performance degradation test is an important reliability work for the verification of high reliability products, because the pseudo failure data can be obtained from the degradation curve which is confirmed by the performance degradation data, then the reliability can be assessed by the pseudo failure data. The two-step method is always used by the traditional reliability assessment based on accelerated performance degradation pseudo failure lifetime data, but the method can not deal with the problem when there is only one sample under each test stress level. According to this situation, an integral estimation method is presented, the method can make the pseudo failure lifetime data from different stress levels an integral one for statistical analysis, and it can reach higher precision than the traditional two-step method do.

2 INTEGRAL ESTIMATION MODEL OF ACCELERATED PERFORMANCE DEGRADATION PSEUDO FAILURE DATA

In an accelerated performance degradation test, the test sample is divided into several groups, each group is tested under a stress level, which is higher than the normal stress level, and then the reliability of products under normal stress level can be obtained by statistical analysis with the pseudo failure data under other stress levels.

Suppose:

1. The failure mechanisms of products under different stress levels remain constant[2];
2. The lifetime of products under different stress levels obeys the same type of distribution $F(t_j)$, $(j = 0, 1, 2, \dots, m)$;
3. The relation between characteristic lifetime of products and stress level S_j following the accelerated model $f(\eta_j) = a + b\varphi(S_j)$, $(j = 0, 1, 2, \dots, m)$, $f(\eta)$ in the equation is a known function of characteristic lifetime η, a and b are the parameters requiring estimation, and $\varphi(S)$ is a known function of stress S;

4. The performance of some samples under each stress level is degraded, the result after dealing with the performance degradation data is (n_j, t_{jk}, m_j), $(j = 0, 1, 2, \ldots, m, k = 1, 2, \ldots, m_j)$, n_j is the total number of samples under stress level S_j, t_{jk} $(t_{j1} < t_{j2} < \cdots < t_{jm_j})$ is the pseudo failure data under stress level S_j, and m_j is the number of the pseudo failure data under stress level S_j.

The integral estimation model of accelerated performance degradation pseudo failure data is:

$$f(t_{jk}) = g[a + b\varphi(S_j)] + \varepsilon_j \tag{1}$$

where the function f in equation (1) is the same as in the accelerated model, and the function g is known.

3 INTEGRAL ESTIMATION OF ACCELERATED PERFORMANCE DEGRADATION PSEUDO FAILURE DATA FROM NORMAL DISTRIBUTION

Suppose the lifetime of products under different stress levels obeys normal distribution, its distribution function is:

$$F(t_j) = \Phi\left(\frac{t_j - \mu_j}{\sigma_j}\right) \quad t_j > 0 \quad j = 0, 1, 2, \ldots, m \tag{2}$$

μ_j in the equation is the location parameter under stress level S_j, and σ_j is the scale parameter under stress level S_j.

The scale parameter of normal distribution reflects the failure mechanism of products, then the suppose (1) of section II can be expressed as:

$$\sigma_0 = \sigma_1 = \cdots = \sigma_m = \sigma \tag{3}$$

Generally, the accelerated test model for products of which lifetime obeys normal distribution is:

$$\mu_j = a + b\varphi(S_j) \tag{4}$$

if $\varphi_j = \varphi(S_j)$, equation (4) can be transformed into $\mu_j = a + b\varphi_j$.

If the pseudo failure data in the suppose (4) of section II comes from normal distribution, the integral estimation model of accelerated performance degradation pseudo failure data is:

$$T_{jk} = a + b\varphi_j + \sigma u_{jk} + \varepsilon_{jk} \tag{5}$$

u_{jk} in the equation is the mean of order statistic of standard normal distribution, it can be obtained:

$$u_{jk} = \mathrm{E}\left(T^0_{jk}\right) = \frac{n_j!}{(k-1)!(n_j-k)!} \int_0^1 \Phi^{-1}(P_k)(P_k)^{k-1}(1-P_k)^{n_j-k}\,\mathrm{d}P_k \tag{6}$$

In equation (5) $\varepsilon_{jk} = \sigma(T^0_{jk} - u_{jk})$, and $\mathrm{E}(\varepsilon_{jk}) = 0$, $\mathrm{Var}(\varepsilon_{jk}) = \sigma^2 \mathrm{Var}(T^0_{jk})$ $\mathrm{Cov}(\varepsilon_{jk}, \varepsilon_{jl}) = \sigma^2 \mathrm{Cov}(T^0_{jk}, T^0_{jl})$, $k \neq l$.

$$v_{jkk} = \mathrm{Var}\left(T^0_{jk}\right) = \frac{n_j!}{(k-1)!(n_j-k)!} \int_0^1 [\Phi^{-1}(P_k)]^2 (P_k)^{k-1}(1-P_k)^{n_j-k}\,\mathrm{d}P_k - u^2_{jk} \tag{7}$$

$$v_{jkl} = \mathrm{Cov}\left(T^0_{jk}, T^0_{jl}\right) = \frac{n_j!}{(k-1)!(l-k-1)!(n_j-l)!} \times \left[\begin{array}{l}\int_0^1\int_0^{P_l} \Phi^{-1}(P_k)\Phi^{-1}(P_l)(P_k)^{k-1} \\ \times (P_l - P_k)^{l-k-1}(1-P_l)^{n_j-l}\,\mathrm{d}P_k\mathrm{d}P_l\end{array}\right] - u_{jk}u_{jl} \quad k < l \tag{8}$$

The best linear unbiased estimator $\hat{a}$, $\hat{b}$, and $\hat{\sigma}$ can be obtained from the pseudo failure data of normal distribution and equation (5):

$$\hat{a} = \bar{t} - \hat{b}\bar{\varphi} - \hat{\sigma}\bar{u} \tag{9}$$

$$\hat{b} = \frac{L_{1x}L_{22} - L_{2x}L_{12}}{L_{11}L_{22} - L^2_{12}} \tag{10}$$

$$\hat{\sigma} = \frac{L_{2x}L_{11} - L_{1x}L_{12}}{L_{11}L_{22} - L^2_{12}} \tag{11}$$

And the covariance of estimator $\hat{\boldsymbol{\theta}} = (\hat{a}, \hat{b}, \hat{\sigma})^{\mathrm{T}}$ is:

$$\mathrm{Cov}(\hat{\boldsymbol{\theta}}) = \sigma^2 (c^{jk})_{3\times 3} \tag{12}$$

in the equation:

$$\bar{t} = \frac{1}{n^*}\sum_{j=1}^{m}\sum_{k,l=1}^{m_j} v^{jkl} t_{jk} \tag{13}$$

$$\bar{\varphi} = \frac{1}{n^*}\sum_{j=1}^{m}\sum_{k,l=1}^{m_j} v^{jkl} \varphi_j \tag{14}$$

$$\bar{u} = \frac{1}{n^*}\sum_{j=1}^{m}\sum_{k,l=1}^{m_j} v^{jkl} u_{jk} \tag{15}$$

$$n^* = \sum_{j=1}^{m}\sum_{k,l=1}^{m_j} v^{jkl} \tag{16}$$

$$L_{11} = \sum_{j=1}^{m}\sum_{k,l=1}^{m_j} v^{jkl}\left(\varphi_j - \bar{\varphi}\right)\left(\varphi_j - \bar{\varphi}\right) \tag{17}$$

$$L_{12} = \sum_{j=1}^{m}\sum_{k,l=1}^{m_j} v^{jkl}\left(\varphi_j - \bar{\varphi}\right)\left(u_{jk} - \bar{u}\right) \quad (18)$$

$$L_{22} = \sum_{j=1}^{m}\sum_{k,l=1}^{m_j} v^{jkl}\left(u_{jk} - \bar{u}\right)\left(u_{jl} - \bar{u}\right) \quad (19)$$

$$L_{1x} = \sum_{j=1}^{m}\sum_{k,l=1}^{m_j} v^{jkl}\left(\varphi_j - \bar{\varphi}\right)\left(t_{jk} - \bar{t}\right) \quad (20)$$

$$L_{2x} = \sum_{j=1}^{m}\sum_{k,l=1}^{m_j} v^{jkl}\left(u_{jl} - \bar{u}\right)\left(t_{jk} - \bar{t}\right) \quad (21)$$

$$(c^{jk})_{3\times 3} = \begin{bmatrix} \sum_{j=1}^{m}\sum_{k,l=1}^{m_j} v^{jkl} & \sum_{j=1}^{m}\sum_{k,l=1}^{m_j} v^{jkl}\varphi_j & \sum_{j=1}^{m}\sum_{k,l=1}^{m_j} v^{jkl}u_{jl} \\ \sum_{j=1}^{m}\sum_{k,l=1}^{m_j} v^{jkl}\varphi_j & \sum_{j=1}^{m}\sum_{k,l=1}^{m_j} v^{jkl}\varphi_j^2 & \sum_{j=1}^{m}\sum_{k,l=1}^{m_j} v^{jkl}u_{jl}\varphi_j \\ \sum_{j=1}^{m}\sum_{k,l=1}^{m_j} v^{jkl}u_{jk} & \sum_{j=1}^{m}\sum_{k,l=1}^{m_j} v^{jkl}u_{jk}\varphi_j & \sum_{j=1}^{m}\sum_{k,l=1}^{m_j} v^{jkl}u_{jk}u_{jl} \end{bmatrix}^{-1} \quad (22)$$

$$(v^{jkl})_{m_j\times m_j} = (v_{jkl})^{-1}{}_{m_j\times m_j}\quad j = 1, 2, \ldots, m \quad (23)$$

The estimator of normal parameters under different stress levels can be obtained from equations (3) and (4):

$$\hat{\mu}_j = \hat{a} + \hat{b}\varphi(S_j) \quad (24)$$

$$\hat{\sigma}_j = \hat{\sigma} \quad (25)$$

The estimator $\hat{a}$, $\hat{b}$, and $\hat{\sigma}$ are shown in equations (9)~(11).

Suppose $j = 0$, the life distribution parameters of products under normal stress level can be obtained from equations (24) and (25), then the reliability of products can also be obtained from the life distribution of products under normal stress level.

4 INTEGRAL ESTIMATION OF ACCELERATED PERFORMANCE DEGRADATION PSEUDO FAILURE DATA FROM WEIBULL DISTRIBUTION

The distribution function of Weibull distribution is:

$$F(t) = 1 - \exp\left[-\left(\frac{t}{\beta}\right)^{\alpha}\right] \quad (26)$$

The equations of integral estimation with accelerated performance degradation pseudo failure data from normal distribution can also be used in the integral estimation with accelerated performance degradation pseudo failure data from Weibull distribution via displacing the mean, variance and covariance of order statistic from normal distribution by the mean, variance and covariance of order statistic from extreme distribution[3], so the data process steps will not be introduced again here.

5 CONCLUSION

1. A general integral estimation model for the accelerated performance degradation pseudo failure data is established.
2. The integral estimation method of accelerated performance degradation and pseudo failure data from normal distribution is discussed in detail in this paper.
3. The method used in this paper can deal with accelerated performance degradation pseudo failure data from different distributions, and also it can deal with the problem when there is only one sample under each test stress level.

REFERENCES

[1] Jia Zhangqiang, Liang Yuying, Cai jinyan, PCB reliability assessment based on accelerated performance degradation testing [J], Radio engineering, 2008, 38(2): 46–50.

[2] Mao Shisong, Wang Lingling, 2000, Accelerated life test, Beijing: Science Press.

[3] Ao liang, Fu huimin, Regression analysis method for interval censored data [J], Journal of aerospace power, 2007, 22(6): 1013–1017.

Engineering Management and Industrial Engineering – Leung (Ed.)
© 2015 Taylor & Francis Group, London, ISBN 978-1-138-02773-2

An improvement of the simplex method

M.X. Shen
Department of Management Engineering, Xijing University, Xi'an, Shaanxi, China

S.W. Zhang
Department of Engineering Technology, Xijing University, Xi'an, Shaanxi, China

J. Li
Department of Management Technology, Xijing University, Xi'an, Shaanxi, China

ABSTRACT: Aimed at the efficiency of solving linear programming concisely and quickly, a new criterion for selecting the pivot element and a new way to yield an initial feasible basis for the popular simplex method of linear programming are presented by developing a new understanding of basis transformations in this paper. And two examples are constructed to demonstrate the conclusions of the improved method. This study helps to enhance the speed of convergence and further clarifies the simplex method in a certain extent.

1 INTRODUCTION

It is well known that the G. B. Dantzig's simplex method is not a good algorithm due to the reason that it is not a polynomial algorithm for use to handle computational complexity; however, it is the popular basic, most effective and widely used way to solve linear programming problems so far. Herewith, various improvements on the simplex procedure have filled many volumes of works. There are two key works that deserve our attention: how to get an initial feasible basis in the form of a unit matrix; how to get the optimal solution as quickly as possible. The basic thought of the simplex method is aimed at the standard form of a linear programming problem:

$$\max\left\{z \middle| z = C^T X,\ A_{m\times n}X = b,\ X \geq 0,\ b \geq 0,\ m \leq n\right\} \tag{1}$$

Improving the optimal value of the object function via a transform from a basic feasible solution to another basic feasible solution start off at the point where a basic feasible solution is given in its feasible region; and then one obtains the optimal solution or arrives at a judgment that no optimal solution exists. This thought is realized by using a basis transformation in reality. The basis transformation is an operation of iteration by choosing a pivot via determining an entrance variable and an extraction variable according to a given rule. In practical uses, we find that the rule (Editorial Group of Textbook of Operations Research. 1990, J. Q. Xue. 1992, J. Z. Zhang, S. J. Xu. 1990, Z. K. Xu. 2000, Editorial Committee of Handbook of Modern Applied Mathematics. 1998, G. H. Wei, J. L. Fu, etc. 1987, Y.W. Niu. 1994, W.E. Duckworth, A.E. Gear, A.G. Lockett. 1978, Eugene L. Lawler, 1976) of maximum positive criterion:

$$\sigma_l = \max_j\left\{\sigma_j \mid \sigma_j > 0,\ j = 1, 2, \ldots, n\right\}$$

could be amended for selecting the entrance variable, and hence the time of iteration can be reduced, and the efficiency of the simplex method is consequently enhanced.

2 ON BASIS TRANSFORMATIONS

For equation (1), let us take

$A = (a_{ij})_{m\times n}$, and $b = (b_1, b_2, \ldots, b_m)^T$,

$c = (c_1, c_2, \ldots, c_n)^T$ and $X = (x_1, x_2, \ldots, x_n)^T$.

Assume

$$z = C^T X = [C_B, C_N]\begin{bmatrix} X_B \\ X_N \end{bmatrix} = C_{\mathbf{B}}^T X_B + C_N^T X_N \tag{2}$$

$$AX = \begin{bmatrix} B & N \end{bmatrix}\begin{bmatrix} X_B \\ X_N \end{bmatrix} = BX_B + NX_N = b \tag{3}$$

Also, assume that B is a basis, and the iteration coefficients A and b are $A=(\bar{a}_{ij})$, $b=(\bar{b}_i)$. By letting $X_N=0$, we obtain the corresponding basic solution (Editorial Group of Textbook of Operations Research. 1990, J. L. Xue. 1992, G. H. Wei, J. L. Fu, etc. 1987, Y. L. Niu, 1994)

$$\bar{X}=\left[\bar{b}_1,\bar{b}_2,\ldots,\bar{b}_m,0,\ldots,0\right]^T=\begin{bmatrix}B^{-1}b\\0\end{bmatrix},$$

the value of the object function $\bar{Z}=C_B^TB^{-1}b$, and the criterion $\sigma=[\bar{\sigma}_1,\bar{\sigma}_2,\ldots,\bar{\sigma}_n]^T$.

If an extraction variable x_k is selected from the basis ($1\le k\le m$ i.e. $k\in I_B$), the entrance variable x_l is chosen from the set of non-basic variables ($m+1\le l\le n$ i.e. $l\in I_N$), and then the pivot $\bar{a}_{kl}\ne 0$ can be obtained. When the basis B is transformed to B' by a basis transformation, the coefficients A, b are correspondingly transformed to $A=(a'_{ij})$ and $b=(b'_i)$, respectively, the criterion σ to $\sigma=(\sigma'_j)$, satisfying that

$$\begin{cases}a'_{kj}=\dfrac{\bar{a}_{kj}}{\bar{a}_{kl}}, \quad (j=1,2,\ldots,n); \quad b'_k=\dfrac{\bar{b}_k}{\bar{a}_{kl}}\\ a'_{ij}=\bar{a}_{ij}-\dfrac{\bar{a}_{kj}}{\bar{a}_{kl}}\bar{a}_{il}, \quad (i\ne k,\ i=1,2,\ldots,m,\ j=1,2,\ldots,n)\\ b'_i=\bar{b}_i-\dfrac{\bar{b}_i}{\bar{a}_{kl}}\bar{a}_{il}, \quad (i\ne k,\ i=1,2,\ldots,m)\end{cases} \tag{4}$$

$$\sigma'_j=\bar{\sigma}_j-\frac{\bar{a}_{kj}}{\bar{a}_{kl}}\bar{\sigma}_l, \quad (j=1,2,\ldots,n) \tag{5}$$

Therefore, we obtain a new basic solution as (6) (Editorial Group of Textbook of Operations Research. 1990, J. Q. Xue. 1992, G. H. Wei, J. L. Fu, etc. 1987):

$$X'=\Big[b'_1,b'_2\ldots,b'_{k-1},\underset{k\text{-th component}}{0},b'_{k+1},\ldots,b'_m,0,\ldots,0,\underset{l\text{-th component}}{b'_k},0,\ldots,0\Big]^T \tag{6}$$

The new value of the object function is as (7) (Editorial Group of Textbook of Operations Research. 1990, J. Q. Xue. 1992, J. Z. Zhang, S. J. Xu. 1990):

$$Z'=\bar{Z}+\sigma_l b'_k=\bar{Z}+\sigma_l\frac{\bar{b}_k}{\bar{a}_{kl}} \tag{7}$$

where $b_k'=\bar{b}_k/\bar{a}_{kl}=\min\{\bar{b}_i/\bar{a}_{il}\mid\bar{a}_{il}>0\}$, $(1\le k\le m)$ and the criteria $\sigma_j=c_j^i-z_j$ such that

$$z_j=\sum_{i=1}^{m}c_ia'_{ij},\ j=1,2,\ldots,n.$$

3 AN IMPROVEMENT OF THE SIMPLEX METHOD

With the guidance of such a general rule that neither miss the basic feasibility of a solution nor infract the target of optimizing the object function Z' from $\bar{Z}$, that is make Z' converge to the optimization based on $\bar{Z}$ as quickly as possible, we define

$$\mu_j=\min_i\left\{\frac{\bar{b}_i}{\bar{a}_{ij}}\mid\bar{a}_{ij}>0,i=1,2,\ldots,m\right\}$$
$$j=1,2,\ldots,n \tag{8}$$

$$\theta_i=\frac{\bar{b}_i}{\bar{a}_{il}},\ (\bar{a}_{il}>0,\ i=1,2,\ldots,m,\text{ for fixed }l) \tag{9}$$

The reason for us to define equation (8) is analogously as that of determining extraction variable x_k by using $\bar{b}_k/\bar{a}_{kl}=\min\{\bar{b}_i/\bar{a}_{il}\mid\bar{a}_{il}>0\}$. It could be understood as to ensure the non-negativity (feasibility) of decision variables in the linear programming.

In order to make Z' converge to the optimal solution based on $\bar{Z}$ as quickly as possible by using equation (7), we select an entrance variable x_l with a maximum $\sigma_l(\bar{b}_k/\bar{a}_{kl})$ for a fixed l, and for a given $i=k$, we get $\bar{b}_k/\bar{a}_{kl}=\mu_l$ from equation (7). After that, we select an entrance variable x_l by using the following rule:

$$\mu_l\sigma_l=\max_j\left\{\mu_j\sigma_j\middle|\mu_j>0,\sigma_j>0\right\} \tag{10}$$

For the selection of an extraction variable, as long as we can ensure the feasibility of the solution (non-negativity of the decision variables), we select an extraction variable x_k by using the following rule:

$$\theta_k=\min_i\{\theta_i\} \tag{11}$$

In fact, this is the usual minimum ratio rule $\bar{b}_k/\bar{a}_{kl}=\min_i\{\bar{b}_i/\bar{a}_{il}\mid\bar{a}_{il}>0\}$. Otherwise, there is a relationship: $\mu_l=\theta_k$.

Based on what is discussed, we can establish a new method to select the entrance variables and the extraction variables in order to improve the simplex method of linear programming as follows.

In the procedure of the simplex method, first select an entrance variable x_l according to equation (10); second, select an extraction variable x_k

according to equation (11); and then the pivot $\bar{a}_{kl}$ is obtained.

Adopting this technique, computational efficiency of the simplex method of linear programming is evidently improved in practical applications. This fact can be shown easily by using the simplex tableau and more expediently as can be observed in below Example 1.

4 ABOUT THE INITIAL BASIS

The big M method and two-stage way are usually adopted to construct a unit matrix as the initial feasible basis by introducing some artificial variables. When we pay attention to the relationship between the essence of the simplex iteration and the elementary transformation of matrices, a way of solving linear programming problems based on elementary row transformations of matrices on the augmented coefficient matrix of the linear programming model of concern appears. Hence, we can present a determinant method (Shen Mao-xing, Xu Jin. 2007) to obtain the initial feasible basis if we do not have any unit matrix as the initial basis:

First, perform a finite series of elementary row transformations on the augmented coefficients matrix $B = \left[A|b\right]_{m\times(n+1)}$ until obtaining its simplest form (including a unit matrix as its sub-matrix) while keeping the resource vector b always positive, and a $r \times r$ initial feasible basis is obtained ($r \le m$). Second, turn to the usual procedure of the simplex method with the simplex tableau by arranging the matrix in its simplest form into the initial simplex tableau.

The advantages of this scheme are not only produces an initial feasible basis (when $r = m$), but also help to find the dependent constrains (when $r < m$) in linear programming models so as to eliminate the steps of finding them, leading to a decreased amount of computation of the simplex method.

It should be noted that we may only use the second and the third class elementary row transformations without the first class elementary row transformation (exchange positions of any two rows) in our method. It appears that the advantages of this new scheme include that it provides a clue for clear thinking, a concise procedure, and requires a less amount of computation in practice, see Example 2 below for more details.

5 EXAMPLES

Example 1. Solves the following linear programming model.

$$\max\ z = 2x_1 + 3x_2$$

$$s.t.\ \begin{cases} x_1 + 2x_2 \le 8 \\ 4x_1 \qquad \le 20 \\ \qquad 4x_2 \le 12 \\ x_1,\quad x_2 \ge 0 \end{cases}$$, The standard form is

$$\max\ z = 2x_1 + 3x_2 + 0x_3 + 0x_4 + 0x_5$$

$$\text{s.}t.\begin{cases} x_1 + 2x_2 + x_3 = 8 \\ 4x_1 + x_4 = 20 \\ 4x_2 + x_5 = 12 \\ x_1, x_2, x_3, x_4, x_5 \ge 0 \end{cases}$$

Three iterations are needed to get the optimal solution by using the ordinary simplex method, which can be observed in most textbooks (Editorial Group of Textbook of Operations Research. 1990.). However, by using our method, we can get the optimal solution in 2 iterations. For instance, let us construct Table 1.

Therefore, the optimal solution of the model is $x^* = (5, 3/2)^T$ and $z^* = \max z = 14.5$.

Example 2. Find the solution of the following model:

$$\min z = -3x_1 + x_2 + x_3$$

$$s.t.\begin{cases} x_1 - 2x_2 + x_3 + x_4 = 11 \\ -4x_1 + x_2 + 2x_3 - x_5 = 3 \\ -2x_1 + x_3 = 1 \\ x_i \ge 0,\quad (i = 1, 2, \ldots, 5) \end{cases}$$

First, let us apply a finite series of elementary row transformations on the augmented coefficient matrix B while keeping the relevant feasibility:

$$\left[\begin{array}{ccccc|c} 1 & -2 & 1 & 1 & 0 & 11 \\ -4 & 1 & 2 & 0 & -1 & 3 \\ -2 & 0 & 1 & 0 & 0 & 1 \end{array}\right] \rightarrow \left[\begin{array}{ccccc|c} 3 & -2 & 0 & 1 & 0 & 10 \\ 0 & 1 & 0 & 0 & -1 & 1 \\ -2 & 0 & 1 & 0 & 0 & 1 \end{array}\right]$$

$$\rightarrow \left[\begin{array}{ccccc|c} 3 & 0 & 0 & 1 & -2 & 12 \\ 0 & 1 & 0 & 0 & -1 & 1 \\ -2 & 0 & 1 & 0 & 0 & 1 \end{array}\right]$$

Then, let us arrange an initial simplex tableau as Table 2.

Therefore, the optimal solution of the model is $x^* = (4, 1, 9, 0, 0)^T$ and $z^* = -2$. It should be noted that the target is to minimizing the object function in this example. So, the rule in equation (10) is changed to the following:

$$\mu_l \sigma_l = \max_j \left\{ \mu_j \left| \sigma_j \right| \left| \mu_j > 0, \sigma_j < 0 \right. \right\} \tag{12}$$

This example problem was solved by using the big-M method with 3 iterations in [2], by using

Table 1. Procedure of solving Example 1.

c_j		2	3	0	0	0		
C_B	X_B	x_1	x_2	x_3	x_4	x_5	b	θ_i
0	x_3	1	2	1	0	0	8	8
0	x_4	[4]	0	0	1	0	20	5
0	x_5	0	4	0	0	1	12	–
σ_j		2	3	0	0	0	0	
μ_j		5	3	–	–	–		
$\mu_j\sigma_j$		10	9	–	–	–		
0	x_3	0	[2]	1	1/4	0	3	3/2
2	x_1	1	0	0	1/4	0	5	–
0	x_5	0	4	0	0	1	12	3
σ_j		0	3	0	–1/2	0	–10	
μ_j		–	3/2	–	–	–		
$\mu_j\sigma_j$		–	9/2	–	–	–		
3	x_2	0	1	1/2	–1/8	0	3/2	
2	x_1	1	0	0	1/4	0	5	
0	x_5	0	0	–2	1/2	1	6	
σ_j		0	0	–3/2	–1/8	0	–14.5	

Table 2. Procedure of solving example 2.

c_j		–3	1	1	0	0		
C_B	x_B	x_1	x_2	x_3	x_4	x_5	b	0_i
0	x_4	[3]	0	0	1	–2	12	4
1	x_2	0	1	0	0	–1	1	–
1	x_3	–2	0	1	0	0	1	–
σ_j		–1	0	0	0	1	2	
μ_j		4	–	–	–	–	–	
$\mu_j\sigma_j$		4	–	–	–	–	–	
–3	x_1	1	0	0	1/3	–2/3	4	
1	x_2	0	1	0	0	–1	1	
1	x_3	0	0	1	2/3	–4/3	9	
σ_j		0	0	0	1/3	1/3	–2	

the two-phase method with 2 iterations in the first phase and 1 iteration in the second phase. Hence, it is clear that our method is more straightforward, simple, and requires only two iterations.

6 CONCLUSION

As a matter of fact, we presented two results in this paper. The first one is the rule in equation (10) that modifies the "maximum criterion rule" of selecting an entrance variable x_l, and enhances the computational efficiency of the simplex method. This result possesses a very good practicability in applications and/or teachings. It should be pointed that the domain $i \in I_B$ of the subscript, the set of the basic variables, or $(i = 1, 2, \ldots, m)$, $l \in I_N$, $j \in I_N$ (the subscript set of the non-basic variables), or $(j = m+1, m+2, \ldots, n)$ are actually contained in equations (8) and (9). The reason why we adopted $j = 1, 2, \ldots, n$ is that the tidiness of formality. The second result presented in this paper is a concise method of obtaining the initial feasible basis as the simplest form as the unit I. It is essentially the same as the simplex method except that it provides a clear way of thinking useful for teaching purposes.

ACKNOWLEDGEMENT

National Natural Science Foundation of China Program number 61473237.

REFERENCES

Duckworth W.E., A.E. Gear, A.G. Lockett. 1978. A Guide to Operational Research. 3rd edition, London: Chapman and Hall Ltd.

Editorial Committee of Handbook of Modern Applied Mathematics. 1998. Handbook of Modern Applied Mathematics: Volume of Operations Research and Optimization Theory. Beijing: Tsinghua University Press.

Editorial Group of Textbook of Operations Research. 1990. Operations Research (3rd Edition). Beijing: Tsinghua University Press.

Eugene L. Lawler. 1976. Combinatorial Optimization: Networks and Matroids. New York: Holt Rinehart and Winston.

Niu Y.W. 1994. Operations Research. Xi'an: Xi'an Jiaotong University Press.

Ru S.F., M.X. Shen, X.F. Xue. 2008. A concise way of determination for LP initial feasible basis of simplex method, Scientia Magna, 4(1): 94–96.

Shen M.X., J. Xu. 2007. Direct way of simplex method for solving linear programming model, Computer Engineering and Applications, 43(30): 94–96.

Wei G.H., J.L. Fu, etc. 1987. Applied Operations Research. Shanghai: Fudan University Press.

Xu Z.K. 2000. Introduction of Mathematical Programming. Beijing: Science Press.

Xue J.Q., 1992. Principle and method of optimization (2nd edition). Beijing: Metallurgical Industry Press.

Zhang J.Z., S.J Xu. 1990. Linear Programming. Beijing: Science Press.

Engineering Management and Industrial Engineering – Leung (Ed.)
© 2015 Taylor & Francis Group, London, ISBN 978-1-138-02773-2

A new method for solving trajectory fusion estimation model based on trust region

T.H. Zhang, W.L. Yin & J. Tao
Department of Information Engineering, Ordnance Engineering College, Shijiazhuang, China

L.H. Guo
Hebei Provincial Department of Science and Technology, Shijiazhuang, China

ABSTRACT: In order to overcome the disadvantage of the improved G-N method, this paper first builds the nonlinear regression model of trajectory fusion estimation based on the exterior trajectory measurement data, and then the method of trust region is used first time for model's parameter estimation; preconditioned conjugate gradient method which is very suitable for large scale calculation is used for solving trust region sub problem. Convergence property analysis of mathematics and practice experiments proves that the method this paper proposes has fast convergence rate, overall convergence property and is very suitable for large scale calculation problems such as trajectory fusion estimation of shooting range.

Keywords: shooting range test; trajectory fusion estimation; trust region; preconditioned conjugate gradient

1 INTRODUCTION

The models that established by Exterior Trajectory estimation method are usually solved by the improved G-N method[1–2]. As the dissatisfaction rank of Jacobi matrix, the improved G-N method converges to a non-stationary point, or terminates prior to the stable point[3]. In order to overcome the disadvantage of the improved G-N method, this paper proposed the trust region method for solving trajectory fusion estimation model of shooting range[4]. Convergence property analysis of mathematics and practice experiments proves that the method this paper proposed has fast convergence rate, and is very suitable for large scale calculation problems such as trajectory fusion estimation of shooting range.

2 NONLINEAR MODEL OF EXTERIOR TRAJECTORY MEASUREMENT FUSION ESTIMATION

Considering the tracking problem to the target within the time period $[t_1, t_m]$, assuming each measuring trace element data have been pretreated, the position parameter of the track is $X(t) = (x(t), y(t), z(t))^{\mathrm{T}}$, and the velocity parameter is $\dot{X}(t) = (\dot{x}(t), \dot{y}(t), \dot{z}(t))^{\mathrm{T}}$ in the reference coordinate system (the reference coordinate system is assumed to transmit coordinates)[5], then:

$$\begin{cases} x(t) = \sum_{j=1}^{N_x} c_j B_j(t, T_x) \\ \dot{x}(t) = \sum_{j=1}^{N_x} c_j \dot{B}_j(t, T_x) \\ y(t) = \sum_{j=1}^{N_y} c_{j+N_x} B_j(t, T_y) \\ \dot{y}(t) = \sum_{j=1}^{N_y} c_{j+N_x} \dot{B}_j(t, T_y) \\ z(t) = \sum_{j=1}^{N_z} c_{j+N_x+N_y} B_j(t, T_z) \\ \dot{z}(t) = \sum_{j=1}^{N_z} c_{j+N_x+N_y} \dot{B}_j(t, T_z), \end{cases} \tag{1}$$

In the formula (1), T_x, T_y, T_z is optimal spline node sequence in x, y, z directions, and $B_j(t, T_\alpha), \dot{B}_j(t, T_\alpha), j = 1, \ldots, N_\alpha, \alpha = x, y, z$ is freedom node standard spline basis function and its derivative function in α direction.

Setting $\beta = (c_1, \ldots, c_{N_x+N_y+N_z})^{\mathrm{T}}$, formula(1) can be expressed as:

$$\begin{cases} X(t) = B(t)\beta \\ \dot{X}(t) = \dot{B}(t)\beta \end{cases} \tag{2}$$

The system error parameters of measuring element are $\alpha = (\alpha_1, \ldots, \alpha_{N_2})^{\mathrm{T}}$. According to the above discussion, the nonlinear model for trajectory parameters can be obtained based on measurement data of different stations as follows:

$$\begin{cases} \tilde{A}_j(t_i^A) = A_j(t_i^A, \beta) + s_j(t_i^A, \alpha, \beta) + \varepsilon_j(t_i^A) \\ \tilde{E}_j(t_i^E) = E_j(t_i^E, \beta) + s_j(t_i^E, \alpha, \beta) + \varepsilon_j(t_i^E) \\ \tilde{R}_j(t_i^R) = R_j(t_i^R, \beta) + s_j(t_i^R, \alpha, \beta) + \varepsilon_j(t_i^R) \\ \tilde{\dot{R}}_j(t_i^{\dot{R}}) = \dot{R}_j(t_i^{\dot{R}}, \beta) + s_j(t_i^{\dot{R}}, \alpha, \beta) + \varepsilon_j(t_i^{\dot{R}}) \end{cases} \tag{3}$$

In the formula (3), s_j is the system error parameters of measuring element, such as the system error of constant value, linear slowly varying error, and refraction correction residuals. In the case of a large amount of data processing, correlation and non-Gaussian have little effect on the estimation results. Therefore, the random error is assumed as Gaussian white noise in the following general assumption fusion model. So, $E\varepsilon_j = 0, E\varepsilon_j\varepsilon_k^T = \delta_{jk}\sigma_j^2 I_{m_l^j}, l = A, E, R, \dot{R}$.

The column vector formed by each measurement element in each time in left side of the formula is denoted as Y, the first term in right side of the formula is denoted as $f(\beta)$, the second term is denoted as $s(\alpha, \beta)$, and the third term is denoted as ε. So

$$Y = f(\beta) + s(\alpha, \beta) + \varepsilon \tag{4}$$

For convenience, noting $F(\alpha,\beta) = f(\beta) + s(\alpha,\beta)$, $\theta = (\alpha^T, \beta^T)^T$, so formula (4) transforming to.

$$Y = F(\theta) + \varepsilon \tag{5}$$

The formula (5) is a non-linear regression model. For convenience, noting $Y - F(\theta) = r(\theta)$, determining the parameter estimates by formula (5), the most commonly used method is the least squares estimation.

$$\min_{\gamma} S(\theta) = \frac{1}{2}\|Y - F(\theta)\|_2^2 = \frac{1}{2}\|r(\theta)\|_2^2 \tag{6}$$

3 THE SOLVING METHOD TO NONLINEAR MODEL

The formula (6) is a least squares estimation problem. We anglicize the calculation in solving the formula (6) first. Assuming that the sampling point numbers for azimuth and elevation angle are $m_A^j (j = 1, \ldots, n_A)$ and $m_E^j (j = 1, \ldots, n_E)$, the sampling point numbers for distance measurement are $m_R^j (j = 1, \ldots, n_R)$, the sampling point numbers for speed measurement are $m_{\dot{R}}^j (j = 1, \ldots, n_{\dot{R}})$, therefore, the number for column vector is $m = \left(\sum_{j=1}^{n_A} m_A^j + \sum_{j=1}^{n_E} m_E^j + \sum_{j=1}^{n_R} m_R^j + \sum_{j=1}^{n_{\dot{R}}} m_{\dot{R}}^j \right) \times 1.$

In the above formula, n_β is the components number for the ballistic parameters spline coefficient vector β, and n_α is the components number for error coefficient α. So, the number for estimating parameters is $p = n_\beta + n_\alpha$. If we want to process the tracking data spanning 100 seconds, assuming there will be a node in two seconds averagely, the spline coefficients number one coordinate direction required is 50 + 3 = 53, and the number for three coordinate directions is 159; if adding the estimated error coefficient, 170 parameters should be estimated at least.

For example, there are four distance measurements (two continuous wave radar, and two single pulse radar), four optical goniometer measurements, and two speed measurements; three will be 20000 measurement data with a sampling rate of 20 frames per second. Because of the enormous amount of calculation, we must find suitable method for solving large-scale computing.

The common method for solving least squares problems is Gauss-Newton method, improved Gauss-Newton method, and Levenberg-Marquardt method. But these methods are more suitable for smaller or medium-sized computing problems[6].

3.1 *Rust region method*

The main idea of trust region methods is to departure from current iteration points and construct a model approaching to the original objective function. As the main structure of the model is based on the information at the point θ_k of the original problem, so this model is a good approximation of the original problem nearby the points θ_k.

In the vicinity of the current iteration point θ_k, using quadratic function approximating the target function, we construct trust sub-domains as:

$$\begin{aligned} \min \quad & m(\delta) = S(\theta_k) + g_k^T \delta + \frac{1}{2}\delta^T G_k \delta \\ s.t. \quad & \|\delta\| \le \Delta_k \end{aligned} \tag{7}$$

In formula (7), $\delta = \theta_{k+1} - \theta_k$, $g_k = \nabla S(\theta_k)$, G_k, is hessen matrix $\nabla^2 S(\theta_k)$, or an approximate to $\nabla^2 S(\theta_k)$, Δ_k is trust region radius. Δ_k is usually adjusted according to fitting degree between the model function $m(\delta)$ and the objective function $S(\theta)$.

In order to solve formula (7), we assume that the decreasing amount of the objective function $Ared_k = S(\theta_k) - S(\theta_k + \delta_k)$ is the actual decreasing amount and note the decreasing amount of model function is $Pred_k = m(0) - m(\delta_k)$. In order to predict the decreasing amount, the following is defined:

$$\rho_k : \rho_k = \frac{Ared_k}{Pred_k}$$

The closer ρ_k is to 1, the better fitting degree of the model function is to the objective function. Δ_k can be increased to expand the trust region. If $\Delta_k > 0$ but not close to 1, then keep the trust region radius Δ_k unchanged. If Δ_k is close to zero or a negative value, the fit degree of m to S is not ideal, and it is necessary to reduce trust domain.

Trust region method steps are as follows:[7]

Step 1. Determine the initial point θ_0. Take $\bar{\Delta} > 0$ as the trust region radius circles, set $\Delta_0 \in (0, \bar{\Delta}], 0 < \eta_1 < \eta_2 < 1, 0 < \gamma_1 < 1 < \gamma_2, \varepsilon \geq 0, k = 0$.

Step 2. Test termination condition. Calculate $g_k = \nabla S(\theta_k)$, if $\|g_k\| \leq \varepsilon$, then $\theta = \theta_k$, method terminates.

Step 3. Solve the trust region sub problem, calculate formula (7) to get δ_k.

Step 4. Correct the trust region radius.

Step 5. If $\rho_k > \eta_1$, then $\theta_{k+1} = \theta_k + \delta_k$, or $\theta_{k+1} = \theta_k, k = k+1$, go to step 2.

The suggestion of parameter selection for the above method is:

$$\eta_1 = 0.01, \eta_2 = 0.75, \gamma_1 = 0.5, \gamma_2 = 2.0, \Delta_0 = 1$$

or: $\Delta_0 = 1/10\|g_0\|$

For the problem of this paper, assuming $J(\theta_k)$ is the Jacobi Matrix of $r(\theta_k)$, so

$$\nabla S(\theta_k) = g_k = J(\theta_k)^T s(\theta_k), \nabla^2 S(\theta_k)$$
$$= G_k = J(\theta_k)^T J(\theta_k) + C(\theta_k).$$

In the above formula, $C(\theta_k) = \sum_j r_j(\theta_k) \nabla^2 r_j(\theta_k)$. According to idea of Gauss-Newton method, consider $J(\theta_k)^T J(\theta_k)$ is a good approximation to $\nabla^2 S(\theta_k)$, so $\nabla^2 S(\theta_k) = G_k = J(\theta_k)^T J(\theta_k), C(\theta_k)$ is ignored. As second derivative for $r(\theta_k)$ is not involved, the computing speed is greatly accelerated.

3.2 *Solving the sub problem*

It can be derived from the structure of the above method that the key step in trust region methods is sub problem solving in step 3. There are two methods for solving the sub problem. One is an exact method, and the other is an imprecise method. The exact method usually involves the calculation for full rank values or eigenvectors. In this paper, we utilize preconditioned conjugate gradient method to solve the sub problem of trust region.

The solving steps are as follows:

Step 0. Assuming $j = 0$; $\delta_0 = 0$; $r_0 = -g_k$; calculate $Mz_0 = r_0$.

Step 1. Calculate $\gamma_j = d_j^T G_k d_j$, if $\gamma_j \leq 0$, go to step 4, or setting $\alpha_j = r_j^T z_j / \gamma_j$.

Step 2. If $\|D_k(\delta_j + \alpha_j d_j)\| > \Delta_k$, go to step 4, or setting $\delta_{j+1} = \delta_j + \alpha_j d_j, r_{j+1} = r_j - \alpha_j G_k d_j$.

Step 3. If $\| D_k r_{j+1} \| \leq \eta \| D_k r_0 \|$, $\delta_k = \delta_{j+1}$, terminate the iteration, or calculate $M_k z_{j+1} = r_{j+1}$, get z_{j+1}, and then calculate $\beta_j = r_{j+1}^T z_{j+1} / r_j^T z_j, d_{j+1} = z_j + \beta_j d_j, j = j+1$, go to step 1.

Step 4. Set $t > 0$, calculate $\|D_k(\delta_j + td_j)\| = \Delta_k$ to get t, set $\delta_k = \delta_j + td_j$. Terminate the iteration.

3.3 *Convergence analysis*

Assuming the function $f(x) = 1/2\sum_{i=1}^{m} r_i(x)^2$ is twice continuously differentiable, and $r(x) = [r_1(x), r_2(x), \ldots, r_m(x)]$ is the nonlinear function of $x \in R^n$, $J(x)$ is the jacobian matrix of $r(x)$, for the given initial point $x^{(0)}$, horizontal collection $L(x^{(0)}) = \{x | f(x) \leq f(x^{(0)})\}$ is bounded, and there is a constant $M_1 > 0, M_2 > 0$

$$\left\|\nabla^2 f(x)\right\| \leq M_1, \quad \forall x \in L(x^{(0)})$$

$$\left\|J(x)^T J(x)\right\| \leq M_2, \quad \forall x \in L(x^{(0)})$$

The point columns generated by the above method $\{x^{(k)}\}$ will terminate in a finite number of iterations, or have a limit point of a subsequence x^* who satisfies the equation $\left\|\nabla f(x^*)\right\| = 0$.

The above theorem shows that the trust region of Gauss-Newton method not only overcomes the problem of non-convergence, but also increases the convergence region. So trust region of Gauss-Newton method has global convergence.

Assuming the function $f(x) = 1/2\sum_{i=1}^{m} r_i(x)^2$ is twice continuously differentiable, for the given initial point $x^{(0)}$, horizontal collection $L(x^{(0)})$ is bounded, the sequence $\{x^{(k)}\}$ generated by trust region of G-N method converges at x^*.

4 EXAMPLE CALCULATION AND ANALYSIS

To verify the improved results of the method, we calculate the measured data of three missions using the Improved G-N method, Levenberg-Marquardt method, and preconditioned conjugate gradient trust region method.

Take the same termination condition $\varepsilon = 10^{-6}$ for three method. Parameters for preconditioned conjugate gradient trust region method is $\eta_1 = 0.01, \eta_2 = 0.75, \gamma_1 = 0.5, \gamma_2 = 2.0, \Delta_0 = 1/10\|g_0\|$. The initial value of spline coefficients is calculated from trajectory parameters, system error coefficients is stetted with value 0. The calculating results are shown in Table 1 to Table 3.

It can be observed from Table 1 that, three methods have the same iteration number, and converge to the same value. Comparing with the time spent for the three methods, there is little difference between the improved Gauss-Newton method and the preconditioned conjugate gradient trust region method, and the time the above two methods spent is far less than the time Levenberg-Marquardt method spent. In order to investigate the iteration character for different estimation method comprehensively, assigning different initial value to the second set of data, the calculating results is showed in Table 4.

Table 1. The results for the first set of data.

Method	S	Num	Time/s
G-N Method	721.6538	4	103.451
L-M Method	721.6538	20	577.255
T-R Method	721.6538	4	106.662

Table 2. The results for the second set of data.

Method	S	Num	Time/s
G-N Method	2346.875	3	865.215
L-M Method	2346.875	12	3561.508
T-R Method	2346.875	3	865.932

Table 3. The results for the third set of data.

Method	S	Num	Time/s
G-N Method	6542.387	5	7895.613
L-M Method	6542.387	16	56310.887
T-R Method	6542.387	5	7887.0181

Table 4. The value of S for different method.

Method	First value	Second value	Third value	Fourth value
G-N	2346.887	2346.887	103254.18	2346.887
L-M	2346.887	2346.887	2346.865	2346.887
T-R	2346.889	2346.888	2346.865	2346.887

5 CONCLUSION

This paper established a model for fusion estimation of the estimated range ballistic, and first used trust region method to solve nonlinear model parameter estimation. Simulation examples show that this method has a faster convergence rate and is not sensitive to initial conditions, so the method this paper proposed is ideally suited for trajectory fusion estimation.

REFERENCES

[1] Q. Gui, J. Zhang. Robust biased estimation and its applications in geodetic adjustments[J]. Journal of Geodesy. 1998 (7–8).

[2] Li-sheng Liu. Exterior ballistic measurement data processing [M]. Bejing: National Defense Industry Press, 2002. 298–300.

[3] Yong Feng, Meisheng Xu, Yan Huang. External Trajectory Measurement Data Processing Based on Wavelet Transform [J]. Journal of Projectiles, Rockets, Missiles and Guidance, 2011, 27(4):98–101.

[4] Bo-ning Ma, Zhi-guang Leng, Xiao-an Tang. Wavelet analysis error controlling method for non-uniform sampling signal [J]. Signal processing, 2012, 28(1): 67–71.

[5] Huai-yu Wu. Time Series Analysis and Synthesis [M]. Wu Han: Wuhan University Press, 2004, 78–82.

[6] Berner K., Wells R.O. Smoothness estimates for soft-threshold denoising via translation-invariant wavelet transforms[J]. Applied and computational harmonic analysis, 2002, 12(1): 1–24.

[7] Xian-qin T.U., Dong-yun Y.I., Ning-li Y.U., et al. The Statistics of Measurement Noise of the Instrumentation Ship [J]. Journal of Spacecraft TT & C Technology, 2006, 25(03): 72–77.

Engineering Management and Industrial Engineering – Leung (Ed.)
© 2015 Taylor & Francis Group, London, ISBN 978-1-138-02773-2

A functional hazard analysis based on Multilevel Flow Modeling

Y. Zhao, J. Jiao & T.D. Zhao
School of Reliability and Systems Engineering, Beihang University, Beijing, China

ABSTRACT: Hazardous element such as dangerous materials or high hazardous energy that must be used in complex system hazards will always exist in the system. In this paper, a new hazard analysis technique is proposed which identifying hazard by deviation influence propagation analysis. Using Multilevel Flow Modeling (MFM), a system can be modeled by a set of high hazardous energy, dangerous materials and safety-related information flow substructures, which represent system functions and control functions. Combining with failure mode classes, possible function achievement changes can be determined. A function deviation causal tree of component is introduced, and the influence propagation upon a series of functions is analyzed. In this way, a series of anomaly functions, which are the trigger or initiator events causing the high hazardous energy to unexpectedly release or dangerous materials to leakage, are determined, and the potentially hazards can be identified. The proposed hazard analysis technique is demonstrated by an example of a fuel system on a certain airplane.

1 INTRODUCTION

With dangerous materials or high hazardous energy must be used in complex system, such as weapons systems, and commercial aircraft, hazards will always exist in the system. System safety is the formal process of identifying and controlling risk (USA 2012). As systems become more complex and more hazardous, more effort is required to understand and manage mishap risk. The key to system safety and effective risk management is the identification and mitigation of hazards (Ericson 2005, T.D. Zhao et al. 2012). In order to ensure the safety of system, it is necessary to study hazard analysis technique, as well as which is meaningful for improving the system design.

To successfully identify hazard, it is necessary to understand hazard. A hazard is an entity that contains only the elements necessary and sufficient to result in an accident. It is comprised of the following three basic components (Ericson 2005, T.D. Zhao et al. 2012): Hazardous Element (HE), Initiating Mechanism (IM), and Target and Threat (T/T). Hazardous element is the basic hazardous resource creating the impetus for the hazard, such as high hazardous energy and dangerous materials. The initiating mechanism is the trigger or initiator event(s) causing the hazard to occur. The target and threat are the person or thing that are vulnerable to injury and/or damage in the accident.

Hazardous element must be used is the fundamental reason that hazard will always exist. In this paper, a new hazard analysis technique is proposed which identifying hazard by modeling the hazardous element flows in system. Using Multilevel Flow Modeling (MFM), a system can be modeled by a set of high hazardous energy, dangerous materials and safety-related information flow substructures, which represent system functions and control functions. Based on the model, the influence propagation analysis upon a series of functions is introduced to recognize the initiating mechanisms, target and threat. Hazard can be identified by using the influence propagation analysis. At last, the proposed hazard analysis technique is demonstrated by an example of a fuel system on a certain airplane.

2 FUNCTIONAL HAZARD ANALYSIS BASED ON MFM

2.1 *Multilevel flow modeling*

MFM was proposed by Morten Lind in 1990 (Lind, M. 1994), and extensive attention has been paid to it in Europe, Japan and other western countries. It has been effectively applied to the alarm analysis, failure mode analysis, sensor fault detection and fault diagnosis (J. Zhan 2007). MFM is a functional modeling technique, which expresses hierarchically the intentions of a system from the viewpoints of goals and functions (Larsson J.E. 2000). The functional aspects of a system are expressed diagrammatically. In MFM, the goals describe the purposes of a system and its subsystems, and the functions describe the system's abilities in terms of flows of mass, energy and information (Jose Luis de la M. & Manuel R. 2010). The MFM enables ones to represent knowledge of a system, by which they can capture designers' intentions of a system and its control systems. The MFM model of a system can express the relations among the behaviors and intentions of system components through causal relations. Figure 1 shows the MFM concepts and the extensions proposed.

Whether there is a flow variable throughout the system is the key of the MFM modeling, which also plays a key role in the goal achievement. So it is important to understand the system fully and get the relationships of the flow variables (M.D. Wang et al. 2012).

2.2 *Functional hazard analysis based on MFM*

Dangerous materials and high hazardous energy flow inside the system based on the design intention, during which safety-related information is produced continuously. The safety-related information, taking many different forms, is transferred, processed and stored among the sub-systems or elements in the system, which can play an important role in observing, guiding and controlling those flows of dangerous materials and high hazardous energy. Safety-related information flows are the process for the acquirement, transformation, process and utilization of information, which is produced during controlling and adjusting the dangerous material and high hazardous energy in a system. The technology proposed in this paper aims to model this process of the flows of dangerous materials, high hazardous energy and safety-related information, and analyzes the possible trigger mechanism and the corresponding threaten/objectives in the system.

Using MFM, a system is modeled by a set of high hazardous energy, dangerous materials and information flow substructures, which represent system functions and control functions. A series of function primitives connected by a flow-line provide the information of causal relation among functions, which contribute to achievement of system goals. The relations among system goals, sub-goals, and system functions to achieve goals/sub-goals are represented by a means-end. Each function primitive involves a cause-effect relation upon input, output, and functional achievement.

Combining with failure mode classes, possible function achievement changes can be determined, and the influence propagation upon a series of functions is analyzed. In this way, a series of anomaly functions, which are the trigger or initiator events causing the high hazardous energy to unexpectedly release or dangerous materials to leakage, are determined, and the hazard can be identified.

A failure mode class may or may not be specific to the physical or logical nature of the components' interactions. The definitions of the failure types identified in the hierarchy are (McDermid J.A. & Pumfrey D.J. 1994.):

- Omission—no service is delivered.
- Value—a service is delivered within the correct interval, but with an incorrect value.
- Timing—the correct value is delivered, but outside the correct interval.
- Emission—a response that is expected, but has either (or both) incorrect time or value.

Propagation is the sequence of effects resulting from an initial deviation or failure. Based on these four types of failures, the casual tree of function deviation propagation and the combinations of different deviations are analyzed.

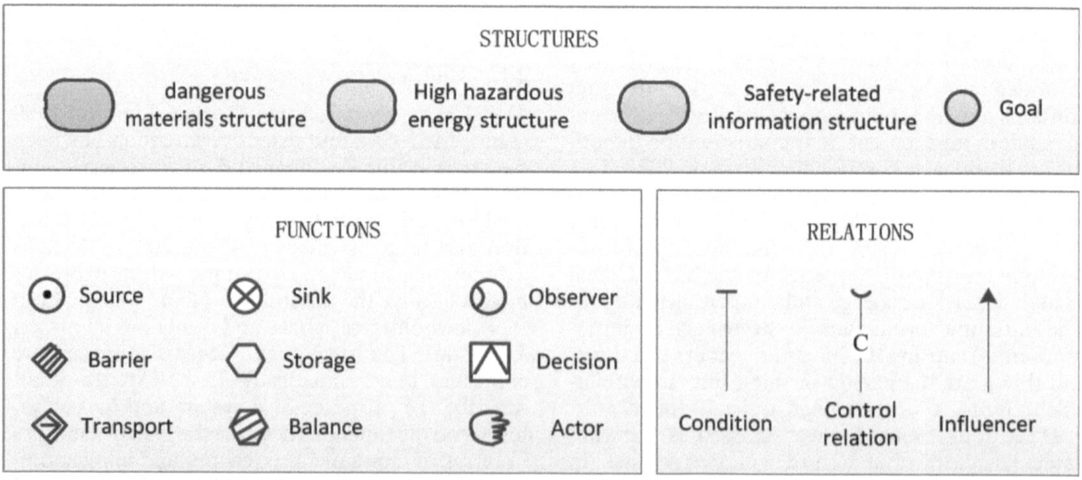

Figure 1. Multilevel Flow Modeling concept and extension.

3 CASE STUDY

In this paper, a preliminary design of aircraft fuel system is taken as a case study for hazard analysis. The fuel system is composed of different subsystems, including fuel supplying subsystem, fuel transmission subsystem, refuel and discharge subsystem, air pressure subsystem and monitoring subsystem. Due to paper space limitations, this paper focuses mainly on the hazard analysis of fuel supplying subsystem to demonstrate the feasible and effective of the technology.

3.1 *Multilevel flow model of fuel system*

The fuel supplying subsystem includes 2 pumps and its pressure sensors respectively, fuel pipeline, manual shut-off valve, and flow sensor. The fuel supplying process inside the fuel system is shown in Figure 2.

The functions of fuel supplying subsystem include the following:

1. Supplying the fuel from fuel tank to engine according to design requirements;
2. Providing the information about fuel flow and fuel exhaust to engine.

The multilevel flow model of the fuel supplying subsystem is shown in Figure 3, whose definitions of each component are listed in Table 1.

There are seven flow nets of M1, I1, I2, E1, E2, E3 and E4, which are defined as follows:

1. M1 is the fuel flow structure in which the fuel flows from tank to engine;
2. I1 is remaining fuel information structure which provide the information to display;
3. I2 is fuel flow information structure providing the operation parameter instructions to pumps and displaying fuel flow speed;
4. E1 is current flow structure which supply power to pumps;
5. E2 is current flow structure which supply power for pump pressure sensors;
6. E3 is the current flow structure supplying power for flow sensor;
7. E4 is the current flow structure supplying power for fuel flow sensor.

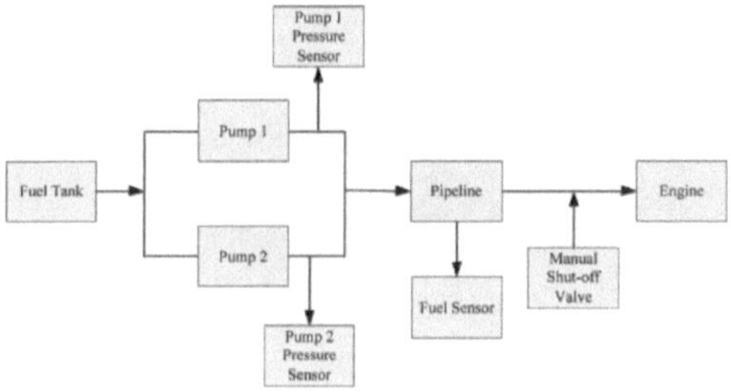

Figure 2. Fuel supplying process of fuel system.

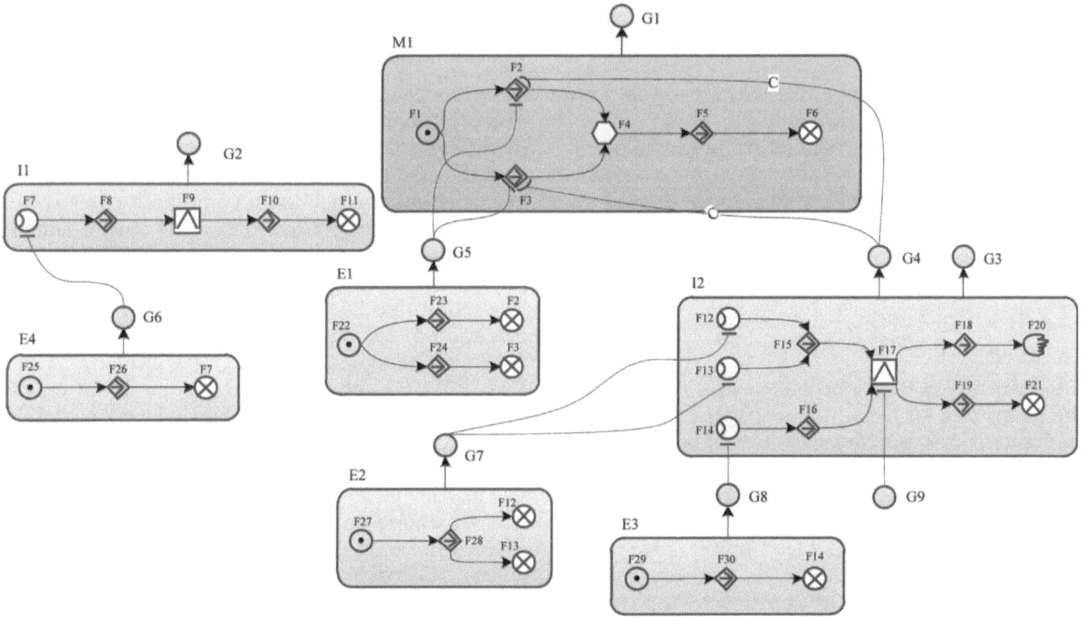

Figure 3. Multilevel Flow Modeling concepts and extensions.

Table 1. Definition of each functional part in the fuel supply subsystem.

No.	Definition	No.	Definition
F1	Fuel tank	F16	Transmission line
F2	Pump 1	F17	Computer
F3	Pump 2	F18	Transmission line
F4	Pipeline	F19	Transmission line
F5	Manual shut-off valve	F20	Pump operation instruction
F6	Engine	F21	Flow speed display
F7	Fuel sensor	F22	Power
F8	Transmission line	F23	Transmission line
F9	Computer	F24	Transmission line
F10	Transmission line	F25	Power
F11	Display	F26	Transmission line
F12	Pump 1 pressure sensor	F27	Power
F13	Pump 2 pressure sensor	F28	Transmission line
F14	Flow sensor	F29	Power
F15	Transmission line	F30	Transmission line

The model contains nine goals of G1, G2, G3, G4, G5, G6, G7, G8 and G9, the representations are as follows:

1. G1: fuel engine according to specific requirements;
2. G2: provide information about fuel exhaust;
3. G3: provide fuel flow information;
4. G4: sub-aim. Supply operation requirement to two pumps;
5. G5: sub-aim. Supply stable current to two pumps;
6. G6: sub-aim. Supply stable current to fuel flow sensor;
7. G7: sub-aim. Supply stable current to two pump pressure sensors;
8. G8: sub-aim. Supply stable current to flow sensor;
9. G9: sub-aim. Provide operation requirements to two pumps.

3.2 *Causal tree of deviation*

This paper takes the pump 1 as an example to build the causal tree for its failure propagations. The fuel feed pump 1 is selected as the root node of the causal tree and all the nodes, which might cause the deviations of the fuel feed pump, are listed according to the logic of the MFM. The function deviation causal tree of the fuel feed pump is shown in Figure 4. The part above F2 is the components that can result in deviation (only shown partly in figure), while the below is the components affected by those deviations. Table 2 shows deviation modes of F2. By the MFM model, it is easy to get the causality relationships between different components. The causal tree provides an effective approach to analyze the cause and effect of deviations.

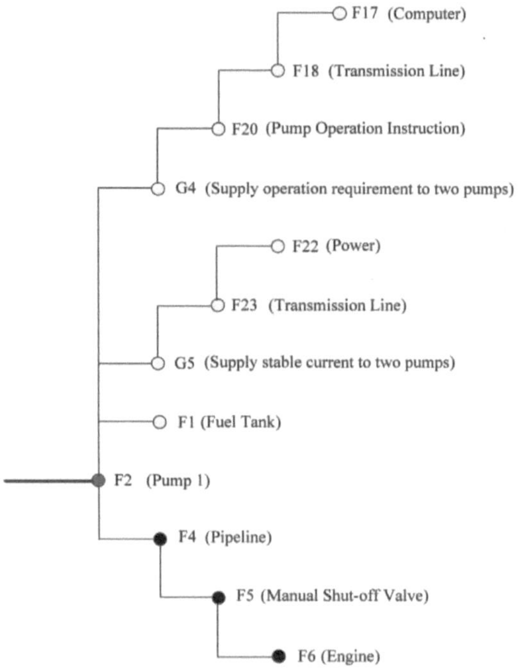

Figure 4. Deviation causal tree of pump 1.

3.3 *Deviation combination analysis*

Based on the analysis of the deviation causal tree of a single component, the deviation combinations of different components are investigated to make sure the effect to dangerous elements and identify hazard. Take the deviation combinations between pump 1 and pump 2 for example. Figure 5 shows the propagation of deviation combination between pump 1 (F2) and pump 2 (F3) resulting from G5. The red dotted line in the figure describes that the deviation of G5 results in the deviation of F3, the red solid line represents the deviations of F12, F15, F17, F18, F20 and G4 result in the deviation of F2, and the black solid line shows the propagation effect resulted from the deviation combinations between F2 and F3.

The deviations of F2 are presented in section 4.2. Based on the same method, the deviations of F3 can be investigated. Table 3 shows only the combination effect of the same deviation modes of F2 and F3.

According to the information listed in Table 3, it can be concluded that the omission and timing

Table 2. Deviation analysis for the deviation of pump 1.

Deviation (failure)	Possible reasons	Consequence
Omission	1. No fuel in tank 2. No stable current supplying to pumps 3. Transmission line open 4. No power 5. No instruction to pumps (instruction omission) 6. Computer omission	1. Omission of pipeline (combining with the deviation of fuel flow sensor can result in hazard) 2. No fuel to engine
Value	1. Fuel in tank too low 2. Unstable current 3. Pump operation instruction deviation 4. Value deviation of computer 5. Value deviation of pump pressure sensor and flow speed sensor	1. Value of pipeline 2. No fuel to engine and cannot meet the engine's requirement
Timing	1. Timing deviation of power supply 2. Timing deviation of operation instruction	1. Timing of pipeline 2. Intermittent fuel to engine
Emission	1. Operation instruction emission	1. Emission of pipeline 2. Cannot supply enough fuel to engine

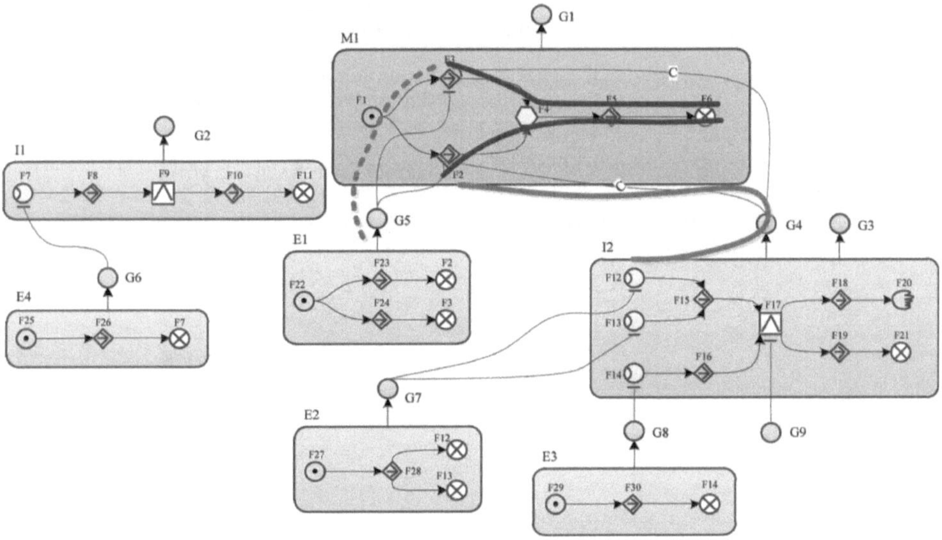

Figure 5. Deviation combination propagation.

Table 3. Effect analysis for the deviation combination of pump 1 and 2 (parts).

Deviation (failure)	Consequence
Omission	1. Omission of pipeline (combining with the deviation of fuel flow sensor can result in hazard) 2. No fuel to engine
Value	1. Value of pipeline 2. No fuel to engine and cannot meet the engine's requirement
Timing	1. Timing of pipeline 2. Intermittent fuel to engine
Emission	1. Emission of pipeline 2. Cannot supply enough fuel to engine

of pump 1 and/or pump 2 can result in the fuel to engine discontinuity or intermittence, which would influence the flight quality directly, and the value of pump can affect certain specific attitudes of aircraft.

From the common cause of two pumps, the deviations of pump power, computer and pressure sensor power are identified, which, as a design weakness, can result in two pumps' common failure. Further analysis can be focused on the deviation combination between pumps and sensors (pressure sensor or flow sensor) to identify possible combination hazards.

4 CONCLUSION

In this study, a functional hazard analysis based on multilevel flow modeling is developed to identify hazard. Using the hazard analysis, high hazardous energy, dangerous materials and information flow substructures are modeled, which made the analysis process more targeted and the analysis results more comprehensive.

Each component of the model may exhibit deviation or may be sensitive to deviations in its interactions, which are initiated by other components. Combine with failure mode classes, possible function achievement changes can be determined. The function deviation causal tree of component is introduced, and the influence propagation upon a series of functions is analyzed to recognize the initiating mechanisms, target and threat. Hazard can be identified by using the influence propagation analysis.

REFERENCES

Ericson, C.A. 2005. Hazard Analysis Techniques for System Safety [M]. Hoboken: Wiley.

Jose Luis de la, M. & Manuel, R. 2010. Accident prevention by control system reconfiguration. Computers and chemical engineering 34: 846–855.

Larsson, J.E. 2000. Knowledge engineering using multilevel flow models. Technical report. Lund Institute of Technology.

Lind, M. 1994. Modeling goals and functions of complex industrial plant. Applied Artificial Intelligence, 8(2).

McDermid, J.A. & Pumfrey D.J. 1994. A Development of Hazard Analysis to aid Software Design, in proceedings of 9th Conference on Computer Assurance (COMPASS '94), pp. 17–25.

USA. 2012. MIL-STD-882E. Department of defense standard practice system safety [S].

Wang, M.D. & G.M. Chen & J.M. Fu et al. 2012. Safety Analysis Approach of MFM-HAZOP and Its Application in the Dehydration System of Oilfield United Station. Procedia Engineering 43: 437–442.

Zhan, J. 2007. Application of Reliability Analysis Based on Multilevel Flow Models in Nuclear Power Plant, Harbin Engineering University.

Zhao, T.D. & J. Jiao & Y. Zhao et al. 2012. Hazard Analysis Techniques for System Safety [M]. Translated. Beijing: National Defense Industry Press.

Manufacturing systems and industrial design

Engineering Management and Industrial Engineering – Leung (Ed.)
© 2015 Taylor & Francis Group, London, ISBN 978-1-138-02773-2

Analysis of torsional vibration characteristics of Doubly Fed Induction Generators connected to grid

A.Y.C. Lou & D. Xie
Department of Electrical Engineering, Shanghai Jiao Tong University, Shanghai, China

B.J.Q. Feng
Department of Electrical Planning, East China Electric Power Design Institute, Shanghai, China

ABSTRACT: The Doubly Fed Induction Generators connected to the grid are likely to increase regional low frequency oscillation and wind turbine shafting torsional vibration. To study the DFIG torsional vibration, a detailed model based on small signal stability analysis is established, including the mechanical system and electrical system. Then, the characteristics of the DFIG torsional vibration is analyzed through the modal analysis method. And the effects of the model parameters on shaft torsional vibration are studied. Results show that the damping ratio can change with some parameters changed, which is of great significance to the torsional vibration suppression and power system stability.

1 INTRODUCTION

In recent years, wind energy has been developed a lot [1–2], which has led to large wind turbines being integrated into electric power grids. Since mechanical failure of large wind turbines occurred frequently. Issues of shaft torsional vibration, the interaction between the mechanical and electrical systems of wind turbines have been receiving wide concerns. The exciting electrical torques related to the quick action of the power electronic devices, harmonics and power system faults introduce a negative damping. If the damping created by the mechanical torques cannot compensate the negative damping of the electrical system, the mechanical torques at torsional frequencies can be amplified, which potentially leads to shaft fault [3]. So it is necessary to study the wind turbine torsional vibration characteristics in deep.

In order to investigate the interaction between the wind turbines and power grid, a variety of wind turbine models have been developed [4]–[7]. [8] simplify the modeling of the mechanical drive train of windmill by using lump model. Fundamental principles of rotational system are described and two-mass shaft model of windmill is developed in [9]. The computational method of shaft system parameters is provided in [10]. Based on that, the corresponding three-mass windmill drive train model, the simplified two-mass model and the single lumped mass windmill drive train are then given. [11–12] presents a decoupled active and reactive powers control of the DFIG wind turbine based on time-domain simulations. [13–14] uses the modal analysis to characterize the small-signal behaviour of DFIG wind turbine.

This paper provides the small-signal stability model of the grid-connected DFIG wind turbine in detail. The modal analysis is used to analyze the torsional vibration characteristics. Finally, the effect of system parameters on the torsional vibration is studied.

2 MODELING OF GRID-CONNECTED DFIG WIND TURBINE

The schematic diagram of the studied DFIG wind turbine is shown in Figure 1.

2.1 *Three-mass shaft model*

The three-mass shaft model is adopted to describe the drive train system, and the corre-

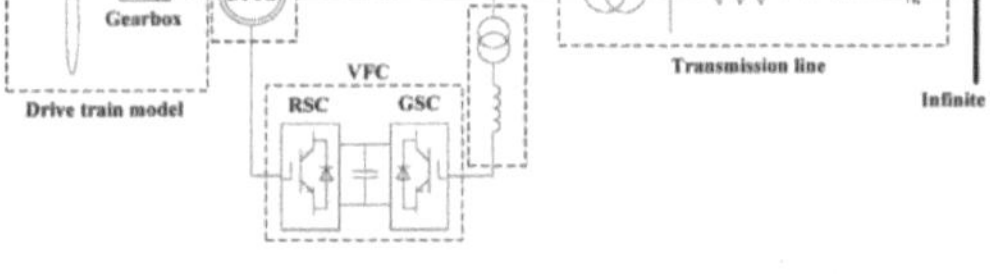

Figure 1. Schematic diagram of the DFIG wind turbine connected to infinite bus system.

sponding standard state equation is presented as follows [15]:

$$\dot{X}_M = A_M X_M + B_M u_M \tag{1}$$

where $X_M = \left[\Delta\theta_1 \quad \Delta\theta_2 \quad \Delta\theta_3 \quad \Delta\omega_1 \quad \Delta\omega_2 \quad \Delta\omega_3\right]^T$

$u_M = \left[\Delta T_W \quad \Delta T_e\right]^T$

$$A_M = \begin{bmatrix} 0 & 0 & 0 & 1 & 0 & 0 \\ 0 & 0 & 0 & 0 & 1 & 0 \\ 0 & 0 & 0 & 0 & 0 & 1 \\ \dfrac{-K_{12}}{M_1} & \dfrac{K_{12}}{M_1} & 0 & \dfrac{-(D_1+D_{12})}{M_1} & \dfrac{D_{12}}{M_1} & 0 \\ \dfrac{K_{12}}{M_2} & \dfrac{-(K_{12}+K_{23})}{M_2} & \dfrac{K_{23}}{M_2} & \dfrac{D_{12}}{M_2} & \dfrac{-(D_{12}+D_2+D_{23})}{M_2} & \dfrac{D_{23}}{M_2} \\ 0 & \dfrac{K_{23}}{M_3} & \dfrac{-K_{23}}{M_3} & 0 & \dfrac{D_{23}}{M_3} & \dfrac{-(D_{23}+D_3)}{M_3} \end{bmatrix}$$

$$B_M = \begin{bmatrix} 0\ 0\ 0\ \dfrac{1}{M_1}\ 0\ 0 \\ 0\ 0\ 0\ \ 0\ \ 0\ -\dfrac{1}{M_3} \end{bmatrix}^T$$

2.2 *Induction generator model*

For modeling the induction generator, A 4th order dynamic model is used to describe as follows [16]:

$$\dot{X}_G = A_G X_G + B_G u_G \tag{2}$$

where $X_G = \left[\Delta\psi_{qs} \quad \Delta\psi_{ds} \quad \Delta\psi_{qr} \quad \Delta\psi_{dr}\right]^T$

$u_G = \left[\Delta u_{qs} \quad \Delta u_{ds} \quad \Delta u_{qr} \quad \Delta u_{dr} \quad \Delta\omega_r\right]^T$

Defining

$$\begin{cases} D = X_s X_r + (X_s + X_r) X_m \\ X_{ss} = X_s + X_m \\ X_{rr} = X_r + X_m \end{cases} \tag{3}$$

Then

$$A_G = \begin{bmatrix} \dfrac{-\omega_b R_s X_{rr}}{D} & -\omega_s\omega_b & \dfrac{\omega_b R_s X_m}{D} & 0 \\ \omega_s\omega_b & \dfrac{\omega_b R_s X_{rr}}{D} & 0 & \dfrac{\omega_b R_s X_m}{D} \\ \dfrac{\omega_b R_r X_m}{D} & 0 & \dfrac{-\omega_b R_r X_{ss}}{D} & -s_0\omega_s\omega_b \\ 0 & \dfrac{\omega_b R_r X_m}{D} & s_0\omega_s\omega & \dfrac{-\omega_b R_r X_{ss}}{D} \end{bmatrix}$$

$$B_G = \begin{bmatrix} \omega_b & 0 & 0 & 0 & 0 \\ 0 & \omega_b & 0 & 0 & 0 \\ 0 & 0 & \omega_b & 0 & \omega_s\omega_b\psi_{dr0} \\ 0 & 0 & 0 & \omega_b & -\omega_s\omega_b\psi_{qr0} \end{bmatrix}$$

2.3 *DFIG converter model*

The DFIG converter consists of a rotor-side converter and a grid-side converter which are connected back-to-back via a dc-link. The circuit of the DFIG converter is shown in Figure 2.

2.3.1 *DC-link model*

The dc-link model can be described as follows:

$$CV_{DC}\dot{V}_{DC} = u_{dg}i_{dg} + u_{qg}i_{qg} - \left(u_{dr}i_{dr} + u_{qr}i_{qr}\right) \tag{4}$$

And its corresponding standard state equation can be obtained:

$$\dot{X}_{DC} = A_{DC} X_{DC} + B_{DC} u_{DC} \tag{5}$$

where $X_{DC} = \left[\Delta V_{DC}\right]$

$u_{DC} = \left[\Delta u_{qg} \ \ \Delta u_{dg} \ \ \Delta i_{qg} \ \ \Delta i_{dg} \ \ \Delta u_{qr} \ \ \Delta u_{dr} \ \ \Delta i_{qr} \ \ \Delta i_{dr}\right]^T$

$A_{DC} = 0$

$B_{DC} = \left[i_{qg0} \ \ i_{dg0} \ \ u_{qg0} \ \ u_{dg0} \ \ -i_{qr0} \ \ -i_{dr0} \ \ -u_{qr0} \ \ -u_{dr0}\right].$

2.3.2 *Rotor-side converter model*

The rotor-side converter is responsible for regulating DFIG active power and terminal voltage based on stator flux vector orientation method. The block diagram of rotor-side converter controller is shown in Figure 3.

Then the standard state equation of rotor-side converter can be obtained as follows

$$\dot{X}_r = A_r X_r + B_r u_r \tag{6}$$

where $X_r = \left[\Delta x_1 \quad \Delta x_2 \quad \Delta x_3 \quad \Delta x_4\right]^T$

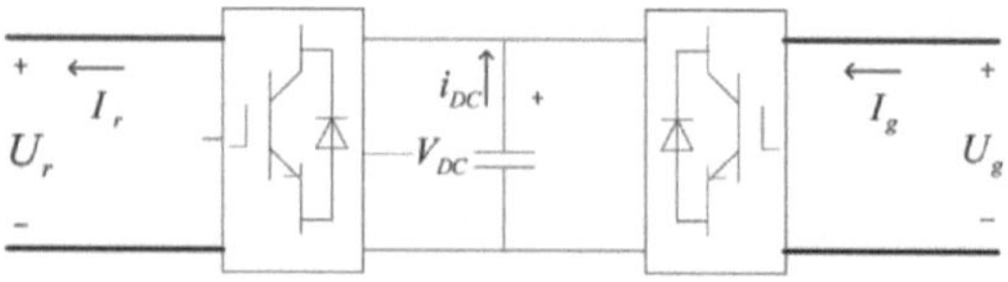

Figure 2. Circuit of the DFIG converter.

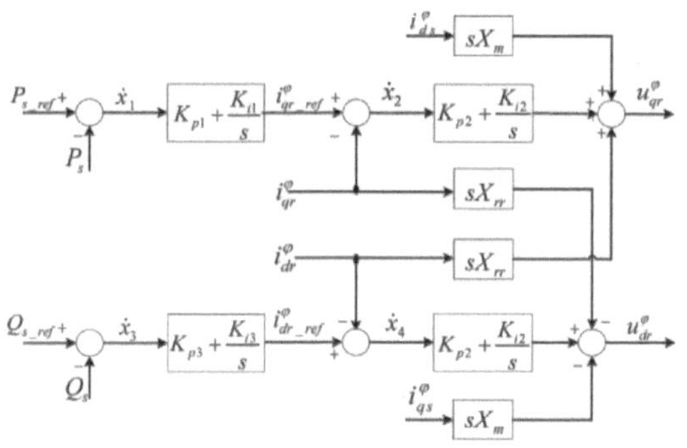

Figure 3. Block diagram of rotor-side converter controller.

$$u_r = \left[\Delta u_{ds}^{\phi} \;\; \Delta u_{qs}^{\phi} \;\; \Delta i_{ds}^{\phi} \;\; \Delta i_{qs}^{\phi} \;\; \Delta i_{dr}^{\phi} \;\; \Delta i_{qr}^{\phi} \;\; \Delta P_{s\_ref} \;\; \Delta Q_{s\_ref}\right]^T$$

$$A_r = \begin{bmatrix} 0 & 0 & 0 & 0 \\ k_{i1} & 0 & 0 & 0 \\ 0 & 0 & 0 & 0 \\ 0 & 0 & k_{i3} & 0 \end{bmatrix}$$

$$B_r = \begin{bmatrix} -i_{ds0}^{\phi} & -i_{qs0}^{\phi} & -u_{ds0}^{\phi} & -u_{qs0}^{\phi} & 0 & 0 & 1 & 0 \\ -k_{p1}i_{ds0}^{\phi} & -k_{p1}i_{qs0}^{\phi} & -k_{p1}u_{ds0}^{\phi} & -k_{p1}u_{qs0}^{\phi} & 0 & -1 & k_{p1} & 0 \\ i_{qs0}^{\phi} & -i_{ds0}^{\phi} & -u_{qs0}^{\phi} & u_{ds0}^{\phi} & 0 & 0 & 0 & 1 \\ k_{p3}i_{qs0}^{\phi} & -k_{p3}i_{ds0}^{\phi} & -k_{p3}u_{qs0}^{\phi} & k_{p3}u_{ds0}^{\phi} & -1 & 0 & 0 & k_{p3} \end{bmatrix}$$

2.3.3 *Grid-side converter model*

The grid-side converter is responsible for controlling the dc-link voltage as a present constant value and DFIG reactive power. The block diagram of GSC controller is shown in Figure 4.

Similar to the rotor-side converter, the standard state equation of GSC can be obtained as follows

$$\dot{X}_g = A_g X_g + B_g u_g \tag{7}$$

where $X_g = \left[\Delta x_5 \;\; \Delta x_6 \;\; \Delta x_7\right]^T$

$$u_g = \left[\Delta V_{DC\_ref} \;\; \Delta V_{DC} \;\; \Delta i_{qg\_ref} \;\; \Delta i_{dg} \;\; \Delta i_{qg}\right]^T$$

$$A_g = \begin{bmatrix} 0 & 0 & 0 \\ k_{idg} & 0 & 0 \\ 0 & 0 & 0 \end{bmatrix}$$

$$B_g = \begin{bmatrix} 1 & -1 & 0 & 0 & 0 \\ k_{pdg} & -k_{pdg} & 0 & -1 & 0 \\ 0 & 0 & 1 & 0 & -1 \end{bmatrix}.$$

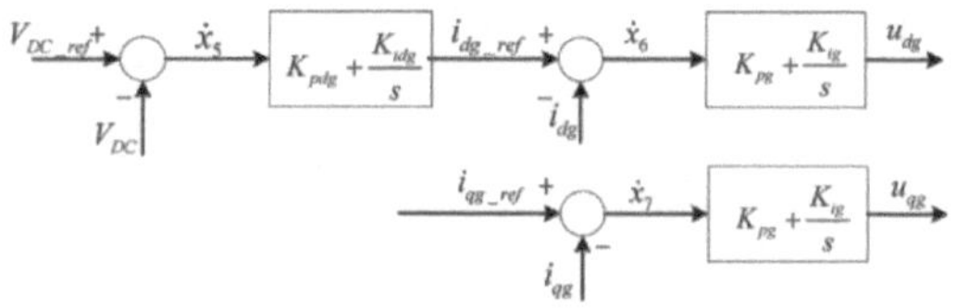

Figure 4. Block diagram of grid-side converter controller.

2.4 *Transmission line model*

The standard state equation of transmission line model is described by

$$\dot{X}_{RLC} = A_{RLC}X_{RLC} + B_{RLC}u_{RLC} \tag{8}$$

where $X_{RLC} = \left[\Delta i_x \;\; \Delta i_y \;\; \Delta u_{cx} \;\; \Delta u_{cy}\right]^T$

$$u_{RLC} = \left[\Delta u_{x1} \;\; \Delta u_{y1} \;\; \Delta u_{x2} \;\; \Delta u_{y2}\right]^T$$

$$A_{RLC} = \begin{bmatrix} -\omega_b r/x & \omega_b & -\omega_b/x & 0 \\ -\omega_b & -\omega_b r/x & 0 & -\omega_b/x \\ -\omega_b/x & 0 & 0 & \omega_b \\ 0 & -\omega_b/x & -\omega_b & 0 \end{bmatrix}$$

$$B_{RLC} = \begin{bmatrix} \omega_b/x & 0 & -\omega_b/x & 0 \\ 0 & \omega_b/x & 0 & -\omega_b/x \\ 0 & 0 & 0 & 0 \\ 0 & 0 & 0 & 0 \end{bmatrix}$$

3 MODAL ANALYSIS OF GRID-CONNECTED DFIG WIND TURBINE

3.1 *The union small-signal stability model*

The union model of grid-connected DFIG wind turbine is obtained as follows:

$$\dot{X} = AX + Bu \tag{9}$$

where $X = \left[X_M \;\; X_G \;\; X_{TL} \;\; X_{DC} \;\; X_r \;\; X_g\right]^T$.

3.2 *Modal analysis*

Based on small-signal stability analysis, the aforementioned wind turbine components are integrated to form a 2 MW, 690 V stall wind turbine model in MATLAB/SIMULINK.

The eigenvalues of the system state matrix are calculated and shown in Table 1. It can be seen that the system is stable since all of the eigenvalues have negative real parts. There are seven oscillation modes. The participation factors which disclose

Table 1. Eigenvalues of matrix A.

	Eigenvalue	Frequency, Hz	Damping
1	$-15.55 \pm 508.40i$	80.914	0.031
2	$-14.83 \pm 125.15i$	19.918	0.118
3	$-36.05 \pm 74.19i$	11.807	0.437
4	$-0.15 \pm 78.08i$	12.427	0.0019
5	$-0.029 \pm 11.09i$	1.765	0.0026

Table 2. Participation factors of torsional modes.

	1	2	3	4	5
$\Delta\theta_1$	0.0000	0.0000	0.0000	0.0002	0.0810
$\Delta\theta_2$	0.0000	0.0000	0.0000	**0.4451**	0.0434
$\Delta\theta_3$	0.0000	0.0000	0.0000	0.0547	**0.3756**
$\Delta\omega_1$	0.0000	0.0000	0.0000	0.0002	0.0810
$\Delta\omega_2$	0.0000	0.0000	0.0000	**0.4451**	0.0434
$\Delta\omega_3$	0.0000	0.0000	0.0000	0.0547	**0.3756**
$\Delta\psi_{qs}$	0.0062	0.0429	**0.1011**	0.0000	0.0000
$\Delta\psi_{ds}$	0.0073	0.0058	0.0371	0.0000	0.0000
$\Delta\psi_{qr}$	0.0004	0.0230	**0.1976**	0.0000	0.0000
$\Delta\psi_{dr}$	0.0013	0.0057	**0.3635**	0.0000	0.0000
Δi_{Lx}	**0.2443**	**0.2294**	0.0556	0.0000	0.0000
Δi_{Ly}	**0.2450**	**0.1929**	0.0440	0.0000	0.0000
$\Delta u_{sc,x}$	**0.2511**	**0.2206**	0.0203	0.0000	0.0000
$\Delta u_{sc,y}$	**0.2514**	**0.2365**	0.0345	0.0000	0.0000
Δi_{gx}	0.0000	0.0000	0.0007	0.0000	0.0000
Δi_{gy}	0.0000	0.0000	0.0002	0.0000	0.0000
ΔV_{DC}	0.0000	0.0000	0.0009	0.0000	0.0000
Δx_1	0.0003	0.0424	**0.2737**	0.0000	0.0000
Δx_2	0.0000	0.0011	**0.1106**	0.0000	0.0000
Δx_3	0.0005	0.0250	**0.1881**	0.0000	0.0000
Δx_4	0.0000	0.0042	0.0604	0.0000	0.0000
Δx_5	0.0000	0.0000	0.0000	0.0000	0.0000
Δx_6	0.0000	0.0000	0.0002	0.0000	0.0000
Δx_7	0.0000	0.0000	0.0022	0.0000	0.0000

the relation between the modes and the variables are shown in Table 2. The factors typed boldly express that the eigenvalues are highly sensitive to the corresponding variables.

As shown in Tables 1 and 2, the frequencies of oscillation mode 1 and 2 are 80.914 Hz and 19.918 Hz. They are highly sensitive to Δi_{Lx}, Δi_{Ly}, $\Delta u_{sc,x}$, $\Delta u_{sc,y}$. The oscillation mode 3, with frequency as 11.807 Hz, is mainly affected by the stator and rotor flux. The oscillation mode 4 is mainly affected by the torsion angle and speed of the low-speed shaft. In addition, it is also affected by the torsion angle and speed of the high-speed shaft to some degree. Accordingly, the oscillation mode 4 is the mechanical oscillation mode. The oscillation mode 5 is mainly affected by the torsion angle and speed of the high-speed shaft. In addition, it is also affected by the torsion angle and speed of the blades and high-speed shaft to some extent. Consequently, the oscillation mode 5 is also the mechanical oscillation mode. So the model in this paper has only two natural frequencies of 12.427 Hz and 1.765 Hz, and the oscillation mode 4 and 5 are the torsional modes.

4 SYSTEM PARAMETERS'S EFFECTS

4.1 *Parameters of series compensation capacitor*

When we keep all other parameters constant and make the series compensation degree k_{com} very, the variations with frequencies and damping ratios of the mechanical oscillation mode 4, 5 are shown in Table 3.

As is shown in Table 3, the magnitudes of the frequency and damping ratio have little change for the modes 4 and 5 as k_{com} increases as a whole. Hence, the observations indicate that the modes have little association with the electrical torque.

It is also true when the value of the flat-wave reactor and the seven PI parameters of converter controllers vary respectively, the frequencies and the damping ratios of mode 4 and 5 remain unchanged. And so is the series compensation capacitor.

4.2 *Parameters of speed controller*

As is shown in Figure 4, the rotor-side converter uses the stator active power as the controlled variable. If we use the rotor speed as the controlled variable respectively, and apply a PI controller in speed control scheme to track the reference speed, the association between rotor speed and PI parameters will be strengthened a lot. The block diagram

Table 3. Effect of series compensation level.

	Mode 4		Mode 5	
k_{com}	Frequency, Hz	Damping	Frequency, Hz	Damping
0.10	12.43	0.0019	1.765	0.0026
0.20	12.43	0.0019	1.765	0.0026
0.30	12.43	0.0019	1.765	0.0026
0.40	12.43	0.0019	1.765	0.0026
0.50	12.43	0.0019	1.765	0.0026
0.60	12.43	0.0019	1.765	0.0026
0.70	12.43	0.0019	1.765	0.0026
0.80	12.43	0.0019	1.765	0.0026
0.90	12.43	0.0019	1.765	0.0026
1.00	12.43	0.0019	1.765	0.0026

of rotor-side converter with speed controller listed below is shown in Figure 5, and the speed controller is marked with dotted lines.

The eigenvalues of the new system state matrix A are calculated and shown in Table 4.

Compared with Table 2, one more oscillation mode with frequency 0.406 Hz and damping ratio 0.009 can be found in Table 4. Given that mode 1~3 have few changes and are not mechanical torsional mode, we just take mode 4~6 into consideration, whose corresponding participation factors are shown in Table 5.

As is shown in Table 4, mode 4 has little changes in its frequency and damping ratio, while mode 5 changes with the frequency from 1.765 Hz to 2.109 Hz and the damping ratio from 0.0026 to 0.02 respectively, which can be explained from Table 5. Mode 4 is mainly affected by low-speed shaft, while mode 5 is affected by high-speed shaft. Since the speed controller only enhances effect of PI parameters on rotor speed (of high-speed shaft), the state variable Δx_8 has more significant effect on mode 5 and less on mode 4. In addition, it also can be obtained from Table 4 that the speed controller can increase system damping of mode 5.

Mode 6 is mainly affected by the angle and speed of blades and high-speed shaft as well as PI controller, which are concerned with not only the mechanical system but also the electrical system.

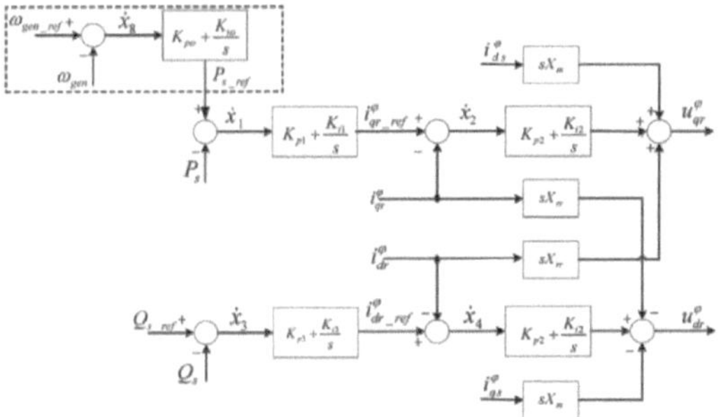

Figure 5. Block diagram of rotor-side converter with speed controller.

Table 4. Eigenvalues of matrix A.

No.	Eigenvalues	Frequency, Hz	Damping
1	$-15.55 \pm 508.40i$	80.914	0.031
2	$-14.79 \pm 125.16i$	19.919	0.117
3	$-36.05 \pm 74.19i$	11.781	0.437
4	$-0.16 \pm 78.13i$	12.434	0.002
5	$-0.27 \pm 13.25i$	2.109	0.020
6	$-0.02 \pm 2.55i$	0.406	0.009

Table 5. Participation factors of torsional modes.

No.	4	5	6
$\Delta\theta_1$	0.0002	0.0398	**0.4600**
$\Delta\theta_2$	**0.4438**	0.0495	0.0065
$\Delta\theta_3$	0.0551	**0.2554**	**0.3102**
$\Delta\omega_1$	0.0002	0.0398	**0.4600**
$\Delta\omega_2$	**0.4439**	0.0495	0.0064
$\Delta\omega_3$	0.0558	**0.4106**	0.0335
$\Delta\psi_{qs}$	0.0001	0.0002	0.0000
$\Delta\psi_{ds}$	0.0001	0.0018	0.0007
$\Delta\psi_{qr}$	0.0001	0.0000	0.0000
$\Delta\psi_{dr}$	0.0004	0.0007	0.0000
Δi_{Lx}	0.0000	0.0000	0.0000
Δi_{Ly}	0.0002	0.0002	0.0000
$\Delta u_{sc,x}$	0.0000	0.0000	0.0000
$\Delta u_{sc,y}$	0.0000	0.0000	0.0000
Δi_{gx}	0.0000	0.0000	0.0000
Δi_{gy}	0.0000	0.0000	0.0000
ΔV_{DC}	0.0000	0.0000	0.0000
Δx_1	0.0003	0.0048	0.0022
Δx_2	0.0001	0.0006	0.0000
Δx_3	0.0001	0.0001	0.0000
Δx_4	0.0000	0.0000	0.0000
Δx_5	0.0000	0.0000	0.0000
Δx_6	0.0000	0.0000	0.0000
Δx_7	0.0000	0.0000	0.0000
Δx_8	0.0005	**0.1550**	**0.3437**

And the corresponding Participation factors are 0.4600, 0.3102 and 0.3437. Therefore mode 6 is the electromagnetic oscillation mode.

When k_{pw} of PI controller in speed control scheme varys, the variations with frequencies and damping ratios of mode 4 and 5 is shown in Figure 8.

As is shown in Figure 6, with the k_{pw} value increasing, the frequencies of mode 4 and 5 remain unchanged, while the damping ratios increase significantly. Therefore the increase of k_{pw} can contribute to the torsional vibration suppression, especially for the torsional mode affected by high-speed shaft.

When k_{iw} of PI controller in speed control scheme varys, the variations with frequencies and damping ratios of mode 4 and 5 is shown in Figure 9.

As is shown in Figure 7, with the addition of k_{iw} value, the frequencies of mode 4 increase slightly. And mode 5 changes with the frequencies from 1.941 Hz to 4.174 Hz and the damping ratios from 0.0226 to −0.0028 respectively. The system of mode 5 becomes unstable. It is shown that the increase of k_{iw} can cause the damping to decrease and the exciting of the torsional vibration will

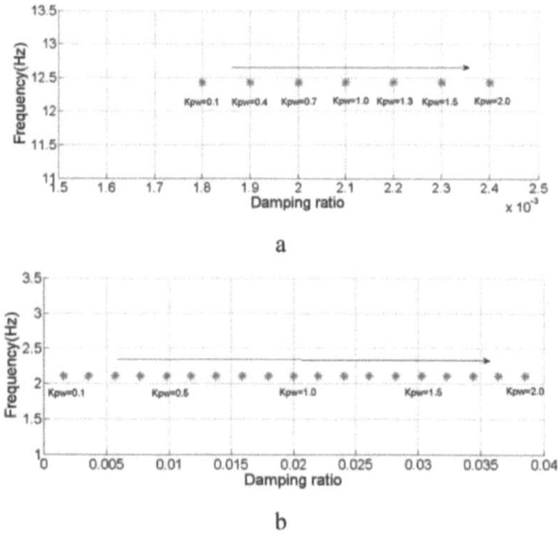

Figure 6. Variations with frequencies and damping ratios of mode 4 and 5 for increasing. k_{pw}.

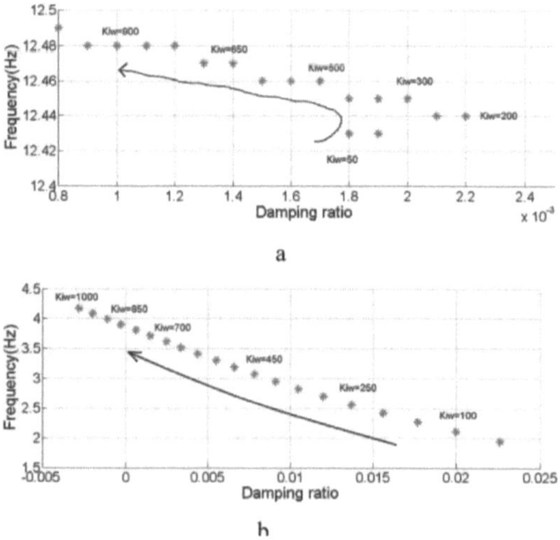

Figure 7. Variations with frequencies and damping ratios of mode 4 and 5 for increasing k_{iw}.

cause the mechanical failure. In addition, the torsional mode affected by high-speed shaft is highly sensitive to k_{iw}.

5 CONCLUSION

This paper analyzes the torsional vibration characteristics of DFIG-based wind turbine. Firstly union model based on small-signal stability analysis is established. Then the modal analysis is applied in the union system and the effects of several parameters on the torsional vibration modes are discussed. The conclusions are drawn as follows.

A Given to the DFIG wind turbine without speed controller, the torsional modes are mainly affected by the high-speed and low-speed shaft. Accordingly, parameters, like the series compensation capacitor, the flat-wave reactor and each PI parameter of converter controllers, do not have significant effect on both mechanical modes.

B The speed controller contributes to the increase of torsional damping, especially for the mode affected by the high-speed shaft. And the torsional damping increases with the addition of the proportion coefficient in speed controller but it decreases for integration coefficient. And other parameters of converter controller have slight effect on the torsional modes.

ACKNOWLEDGMENT

This work has been funded by the National Natural Science Foundation of China (51277119).

REFERENCES

[1] Li S.Y., Sun Y. Wu T. et al. 'Analysis of Small Signal Stability of Grid-Connected Doubly Fed Induction Generators'. Power and Energy Engineering Conference (APPEEC), Asia-Pacific, 2010, pp. 1–4.

[2] Luna A., Lima F.K.A., Santos D., et al. 'Simplified modeling of a DFIG for transient studies in wind power applications'. IEEE Transactions on Industrial Electronics, 2011, 58, (1), pp: 9–20.

[3] I.M. Canay, 'A Novel Approach to the Torsional Interaction and Electrical Damping of the Synchronous Machine Part I: Theory,' IEEE Transactions on Power Apparatus and Systems, vol. PAS-101, no. 10, pp. 3630–3638, 1982.

[4] V. Akhmatov and H. Khudsen, "Modeling of windmill induction generators in dynamic simulation programs," in Proc. IEEE Power Tech. Conf., Budapest, Hungary, Aug./Sept. 29–2, 1999, BPT99-243-12.

[5] E. Welfonder, R. Neifer, and M. Spanner, "Development and experimental identification of dynamic models for wind turbines and their fluctuating power generation," in Proc. Vol. From the IFAC Symp., Dec. 6–8, 1995, pp. 73–81.

[6] Ulas Eminoglua, "A New Model for Wind Turbine Systems," Electric Power Components and Systems, Vol. 37, no. 10, pp. 1180–1193, Sep. 2009.

[7] S.M. Muyeen, Mohd. Hasan Ali, Rion Takahashi, etc. "Damping of Blade-shaft Torsional Oscillations of Wind Turbine Generator System," Electric Power Components and Systems, Vol. 36, no. 2, pp. 195–211, Jan. 2008.

[8] S.K. Salman and I.M. Rida, "Investigating the impact of embedded generation on relay settings of utilities' electrical feeders," IEEE Transactions on Power Delivery, vol. 16, pp. 246–251, Apr. 2001.

[9] S. Santoso, L. Ha Thu, "Fundamental time-domain wind turbine models for wind power studies," Renewable Energy, vol. 32, no. 14, pp. 2436–2452, November 2007.
[10] S.K. Salman, A.L.J. Teo. "Windmill modeling consideration and factors influencing the stability of a grid-connected wind power-based embedded generator," IEEE Transactions on Power Systems, vol. 18, no. 2, pp. 793–802, May, 2003.
[11] Ghennam, T., E.M. Berkouk, and B. Francois. 'Modeling and control of a doubly fed induction generator (DFIG) based wind conversion system'. International Conference on Power Engineering, Energy and Electrical Drives, Lisbon, 2009, pp. 507–512.
[12] Kayikci M, Milanovic J.V. 'Assessing transient response of DFIG-based wind plants-The influence of model simplifications and parameters'. IEEE Transactions on Power Systems, 2008, 23, (2), pp. 545–554.
[13] Mei, F, Pal, B. 'Modal analysis of grid-connected doubly fed induction generators'. IEEE Transactions on Energy Conversion, 2007, 22, (3), pp. 728–736.
[14] Lingling Fan, Chanxia Zhu, Zhixin Miao, et al. 'Modal analysis of a DFIG-based wind farm interfaced with a series compensated network'. IEEE Transactions on Energy Conversion, 2011, 26, (4), pp. 1010–1020.
[15] Zhang Y, Xie D, Feng J, et al. 'Small-signal Modeling and Modal Analysis of Wind Turbine Based on Three-mass Shaft Model'. Electric Power Components and Systems, 2014, 42, (7), pp: 693–702.
[16] Prabha Kunder. 'Power system stability and control'. McGraw-Hill Professional, Canada, 1994.

Engineering Management and Industrial Engineering – Leung (Ed.)

 ISBN 978-1-138-02773-2

Development of a Wireless Indoor Electric Appliance Monitor for energy report

W. Kiatruangkrai & E. Leelarasmee
Chulalongkorn University, Bangkok, Thailand

ABSTRACT: Energy is very important in today's world, electricity from petroleum is facing problems due to the prediction of its scarcity in the future. In recent year, many devices were developed to support their users to save energy. One of the most popular devices is the In-House-Display (IHD) that is supported by smart meters. However, this method is still having some obstacles. The purpose of this paper is to find a new way to help users save energy by developing a new device to increase users' awareness, and enables him/her to change in consumption behavior that leads to energy waste reduction. The image matching technique was selected to develop this new device called Wireless Indoor Electric Appliances Monitor (WIEAM). WIEAM implements technique that captures image of appliances by using digital camera, detect their on/off status with algorithm of similarity and calculate energy consumption. Compared with other devices, WIEAM is simple, does not require highly skill for installation, and accurate with the estimation error within 10%.

1 INTRODUCTION

In today's world, energy is playing a very crucial role. Many aspects of society such as household, industrial, agriculture and transportation rely on energy in order to ensure the life styles and production of people. Most of energy productions are from petroleum such as oil, coal and gas. Based on the energy reserve survey and consumption ratio, it is found that the world has oil reservation for 25 years in estimate (EIA, 2013).

So energy saving is a very important issue that has recognized by people around the world. Many methods have been developed to help users to save energy. One of the most popular methods is energy consumption feedback. This method provides information of energy usage pattern to users as shown in Figure 1. This information helps users to know about their energy consumption, which will increases their awareness and leads to energy usage reduction (Darby, 2001).

In order to collect, record and provide energy consumption feedback to users, many devices have been developed and one of the most popular device is the so called In-House-Display (IHD) (Allen et al., 2006, Faruqui et al., 2010 & Martinez et al., 2010) supported by smart meter. In household, one meter is installed outside the house and connected to the utility interface. With this approach, only total energy consumption in the house can be collected. If consumption for each appliance is required, additional meter has to be installed with each appliance.

The information that IHD provides can be categorized as 2 groups, i.e. total consumption in the house and consumption of each appliance in the house. The first group requires one smart meter installed outside the house. The second group requires smart meter installed at each appliance as shown in Figure 2.

The benefit that users can obtain from IHD is the display of the energy usage pattern, which can be compared with reference energy use to increase awareness and change their energy consumption behavior (van Houwelingen & van Raaij, 1989).

However, there are few barriers that discourage many users from using IHD. First is the cost of devices or smart meters. The more energy consumption details are needed, the more devices have to be installed. The reason that users should acquire the energy usage in details, i.e. per each appliance or each point of use, is that it can provide specific behavior or action which can influence the users to achieve behavior changing more effectively (Wood & Newborough, 2003). However, more devices also mean a huge investment for users.

Second, the proper installation of IHD requires skillful technician to operate. So it is not convenient for users and costly to find technician to do the job. Furthermore, for old buildings which have a complex design for electricity cable, it would take a long time to find power line for each appliance.

Third, the installation of smart meter to support IHD cannot proceed without landlord approval for the users who rents the house because it has

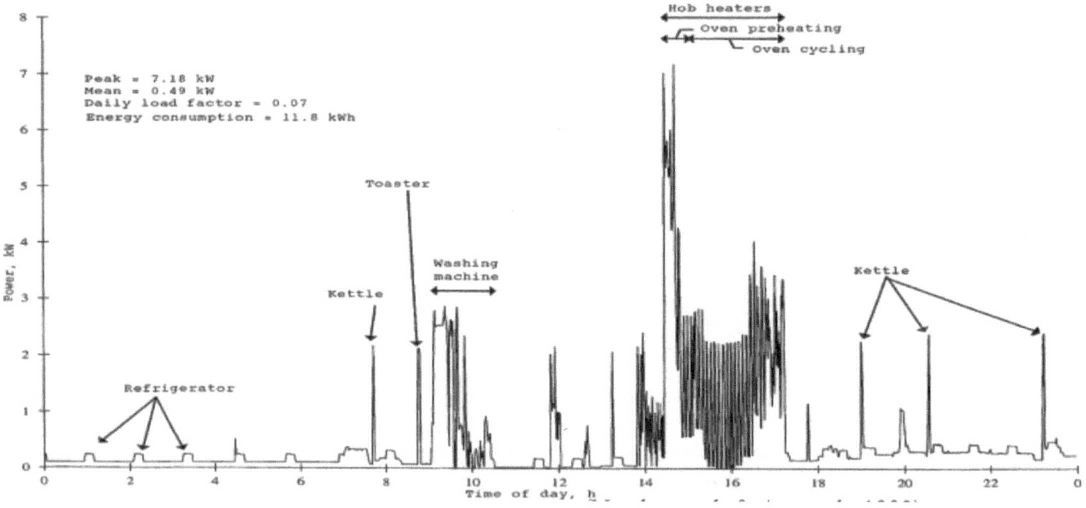

Figure 1. Example of energy usage pattern in household. (Newborough & Augood, 1999).

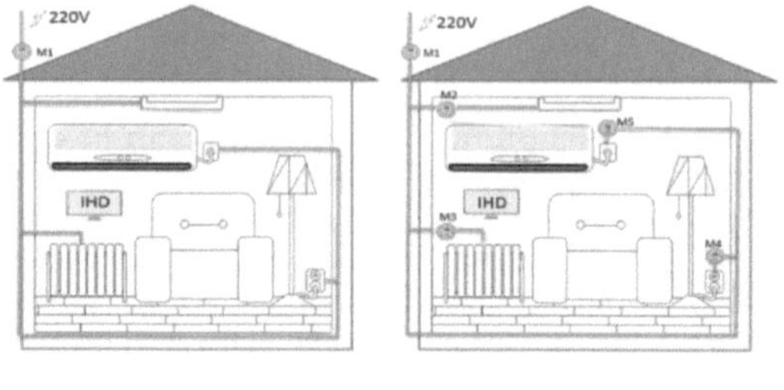

Figure 2. Two groups of IHD installation in house, Mx are the smart meters installed for energy monitoring.

to be connected with power line. This means users have to relocate and rearrange the power lines in the place.

The purpose of this paper is to find a new way to acquire energy usage of each appliance for users in order to assess their usage and change their behavior by using non-intrusive approach. While this approach has been studied by using several types of sensors such as lighting, thermal and acoustics in the past in order to assess the wastes energy usage (Richman et al., 1994, Maniccia et al., 1999 & Maniccia et al., 2001), several downsides have been found. First, the combination of sensors has to be installed in the same room in order to detect and acquire energy usage. For example, photo-sensor is used to detect luminaire and thermal sensor to detect Heating, Ventilation and Air Conditioning (HVAC). Second, larger room requires more than one sensor of each type in order to acquire accurate information. Third, uncontrollable factors, such as light or heat from outside windows, may cause malfunction of sensors and inaccurate energy usage acquisition. In this paper, one type of sensor will be used to detect luminaire and HVAC in a room by installing a digital camera as a sensor. The finished device is called Wireless Indoor Electrical Appliance Monitor (WIEAM) and it is described in the following. The concept and prototype design will be described in section 2, followed by software development in section 3 and conclusion in section 4.

2 CONCEPT AND PROTOTYPE DESIGN OF WIRELESS INDOOR ELECTRIC APPLIANCES MONITOR (WIEAM)

The conceptual design of this WIEAM is shown in Figure 3 and described as follows. A digital camera is used to capture picture of the room or office, a computer board (CPU) will receive the image that captured by digital camera to its memory and process for the on/off status of each electronic appliance detection in the room. The detection result will be sent via Wireless Fidelity (Wi-Fi) component to a computer server in order to calculate and record energy usage for users who can directly acquire information from this computer or via internet.

A prototype of the WIEAM is shown in Figure 4. It contains 3 main components. They are as follows:

1. A CMOS digital camera module shown at the bottom left of the figure. This work uses a CMOS chip number OV5647 with resolution of 5 megapixels.
2. A computer board made by Raspberry-Pi. It employs a processor BCM2835 that combines

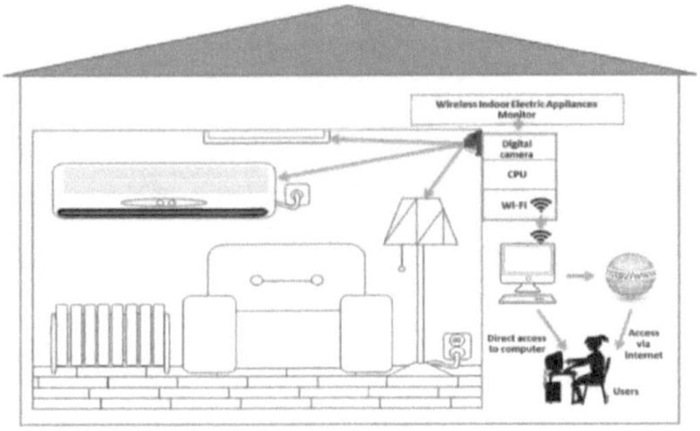

Figure 3. Conceptual design of the WIEAM.

Figure 5. Example of image taken periodically.

Figure 4. Component of the WIEAM.

Figure 6. Detection tape installed at air conditioner.

an ARM11 700 MHz processor and a Graphic Processor Unit in one chip, 512 MB of RAM for storing images and runs under Linux based operating system called Raspbian. Application programs are written in C# and Python programming. This computer is powerful enough to perform image processing algorithm for detecting the status of each appliance.

3. A Wi-Fi communication module situated at the top right of the figure. It is connected with the computer board via USB at 150 Mbps. With the Wi-Fi module, the WIEAM can transfer detected data to other devices such as notebook directly without attaching any module to the receiving end.

The WIEAM will be installed and fixed inside the room or office. Its location should be at the place where digital camera can capture the image of the whole room including all lamps and air conditioners. The digital camera will take image of the room every fixed time intervals such as 1 minute. The image will be collected and recorded in order to compare with image from the initial setting and detect status on/off of each appliance as shown in Figure 5.

In addition, the detection tape will be attached to the air conditioner in order to facilitate its status detection accurately as shown in Figure 6.

3 IMAGE MATCHING SOFTWARE DEVELOPMENT

Image matching software development will be described in this section, including algorithm to detect on/off status of each appliance. Few examples will be used to describe how the software will be developed to make it suitable for each energy usage situation. Figure 7 shows a room in which the on/off status of each corresponding appliance is different. This example indicates that the software must implement algorithm to detect on/off status for each appliance by using the image taken from the room.

Another example from Figure 8 shows that uncontrollable factors such as light from window may cause inaccurate detection. So the software must have algorithm to eliminate those factors from detection. This is done by exploiting as follows.

First the location of WIEAM installed in the room must be fixed before it can be operated.

Figure 7. Example of images which each appliance status on/off appears different.

Figure 8. Example of uncontrollable window light.

Users then use the software to make initial setting by capturing 2 images, one when all appliances are turned on and the other when they are turned off. These images will be saved in the database as reference images and used for matching comparison. To save computation, users have to select regions in the image that contain each appliance to avoid influence from outside factor and enable software to detect status of appliance separately as shown in Figure 9. Regions outside these will not be considered.

During the initial setting, users have to set the similarity level for the software to decide whether the appliance status is on or off by using algorithm of image matching percentage, too low similarity level may lead to false detection so it is important for users to test the similarity level several times as the environment may vary by each room. Details of each appliance are also need to be recorded in the database in order to calculate for energy usage and more information such as payment and CO_2 saving.

When initial setting is completed and original image of status on/off is saved in database, the WIEAM can operate by taking image of the room every minute and detect status by comparing present image with the original image, calculating the similarity level and providing result of appliance status.

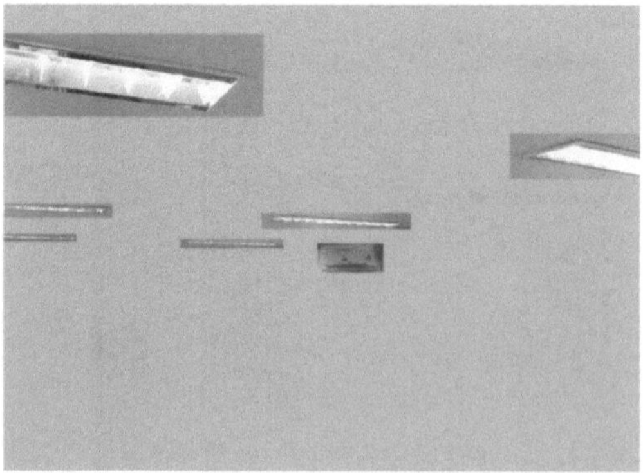

Figure 9. Example of selected area for energy usage detection.

After detection is finished, the results will be calculated by using each appliance details that is recorded at the initial setting. All calculation will be collected and saved in the database so users can search by query and get the energy usage report as per his/her requirements.

4 CONCLUSIONS AND DISCUSSION

The concept of image matching technique is briefly described and implemented in the prototype Wireless Indoor Electrical Appliance Monitor (WIEAM). Preliminary tests have found that it has a good performance when used to detect status on/off for luminaires such as light bulb and LED and provides users with useful information of energy consumption both as a total sum and per appliance with result within 10% error.

The future research for the WIEAM can be considered with several concepts. First, the hardware development can be improved to cost less and consume low power. Second, for many rooms or offices which already have CCTV or IP camera system installed for security reason, it is very interesting if this concept can be applied to those existing camera systems to save cost and time. Third, a combination of the WIEAM and motion sensor can be used to develop appliance automation control to replace the combination of photo sensor and motion sensor which are used currently to make positive energy control instead of negative energy control which relied to users awareness and behavior only.

ACKNOWLEDGEMENT

This work is supported by grants from Energy Policy and Planning Office, Ministry of Energy, Thailand.

REFERENCES

Allen, D., & Janda, K. (2006), *The Effects of Household Characteristics and Energy Use Consciousness on the Effectiveness of Real-Time Energy Use Feedback: A Pilot Study*, 2006 ACEEE Summer Study on Energy Efficiency in Buildings.

Darby, S. (2001), *Making it obvious: designing feedback into energy consumption*. Proceedings, 2nd International Conference on Energy Efficiency in Household Appliances and Lighting. Italian Association of Energy Economists.

Ehrhardt-Martinez, K., Donnelly, K., & Laitner, J. (2010), *Advanced metering initiatives and residential feedback programs,* E105, US Council for an Energy Efficient Economy.

Faruqui, A., Sergici, S., & Sharif, A. (2010), *The impact of informational feedback on energy consumption-A survey of the experimental evidence*, Energy 35 (2010) 1598–1608.

Maniccia, D., Rutledge, B., Rea, M., & Morrow, W. (1999), *Occupant use of manual lighting controls in private offices*, Journal of the Illuminating Engineering Society, 28, 42–56.

Maniccia, D., Tweed, A., Bierman, A., & Neida, B.V. (2001). *The effects of changing occupancy sensor timeout settings on energy savings, lamp cycling and maintenance costs*. Journal of Illuminating Engineering Society, 30, 97–110.

Newborough, M., & Augood, P. (1999), *Demand-side management opportunities for the UK domestic sector*, IEE Proceedings of Generation Transmission 146 (3) (1999) 283–293.

Richman E.E., Dittmer A.L., & Keller J.M. (1994), *Field Analysis of Occupancy Sensor Operation*. Prepared for the U.S. Department of Energy, Pacific NW Lab, Richland, Washington.

U.S. Energy Information Administration (EIA), (2013), [ONLINE] *International Energy Outlook 2013*, at http://www.eia.gov/forecasts/aeo/pdf/0383%282013%29.pdf, (accessed 15 July 2014).

Van Houwelingen, J., & Van Raaij, W. (1989), *The Effect of Goal-Setting and Daily Electronic Feedback on In-Home Energy Use*, Journal of Consumer Research, Vol. 16, 98–105.

Wood, G., & Newborough, M. (2003), *Dynamic energy-consumption indicators for domestic appliances: environment, behaviour and design*, Energy and Buildings 35, 821–841.

Engineering Management and Industrial Engineering – Leung (Ed.)
© 2015 Taylor & Francis Group, London, ISBN 978-1-138-02773-2

The application of Cascade Integrator Comb filter on quasi-continuous wave system

J. Bai, Y.J. Zhang, Q.H. Zhang & X.F. Zhang
Science and Technology on Millimeter-Wave Laboratory, Beijing, China

ABSTRACT: High PRF and high duty-cycle quasi-continuous wave system is studied in this paper. Cascade integrator comb filter instead of FIR filter is used as anti-aliasing digital narrowband filter. This paper analyzes the amplitude-frequency characteristic of cascade integrator comb filter and points out its shortcomings and improved methods. And this paper realizes the simulation of quasi-continuous wave system. Finally SNR loss of quasi-continuous wave system is calculated.

1 INTRODUCTION

The narrow band FIR filter is based on multi-rate signal processing. Specifically, this theory is the study of multistage realization of extraction, interpolation, equivalent transformation and sampling rate conversion. The narrow band filter whose filtering characteristic largely determines performance of system is considered as the important part of quasi-continuous wave system. How to design that high sampling frequency system has both superior filter performance and small footprint is the urgent problem. This paper uses cascade integrator comb filter to design efficient narrowband filter, which can save hardware resources while ensuring the filter characteristics. Quasi-continuous wave system using cascade integrator comb filter is simulated. Moreover, compared with the technology of wideband matching, SNR loss in quasi-continuous-wave receiver is deduced and calculated in this paper[1].

2 DESIGN OF CASCADE INTEGRATOR COMB FILTER

Cascade Integrator Comb filter (CIC) which is widely used in the multi-rate signal processing is an efficient interpolation and extraction filter. The essence of CIC filter is a FIR filter with pole-zero cancellation. CIC filter just needs adders without multipliers and system storage in hardware realization. And all of its processing coefficients are 1. So CIC filter can substantially decrease hardware resources.

The system function of CIC filter is formula 1:

$$H(z)=\left[\frac{1}{1-z^{-1}}\cdot\left(1-z^{-DM}\right)\right]^{N} = [H_1(z)\cdot H_c(z)]^{N} \quad (1)$$

D is the extracting multiple and M which is 1 or 2 in practice is the delay factor. In formula 1 $H_1(z)=1/1-z^{-1}$ is integrator and $H_c(z)=1-z^{-DM}$ is comb filter. So CIC filter is the multistage realization of the integrator and the comb filter[2].

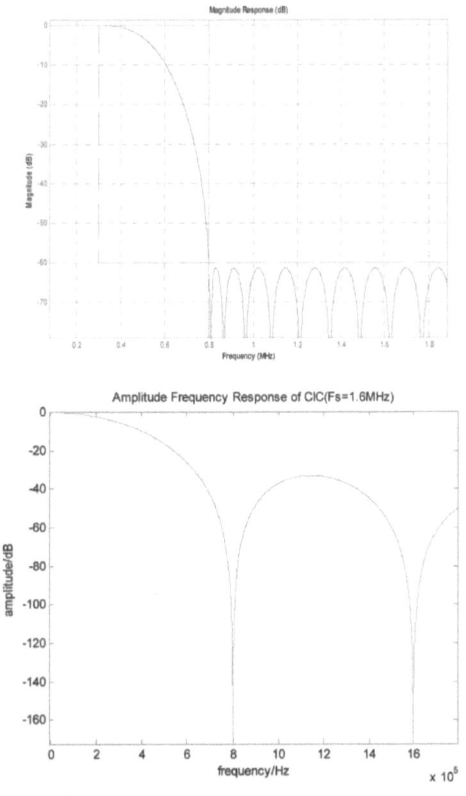

Figure 1. Amplitude-frequency characteristics of filters.

Table 1. Design parameters of filter.

Filter parameter	
Fs	1.6 MHz
Fp	800 KHz
Fst	300 KHz
Extracting multiple	25

Furthermore, the growth of the significant bit caused by CIC filter is $N \log_2 DM$.

The design of CIC filter starts from practical amplitude-frequency characteristics. Then base parameters of CIC filter such as N, D and M are determined by system requirements. Significant bits of required registers are also calculated. Finally, according to the basic structure of CIC filter, satisfactory filter is designed. This paper simply uses one of filters in this system as an example. The result is shown in Figure 1. The result of CIC is in the left and the result of FIR is in the right. The design parameters are shown in Table 1.

The results of CIC and FIR both meet the needs of this system. But the order of FIR is 265 and the other is 125, CIC filter can really save hardware resources. For the CIC filter, bigger pass-band attenuation is an inevitable problem. To compensate for this, Sharpened CIC filter is put forward. Moreover, interpolation function of second-order polynomial can also improve its performance. Behind CIC filter interpolation function of second-order polynomial is used to show an upward trend in passband and increase the passband width. Meanwhile better stop-band attenuation isn't heavily influenced. It can meet the requirements of the wide pass-band system design[3–5].

3 THE SIMULATION OF QUASI-CONTINUOUS WAVE SYSTEM

In this paper, designed CIC filter is applied to a high PRF and high duty-cycle 8 MM radar system. System parameters are as follows: center frequency is 60 MHz, and the first sampling frequency is 80 MHz which is 800 KHz after twice extracting. 8192-point FFT is performed. So the resolution of Doppler filter group is $PRF/N \times k = 800 \times 10^3/8192 \times k = 97.65625 \times k$. Meanwhile, in order to ignore straddling losses and mismatching losses, the target echo matches exactly Doppler filter group and the maximum point of the target echo is at a sample point. So the distance and the speed of designed target are 815.5 m and 998.2 m/s respectively (Doppler frequency is about 234 KHz). And SNR of target is 30 dB. Figure 2 is the block diagram of system simulation.

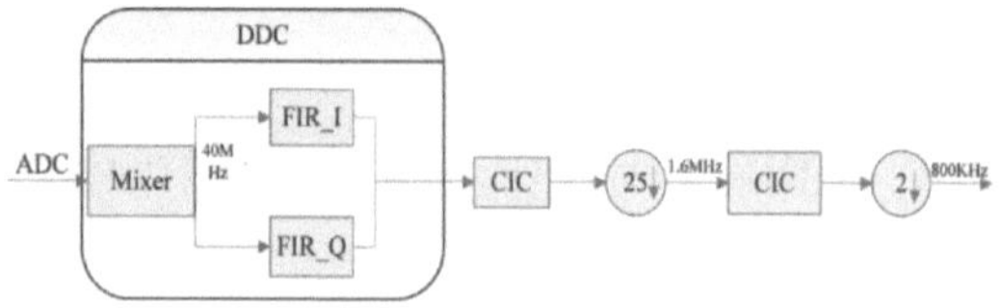

Figure 2. The block diagram of system simulation.

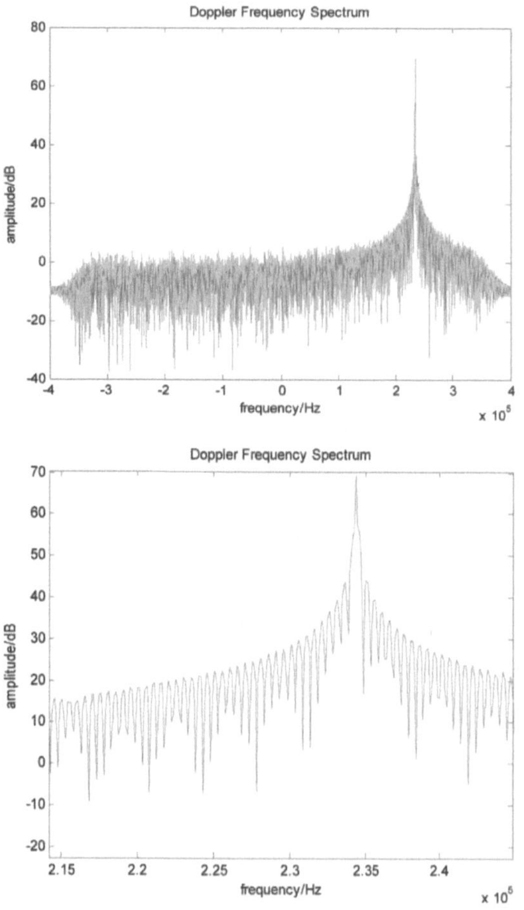

Figure 3. Simulation results.

CIC filter in practical applications causes cumulative overflow. The accumulator in FPGA is used to achieve the integrator of CIC filter. So there would be inevitably causing overflow as a result of DC component. Even if there is no DC component in the input signal, sinusoidal signal not doing integrals from zero also contains a DC component. When the DC component of the input signal is generated after the first stage accumulator, it will lead to overflow of the second stage accumulator.

Therefore, it is necessary to solve the problem in project implementation. In the case of ensuring the accuracy, the first solution is to truncate its valid bit which is the most common approach. Rational design of valid bit and rational planning in register is the second solution to not lose accuracy. And the simulation results are shown in Figure 3.

4 ANALYSIS OF SNR LOSS

In case to meet the maximum Doppler frequency of searching, this paper uses a multi-stage extraction to decline the sampling rate to PRF. And the last narrow filter only filters out the peak of pulsed Doppler train in frequency spectrum. The rest of the energy in frequency spectrum is lost which results in the loss of the SNR. High extraction rate and extremely narrow filter bandwidth cannot be blindly pursued at the same time in this system. Process diagram is shown in Figure 4.

This paper uses the above system parameters for example to derive SNR loss in quasi-continuous-wave receiver assuming PRF is 800 KHz and duty-cycle is 0.4. Compared to retain all the energy of pulsed Doppler train, the SNR loss is derived as follows:

First, for Doppler radar system, it is generally assumed that interference is from the white noise. At this time when the Doppler frequency is equal to sampling frequency of DFT, Discrete Fourier transform is the matched filter of ideal and athletic target with a fixed radial velocity. Time domain representation of pulse train is provided by formula (2) and its spectrum is formula (3) where N is the number of sub-pulses, T_r is the pulse interval, T is the pulse width.

$$u(t) = \frac{1}{\sqrt{TN}} \sum_{n=0}^{N-1} u_1(t - nT_r) \tag{2}$$

Among them, $u_1(t) = \begin{cases} 1, & 0 < t < T \\ 0, & \text{else} \end{cases}$

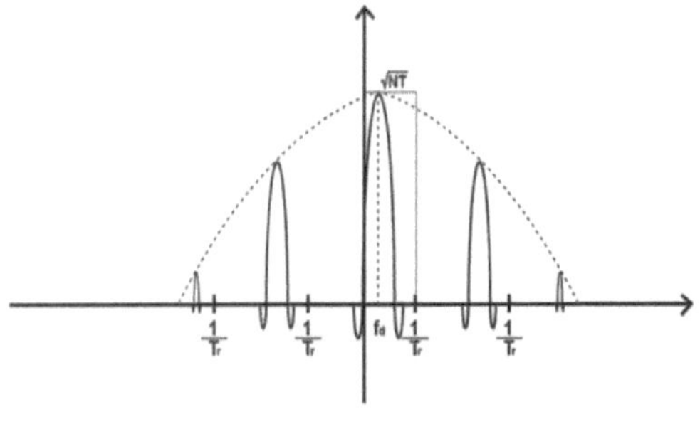

Figure 4. Process diagram.

$$U(f) = \sqrt{NT}\,\sin c(fT)\sin c(NfT_r)e^{j\pi f[(N-1)T_r + T]} \tag{3}$$

Second, the sampling frequency is reduced to pulse repetition frequency and the cutoff frequency of matching CIC anti-aliasing filter is generally half of the extracted sampling frequency. Frequency interval of pulse train in frequency spectrum is equal to pulse repetition frequency. So CIC filter just filters out its peak of target while lose other energy.

For the spectrum of the reserved peek, in this paper it is approximated as a function of sinc in frequency domain and rectangular pulse in time domain which are shown in Formula (4).

$$\sqrt{NT}\,\sin c(f \cdot NT_r) \leftrightarrow \frac{\sqrt{NT}}{NT_r} rect\left(\frac{t}{NT_r}\right) \tag{4}$$

Compared to retain all the energy of pulsed Doppler train, the processing bandwidth of white noise is always the bandwidth of Doppler filter group $1/NT_r$ which is the resolution of FFT. While the power of target changes to peak power of the approximately rectangular pulse from average power of pulse train. Suppose SNR of pulse train after matched filter is S^0/N^0 and SNR of the approximately rectangular pulse is S'/N' which is obtained by Formula (5). Among them, L is a coefficient. Specifically numerator of L is peak power of the approximately rectangular pulse and denominator is average power of pulse train.

$$\frac{S'}{N'} = L \cdot \frac{S^0}{N^0} = \frac{\left(\frac{\sqrt{NT}}{NT_r}\right)^2}{\left(\frac{1}{\sqrt{NT}}\right)^2 \cdot \frac{T}{T_r}} \cdot \frac{S^0}{N^0} = \frac{T}{T_r} \cdot \frac{S^0}{N^0} \tag{5}$$

By simplifying Formula (5), this paper assumes PRF is 800 KHz and duty-cycle is 0.4. Compared to retain all the energy of pulsed Doppler train, SNR loss is 4 dB after narrow-band filter.

5 CONCLUSIONS

This paper in digital down-converter uses CIC filter which has superior filter performance and small footprint to meet quasi-continuous-wave receiver. The simulation results are obtained by matlab. And this paper also puts forward the solutions to solve the problem of cumulative overflow and pass-band attenuation. The SNR loss in quasi-continuous-wave receiver is deduced in detail.

REFERENCES

1. Hogenauer E.B. An economical class of digital filters for decimation and interpolation [J]. IEEE Transactions on Acoustic, Speech and Signal Processing, 1981, 29(4):155–162.
2. Wan Hong xing, Chen He. VLSI architecture of a high speed parallel FFT processor [J].Application of Electronic Technology, 2005(5):1–6. (In Chinese).
3. Stephen G, Stewart R.W. Sharpening of partially non-recursive CIC decimation filters [J]. Signal, System and Computer, 2004(2):2170–2173.
4. Jouny I. Classification of clutter using the spectrum [A]. IEEE Signal Processing Workshop [C], 245–249, 1993.
5. Gao Zhenbin, Chen He, Han Yueqiu. Design and implementation of a pipeline processor for various size FFTS [J].Transactions of Beijing Institute of Technology, 2005, 25(3):268–271. (In Chinese).

Engineering Management and Industrial Engineering – Leung (Ed.)
© 2015 Taylor & Francis Group, London, ISBN 978-1-138-02773-2

Shipboard flow field simulation for helicopter-ship dynamic interface

H.R. Ma & W.S. Sun
Naval Aeronautical Engineering Academy, Qingdao, China

X.X. Gao
Qingdao University, Qingdao, China

ABSTRACT: Dynamic interface of helicopter/ship is one of the most important phases to insure onboard helicopter flight safety. This paper describes the shipboard flow field simulation techniques that demands for dynamic interface simulation. Based on the analysis of the factors involved in dynamic interface, a simplified plan is proposed to solve the coupling flow field under the rotor and above the shipboard. The key techniques in shipboard flow field simulation are described, including shipboard flow simulation with CFD, reconstruction of the flow field for dynamic interface, and the typical trajectory of helicopter landing is also described. Finally, conclusions are made to find methods of shipboard flow field simulation for helicopter/ship dynamic interface, providing references for relevant researchers.

1 INTRODUCTION

The Dynamic Interface (DI) of helicopter and ship is the process of launching and recovering of the helicopter shipboard under certain severe circumstances such as turbulent current, ship movement, obstacles at shipboard, and the poor vision of the pilot. The typical project was JSHIP[1] (Joint Shipboard Helicopter Integration Process), chartered and sponsored by the Office of Secretary of Defense to improve the interoperability of joint forces with U.S. Navy air capable ships, trying to define a process for expanding Wind Over the Deck (WOD) flight envelopes for any ship/helicopter combination using modeling and simulation. With the rapid progress of simulation technology and turbulence theory, CFD has found its popularity in shipboard flow field simulation for DI. On the other side, new achievements in fight simulation are available to us in computers. Several fight simulation softwares, FLIGHTLAB[2][3] or GENHEL, for instance, have been tested by a good application in this area. To sum up the previous research, it can be found that there are two points being focused on. One is the importance of ship airwake fidelity[4–8], and the other is the real time simulation of DI course[9][10].

2 INFLUENCE FACTORS IN DYNAMIC INTERFACE

The complexity to precisely simulate DI course comes from two aspects. One is that flow field on deck, when helicopter landing on or launching from the ship, is time-vary and unsteady, and the other is that influence factors in the course is complex and strongly coupled with each other. Therefore, the factors and their relationship with each other should be analyzed, and on the basis of that, the module of the programme modeling and simulating DI should be simplified.

The module units in DI simulation include wind at-sea, ship's physical model, sea wave, ship's motion model and flight dynamic model of the helicopter. Their relationship is showed in Figure 1.

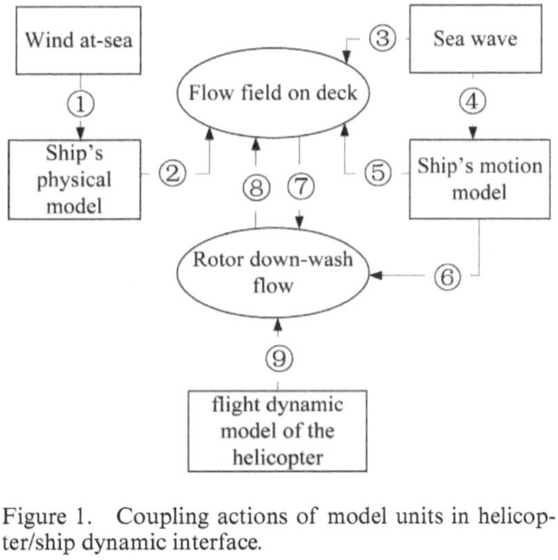

Figure 1. Coupling actions of model units in helicopter/ship dynamic interface.

As you can see in Figure 1, the bridge by which model units have impact on others is the combination with flow field on deck and rotor down-wash flow. Also, each one of them has an effect on the other one, which can be the most important relationship and which DI simulation should carefully deal with. Here are some analyses on the 9 items listed in the figure.

2.1 *Flow field on deck*

Both Items 1 and 2 describe the simulation of shipboard flow field. It has been widely researched from all over the world. Item 3 describes how to deal with the sea level as the boundary condition in shipboard flow field simulation. Generally, the viscosity of sea water can be neglected, and the sea level can be supposed as a mirror surface. Item 4 describes the calculation of ship's movement on the basis of sea condition. Movement of ship is very important, but we generally get the ship's law of motion as the known condition in DI simulation. Item 5 describes the perturbation of the ship movement to flow field on deck. For this movement has a long period and low frequency, the perturbation can be neglected.

2.2 *Shipboard effect*

In the final stage of landing mission or the early stage of launching mission, the helicopter is very close to the ship. Therefore, the down flow of main rotor can easily alter flow field on deck, as described in item 8. On the other hand, the shipboard has a reverse effect on down-wash flow, changing the image of the down-wash flow field. This is similar to the well-known "Ground Effect", and is called "Shipboard Effect", as described in item 7. Owing to the movement of ship (item 6), this "Shipboard Effect" will be much more complex, thus can be called "Dynamic Shipboard Effect".

2.3 *Fully-coupled flow field*

As is mentioned above, flow field on deck and rotor down-wash flow have significant influence on each other. Therefore, how to precisely simulate the fully-coupled flow field is a major challenge. Current tools rely too much on one-way coupled simulations. In this case, the rotorcraft experiences disturbances due to a pre-calculated Computational Fluid Dynamics (CFD) ship airwake, but the ship airwake CFD solutions are unaffected by the downwash of the rotors. However, the unsteady and nonlinear nature of both ship and rotor wakes invalidates superposition and thus the uncoupled solutions are somewhat questionable. There has been a research that develops A fully-coupled simulation tool. An existing computational fluid dynamics code and a flight dynamics simulation model (item 9) have been coupled for this purpose. The flight dynamics model provided the location and orientation of the helicopter and the main rotor as well as the blade loads while the computational fluid dynamics code provided local air velocities. This approach, however, is costly and demanding in computer requirements, and the one-way coupled simulations has been necessary to achieve real-time execution in flight simulators since the computational requirements of CFD make it unfeasible to perform real-time solutions.

3 SIMULATION OF SHIPBOARD FLOW FIELD

The first step is to establish the physical model of the target ship. These studies used generic ship geometry to generate the CFD airwake model. Previous research has proved that geometry fidelity of the ship has significant influence on its CFD result of shipboard flow field. Figure 2 shows the simplification to the geometry, maintaining the main characteristic of the ship.

Then, the external flow field volume was determined and grid division was performed with ICEM CFD, as is shown in Figure 3.

The commercial CFD code, FLUENT, was introduced for the current study. Second order

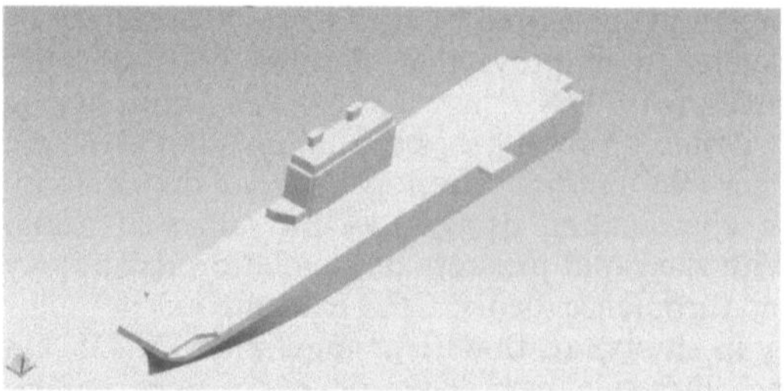

Figure 2. Simplified geometry of the ship.

Figure 3. Grid division of the external flow field.

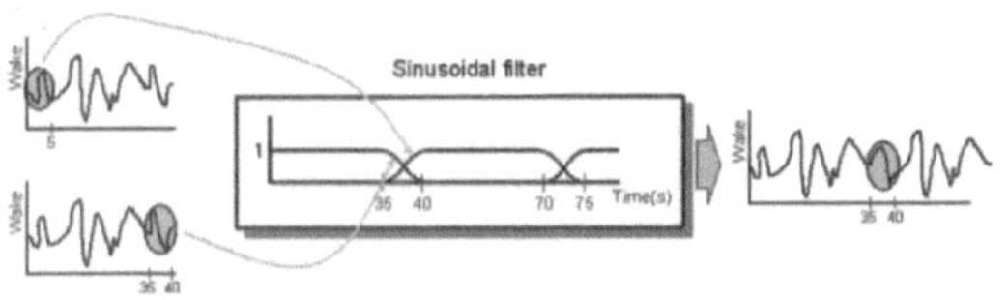

Figure 4. Sine function method used in flow data extension[4].

discretisation was given in time and space, and a blended upwind/central differencing scheme is also used for the convective terms. With the use of the PISO scheme, Pressure-velocity coupling was performed. In this study, the Detached-Eddy Simulation rather than Reynolds Averaged Navier-Stokes (RANS) methods was used to deal with the turbulence modeling.

4 APPROACH OF THE OVERLAPPED TIME HISTORY OF AIRWAKE

Shipboard flow field is time-varying and usually has some certain period, so a result of flow field in 40 s was figured out in the step before. 40 seconds, however, is not long enough for the helicopter finishing its landing/launching mission. A feasible method to solve the problem is to extend the result by period. But a following problem is how to deal with the difference between the result at 0 s and that at 40 s. So the sine function filter method was used to prevent the flow field result break occurring at the end of 40 s, with the combination with result in 35–40 s and result in 0–5 s, as is shown in Figure 4.

5 DEADLINE

The influencing factors of dynamic interface and their effects on others are analyzed. Then process and methods in shipboard flow field CFD simulation are described. For DI simulation, the approach of the overlapped time history of airwake is introduced. From what has been discussed, we can make some useful conclusion:

DI simulation is less costly than sea trails, but is still very complicated. Simplification is necessary. On the basis of that, the interaction between flow field on deck and rotor down-wash flow should be focused on. 'Shipboard Effect' should be involved to precisely calculate the fully-coupled field. For the finite data we catch in the flow field, overlapping time history of airwake serves as a necessary step before flight dynamic simulation. The above material should be with the editor before the deadline for submission. Any material received too late will not be published.

REFERENCES

[1] Roscoe M.F., Wilkinson C.H. DIMSS-JSHIP's modeling and simulation process for ship/helicopter testing and training, AIAA Modeling and Simulation Technologies Conference and Exhibit. Monterey CA: AIAA, 2002.

[2] Roy C. Wagner. Procedural guide for modeling and analyzing the flight dynamics of the SH-60B helicopter using FLIGHTLAB[D]. Monterey California: Naval Postgraduate School, 1995.

[3] Geoffrey J.J., Kristopher S., Christopher B. Combining simulink and flightlab for handling qualities and crew station prototyping[C]//64th AHS. Montréal, 2008.

[4] Dooyong L., Nilay S. Simulation of helicopter shipboard launch and recovery with time-accurate airwakes[C]//AIAA Atmospheric Flight Mechanics Conference and Exhibit. Rhode Island: AIAA, 2004.

[5] Wilkinson C.H., Roscoe M.F. Determining fidelity standards for the shipboard launch and recovery task[C]//AIAA Modeling and Simulation Technologies Conference and Exhibit. Montreal: AIAA, 2001.

[6] Advani S.K., Wilkinson C.H. Dynamic interface modelling and simulation-a unique challenge[C]// The Royal Aeronautical Society Conference on Helicopter Flight Simulation, London, 2001.

[7] Wilkinson C.H., Vander G.M. Modeling and simulation of the ship-helicopter environment[C]//AHS 56th. Virginia: AIAA, 2000.

[8] Lincoln P.E. A preliminary study of the airwake model used in an existing SH-60B/FFG-7 helicopter/ship simulation program[R]. Australia: Department Of Defence Science and Technology Organisation, 1994: AR-008–644.

[9] Forrest J.S., Owen I. Detached-eddy simulation of ship airwakes for piloted helicopter flight simulation[C]//1st International Aerospace CFD Conference. Paris, 2007.

[10] Forrest J.S., Steven J.H. Towards fully simulated ship-helicopter operating limits: the importance of ship airwake fidelity[C]//64th AHS. Montréal, 2008.

[11] Itoga N., Nagashima T. Numerical analysis of ground effect for a lifting rotor hovering above close proximity to inclined flat surface[C]//Proceedings of AHS International Meeting on Advanced Rotorcraft Technology. Japan, 1998.

Engineering Management and Industrial Engineering – Leung (Ed.)
© 2015 Taylor & Francis Group, London, ISBN 978-1-138-02773-2

Research on Vortex ring State boundary of helicopter tail rotor

W.S. Sun & H.R. Ma
Naval Aeronautical Engineering Academy, Qingdao, China

X.X. Gao
Qingdao University, Qingdao, China

ABSTRACT: It is important to research on the occurring time and mechanism of Vortex-ring State because of its serious influence on helicopter flight safety and performance. The characteristics of vortex ring state for helicopter tail rotor are analyzed. According to Gao-Xin criterion, Vortex-ring State boundary for helicopter tail rotor is calculated, the critical rates of helicopter tail rotor vortex ring are figured out and boundary curve of tail rotor vortex ring is drawn. Analysis on the results shows that flight height of helicopter as well as temperature on ground have an effect on the critical rates, and the helicopter should be prevented from turning into tail rotor vortex ring state in lateral flight or turning.

1 INTRODUCTION

When a helicopter is descending, it can easily enter a state called "Vortex-ring State" (VRS). This is a condition of powered flight where the helicopter settles into its own downwash. This unsteady flight condition can be characterized by severe thrust fluctuations, vibrations, rapid rates of descent, sluggishness and vibrations of the controls, and can result in a temporary loss of helicopter control. The flow pattern of helicopter's main rotor in VRS is shown in Figure 1.

Similar to the main rotor, tail rotor may be threatened by VRS in some cases, such as turning around when hovering, hovering in lateral winds or sideward flight. Figure 2 indicates its flow pattern.

Once a helicopter goes into the VRS, no matter by its main rotor or tail rotor, the pilot is usually encountered with a great challenge to save the aerobat from crash. The VRS can cause severe accidents without correct treatment. For the purpose of studying the essential characteristics of the tail rotor inflow in VRS, by defining the occasion of entering tail rotor VRS and providing

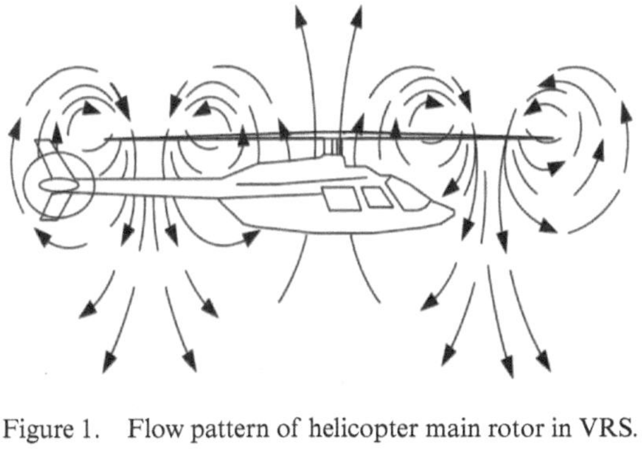

Figure 1. Flow pattern of helicopter main rotor in VRS.

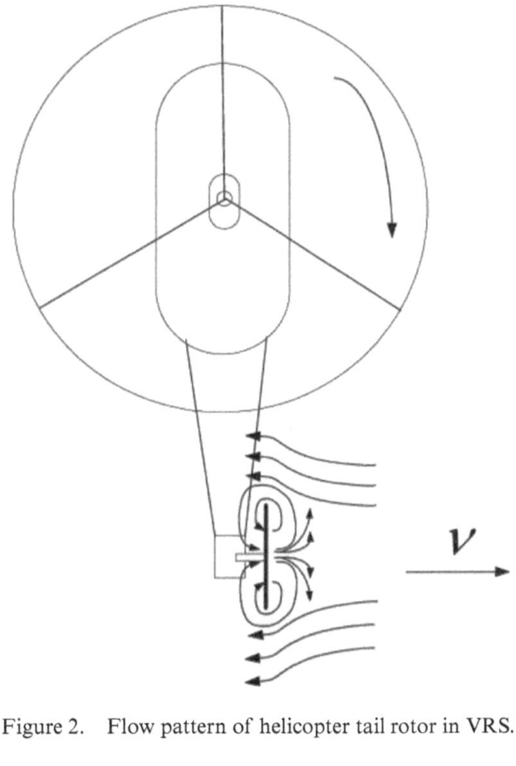

Figure 2. Flow pattern of helicopter tail rotor in VRS.

early warning information to prevent helicopter accidents from being caused by VRS, works from several aspects are conducted as follow. Criterion theories for helicopter VRS are introduced and studied. Method of calculating boundaries of tail rotor VRS is studied and used in a specific helicopter. Helicopter critical velocity entering tail rotor VRS are figured out under a series of conditions, and the VRS boundary charts of the tail rotor are drawn.

2 CRITERIA FOR VRS

After decades of research on this serious issue, researchers as well as pilots and designers of helicopter have formed a general impression on the cause of formation and characteristics of helicopter Vortex-ring State. Furthermore, a lot of works have been done to indicate when and how the dangerous phenomenon happens. This theory defines the criteria of helicopter in this state. The so-called "Vortex-ring State Criteria" was developed from Wolkovitch, Peters to Gao-Xin.

Compared to "Wolkovitch Criteria" and "Peters Criteria", Gao-Xin Criteria is more suitable for tail rotor VRS: When the projection of relative incoming flow velocity vector on the reverse direction of rotor induced airflow velocity vector surpassing a critical number, the rotor enters VRS. This critical vertical velocity can be expressed as.

$$V_{cr}(90^\circ) = 0.28 \cdot v_h \tag{1}$$

where v_h is equivalent hovering induced velocity. To vanish v_h, a dimensionless criteria equation is obtained.

$$\frac{\tilde{V}_x^2 + (\tilde{v} - \tilde{V}_y)^2 - \tilde{v}(\tilde{v} - \tilde{V}_y)}{\sqrt{\tilde{V}_x^2 + (\tilde{v} - \tilde{V}_y)^2}} = -0.28 \tag{2}$$

where $\bar{v}$ is induced velocity of helicopter rotor, $\bar{V}_x$ is horizontal component of helicopter velocity, $\bar{V}_y$ is vertical component of helicopter velocity.

3 CALCULATION OF TAIL ROTOR VRS BOUNDARY

3.1 *Equivalent induced velocity*

Parameters of a helicopter in some certain condition are given: radius of tail rotor R_{tr} (m), rotation rate of tail rotor Ω_{tr} (rad/s), the horizontal distance between center of tail rotor and origin of body coordinates L (m), radius of main rotor R_m (m), rotation rate of main rotor n_m (rpm), power in steady hover P (kw), blade tip lose coefficient K, as well as flight height H (m), local height above sea level H_loc (m) and local temperature t (°C). Equivalent induced velocity can be figured out as follow:

temperature at the height H

$$T = t - 0.0065 \times (H - H\_loc) \tag{3}$$

density of atmosphere

$$\rho = 1.225 \times \left(\frac{273.15 + T}{288.15}\right)^{4.2559} \tag{4}$$

torque of rotor axis

$$Q_m = \frac{1000P}{2\pi \times n_m / 60} \tag{5}$$

thrust coefficient of tail rotor

$$C_{Ttr} = \frac{T_{tr}}{\frac{1}{2}\rho(\Omega_{tr}R_{tr})^2 \pi R_{tr}^2} = \frac{Q_m / L}{\frac{1}{2}\rho(\Omega_{tr}R_{tr})^2 \pi R_{tr}^2} \tag{6}$$

equivalent induced velocity of tail rotor

$$v_{htr} = \frac{1}{2}\sqrt{C_{Ttr}(\Omega_{tr}R_{tr})^2} \tag{7}$$

3.2 *Critical velocity*

Critical lateral velocity represents the minimal velocity that tail rotor enters VRS in lateral flight, equal to $0.28v_{htr}$. So it can be easily figured out in different conditions in terms of height and temperature, as listed in Table 1. The tendency of critical lateral velocity growing in accordance with height and temperature is obvious.

3.3 *Tail rotor vortex ring boundary curve*

At the initial time of the VRS occurring, calculation of induced velocity can be based on

Table 1. Critical lateral rates for tail rotor VRS of a certain type helicopter (km/h).

	H/m					
T/°C	50	100	200	400	800	2000
–10	24.5	24.6	24.7	25.0	25.5	27.3
0	22.7	22.7	22.8	23.1	23.6	25.1
10	21.0	21.0	21.1	21.4	21.8	23.1
20	19.5	19.5	19.6	19.8	20.2	21.4
35	17.5	17.6	17.7	17.8	18.1	19.2

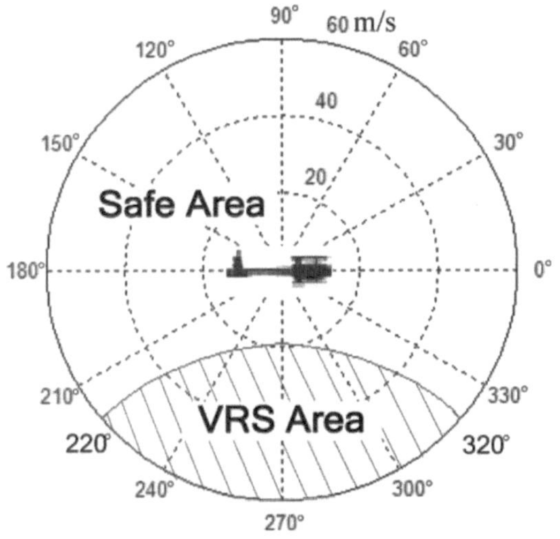

Figure 3. Tail rotor vortex ring boundary curve.

Momentum Theory, when the supplementary equation is expressed as

$$\tilde{v}^2 \cdot (\tilde{V}_x^{\,2} + \tilde{V}_y^{\,2} - 2\tilde{v}\tilde{V}_x + \tilde{v}^2) = 1 \tag{8}$$

Then, equation with two unknowns, $\tilde{V}_x$ and $\tilde{V}_y$, which represent forward speed and lateral speed in tail rotor VRS, are solved. $\tilde{V}_x$ is plotted vs. $\tilde{V}_y$ in the platform of helicopter. Tail rotor vortex ring boundary curve is obtained, and Figure 3 shows one curve of a certain type helicopter in some condition, where latitudes represent helicopter's relative velocity to wind, while longitudes represent orientation angle of the helicopter.

4 DEADLINE

The features of tail rotor in VRS are discussed and tail rotor VRS boundary is figured out. The critical speed at which the helicopter entering tail rotor VRS is heavily influenced by flight height and local temperature. For right-whirling helicopter, the VRS area is located at its right part in the platform of helicopter. Pilots should be careful on the tail rotor VRS, especially when hovering in right-hand winds, lateral flight to right side, and turning left. Further research can be used into development of onboard VRS warning system.

REFERENCES

[1] Wang Shi-cun. Helicopter Aerodynamic [M]. Nanjing, 1985:141–142.

[2] Xin Hong, Gao Zheng. An Experimental Investigation on the Boundary of Helicopter Vortex-Ring State [J]. Transactions of Nanjing University of Aeronautics & Astronautics. 1995, 27(4):439–443.

[3] Tang Lian-gang, Zhu Yu, Li Xian-yao. Flight Test Research on Helicopter Tail Rotor Vortex Ring[J]. Flight Dynamics, 2008, 26(5):63–66.

[4] Xin Hong. Theoretical and Test Research on Unsteady Aerodynamic Characteristics of Rotor [D]. Nanjing: Nanjing University of Aeronautics & Astronautics, 1995.

Engineering Management and Industrial Engineering – Leung (Ed.)
© 2015 Taylor & Francis Group, London, ISBN 978-1-138-02773-2

FPGA-based image edge detection IP core design

S.H. Wang, X.F. Lu & Y.D. Wu
Shanghai University, Shanghai, China

ABSTRACT: Edge detection is the foundation of image processing. Classic edge detection algorithms contain nonlinear operations, which causes high algorithm complexity. Combining processing ability with parallel structure of FPGA, we propose an accelerated edge detection hardware architecture on FPGA. Considering edge detection has been applied into many image processing algorithms, a reusable edge detection IP core is proposed in this paper. Experiments show the FPGA-based edge detection IP core not only implements the edge detection function but also protects the intelligence property of the IP designer.

1 INTRODUCTION

Edge detection extracts the contour information of the image, so it is the first step of many complicated image processing algorithms. Many researchers have been devoted to improve the accuracy and efficiency of edge detection algorithms since it was first proposed in 1959 [1]. Han Yong [2] uses edge detection to measure the accuracy for out-of-gauge goods on the railway. Zhang Yinshi [3] applied edge detection into face and feature recognition and proposed a cartoon avatar recognition method. Akinsowon [4] applied edge detection into biometric recognition to implement the palm-print identification.

Classic edge detection algorithms contain derivate and evolution operations. To meet the demands of digital image processing, convolution masks are designed to simplify the algorithms. FPGA parallel structure improves the algorithm efficiency and real-time performance further. In this paper, we propose a FPGA-based edge detection hardware architecture.

As mentioned above, edge detection is the foundation of complex image processing algorithms, so it is very meaningful to design a reusable edge detection Intelligence Property (IP) core.

As a programmable logic device, FPGA uses Verilog HDL to describe the hardware architecture of digital circuits. There are many functions can be called in high-level programming language, while Verilog HDL does not provide such functions library. The appearance of IP core solves this problem.

IP core is a reusable circuit module. Specific circuit function can be implemented by reusing IP core. There are three kinds of IP cores. They are soft, hard and firm IP core. Soft IP core uses RTL netlists to describe the hardware circuits. It is flexible, but it can be randomly modified during transmission. Hard IP core is a physical circuit module with specific circuit function. It is reliable, but the fixed size and port location makes it restrictive. Firm IP core uses gate-level netlists to describe the hardware circuit. Although it cannot be modified, it can be generated and reused on the software platform. Therefore, it is more reliable than soft IP core and more flexible than hard IP core. In this paper, we design the image edge detection IP core as a firm IP core.

Altera is one of the most competitive FPGA manufacturers. It provides lots of reliable and reusable IP cores for FPGA designers, but most IP cores focus on the communication protocol. Considering the wide use of FPGA in image processing, lots of researchers have been devoted to FPGA-based image processing IP core designs. Ahmad [5] designed a two-dimensional Haar wavelet transform IP core and dynamic partial reconfiguration IP core and implemented the face recognition system. Chandrakanth [6] designed a configurable Fast Fourier Transform IP core, which improved the IP core reusability greatly. Bendaoudi [7] designed a stereovision IP core and implement the obstacle detection system. Di Carlo [8] used FPGA real-time performance to accelerate the image denoising speed and designed an adaptive image denoising IP core.

In this paper, we propose a FPGA-based edge detection IP core. Experiments show the generated IP core not only implements the edge detection function, but also protects the intelligence property of IP designer.

2 EDGE DETECTION IP CORE DESIGN

2.1 *Image edge detection algorithms*

Edge detection aims at identifying points at which image brightness changes sharply. In this section, we introduce four classic edge detection algorithms. They are divided into two methods, first-order differentiator that determines the edge as the maximal gradient, second-order differentiator that determines the edge as second order zero crossing.

We define $I(x,y)$ as the input image signal. Its first partial derivatives in orientations of x,y are shown as follows:

$$I_x(x,y)=\frac{\partial I(x,y)}{\partial x} \tag{1}$$

$$I_y(x,y)=\frac{\partial I(x,y)}{\partial y} \tag{2}$$

The gradient of $I(x,y)$ is shown as follows:

$$\nabla \vec{I}(x,y)=\left[\frac{\partial I}{\partial x}\ \ \frac{\partial I}{\partial y}\right]^T \tag{3}$$

The value and orientation of $\nabla \vec{I}(x,y)$ are shown as follows:

$$\left|\nabla \vec{I}\right|=\sqrt{\left(\frac{\partial I}{\partial x}\right)^2+\left(\frac{\partial I}{\partial y}\right)^2} \tag{4}$$

$$\varphi=\arctan\left(\frac{\partial I}{\partial y}\Big/\frac{\partial I}{\partial x}\right) \tag{5}$$

The orientation of contour θ is shown as follows:

$$\theta=\pi/2-\varphi \tag{6}$$

Robert [9], Sobel [10] and Prewitt [11] edge detection operators meet the demands of digital image processing. The convolution masks of Robert, Sobel and Prewitt are shown in Figure 1. The left part of the each figure is horizontal edge operator and the right part is the vertical edge operator.

1	0		0	-1
0	-1		1	0

Robert

-1	-1	-1		1	0	-1
0	0	0		1	0	-1
1	1	1		1	0	-1

Prewitt

-1	-2	-1		-1	0	1
0	0	0		-2	0	2
1	2	1		-1	0	1

Sobel

0	-1	0		-1	-1	-1
-1	4	-1		-1	8	-1
0	-1	0		-1	-1	-1

Laplacian

Figure 1. Convolution masks of edge detection algorithms.

The second partial derivatives in orientations of x,y are shown as follows:

$$I_{xx}(x,y)=\frac{\partial^2 I(x,y)}{\partial x^2} \tag{7}$$

$$I_{yy}(x,y)=\frac{\partial^2 I(x,y)}{\partial y^2} \tag{8}$$

Laplacian [12] operator is shown as follows:

$$\nabla^2 I=\frac{\partial^2 I}{\partial x^2}+\frac{\partial^2 I}{\partial y^2} \tag{9}$$

The convolution mask of Laplacian is shown in Figure 1.

2.2 *FPGA-based image edge detection hardware architecture*

The block diagram of FPGA-based image edge detection is shown in Figure 2. It includes RGB to grey, shift register delay, edge detection, VGA synchronizing signal delay, RGB delay and display blocks.

The size of the input image is 800 × 525 with blanking pixels.

R, G and B signals are 8 bits. We extend the R, G, and B signals to 10 bits. The highest bit and the second highest bit are zero and zero, respectively. The other bits are the original 8-bit R, G, and B signals. R, G and B signals are transformed into Grey as follows:

$$\mathrm{Grey}=\mathrm{R}\times 306+\mathrm{G}\times 601+\mathrm{B}\times 117 \tag{10}$$

The Grey signal is input into shift register delay block. Since each edge detection convolution mask has three columns and three rows, we need to construct a Grey matrix with three columns and three rows. We use three shift registers to store three rows of Grey signals. Each register occupies 800 × 10 bits. At the same clock tick, three registers output three Grey signals in the same column.

In the edge detection block, the Grey matrix does convolution operation with the convolution masks mentioned in Section 2.1. We use a button to switch Robert, Prewitt, Sobel and Laplacian edge detection algorithms. Since the VGA display scans line by line, when the pixels of the third row come, the edge detection results of the first row are calculated and output. It causes the time delay of two rows, which means 800 × 2 clock ticks. Therefore, the synchronizing signals and the

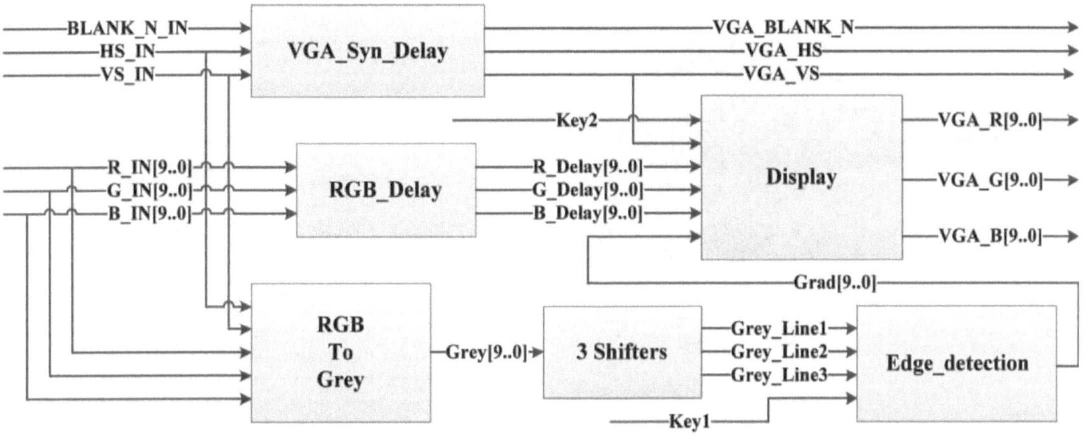

Figure 2. Block diagram of FPGA-based image edge detection hardware architecture.

original R, G, and B signals should be delayed by two rows as well.

Input the edge detection results, delayed R, G, and B signals and synchronizing signals into display block. Then, operate the key to control the dynamic display of the original image and the edge detection results.

2.3 *Generate and reuse the FPGA-based image detection IP core*

We generate and reuse the FPGA-based image edge detection IP core by the QXP file on the Quartus II 11.0.

First, build a new project on Quartus II. In this project, we implement the edge detection hardware architecture mentioned in Section 2.2. Then, compile the project. Finally, we use QXP file to encapsulate the RTL netlist of the whole project. The FPGA-based image detection IP core is generated as a black box. Only its input and output ports can be viewed, while the internal RTL netlist or the Verilog code cannot.

We reuse the edge detection IP core in a video capture and display project on Quartus II. First, add the QXP file to this project. Then, describe the connection between the IP core and the current project by Verilog HDL. Finally, generate the internal reused block and implement the edge detection function.

3 EXPERIMENTAL RESULTS

3.1 *Hardware platform*

We design a FPGA-based image processing hardware platform to accomplish the following experiments. The hardware platform is shown in Figure 3, where number 1 to number 6 are FPGA core board, power supply module, key and led module, serial communication module, video-in module and video-out module, respectively.

The FPGA-based image processing system is shown in Figure 4. Number 1 to number 3 are CCD camera, FPGA-based image processing hardware platform and VGA display, respectively. CCD camera input CVBS analog video signal into hardware platform to do image processing operations. The VGA display shows the image processing results.

3.2 *Qualitative results of the FPGA-based image edge detection IP core*

3.2.1 *Edge detection results*

The edge detection results are shown in Figure 5. In the generated IP core, we implement the hardware architecture of four classic edge detection algorithms. We reuse the generated IP core as an image processing module embedded in the video capture and display project on Quartus II. The experiment results show the generated IP core implements detection function.

3.2.2 *RTL viewer of the reused IP core block*

We reuse the edge detection IP core and generate the internal reused block in video capture and display project on Quartus II. Only the input and output ports are shown in the RTL viewer, while the internal digital sequence and logic circuits are not accessible as a black box. Experiments show the proposed FPGA-based image edge detection IP core not only implements the edge detection function, but also protects the intelligence property of IP designer.

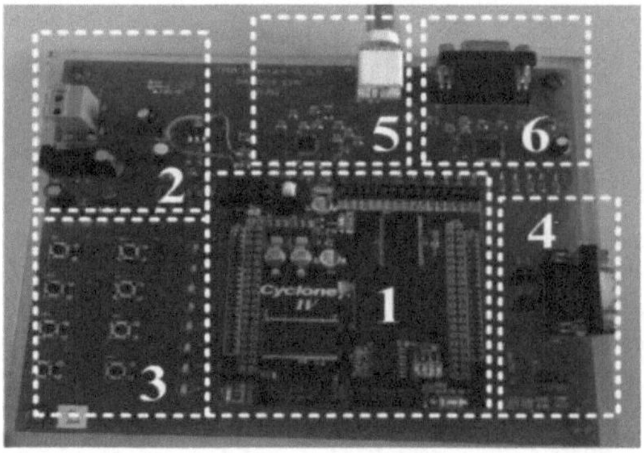

Figure 3. FPGA-based image processing hardware platform.

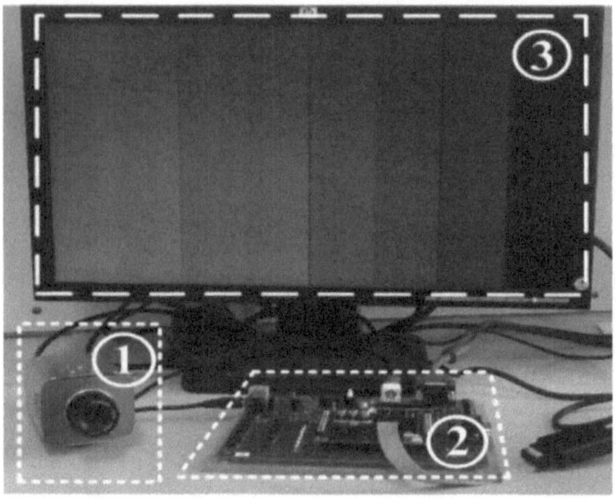

Figure 4. FPGA-based image processing system.

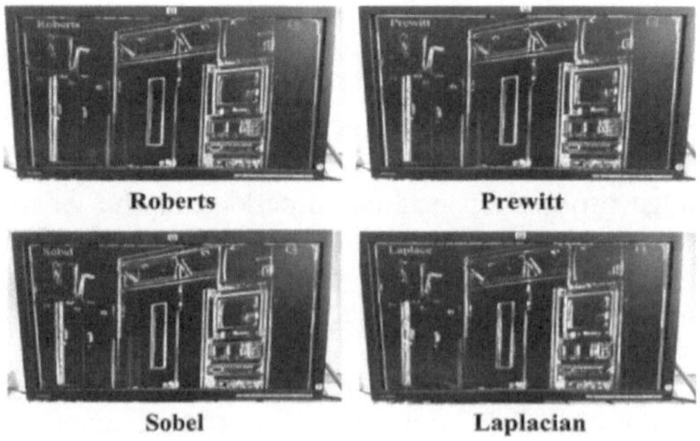

Figure 5. Edge detection results.

3.3 *Quantitative results of the FPGA-based image edge detection IP core*

Table 1 and Table 2 show the power dissipation and resource consumption of FPGA-based image edge detection IP core with four edge detection algorithms respectively. Experiments show the generated IP core is reasonable and efficient.

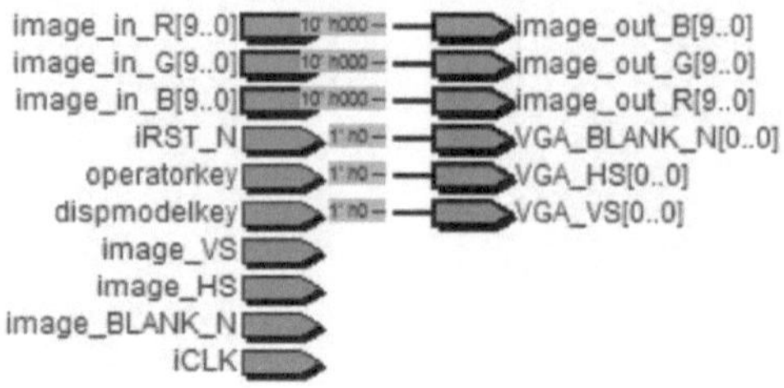

Figure 6. RTL viewer of the reused IP block.

Table 1. Power dissipation of FPGA.

Sort	IP core dissipation (mW)
Total thermal power dissipation	147.06
Core dynamic thermal power dissipation	0
Core static thermal power dissipation	98.60
I/O thermal power dissipation	48.46

Table 2. Resource consumption of FPGA.

Resource consumption	Available	Used
Combinational LE with no register	114,480	913
Sequential LE	114,480	29
Combinational LE with a register	114,480	236
Dedicated logic registers	114,480	265
LABs	7,155	95
M9Ks	432	59

4 CONCLUSION

In this paper, we implement the FPGA-based image edge detection hardware architecture, generate its IP core and reuse it in the video capture and display project on Quartus II. Experiments show the generated IP core implements the edge detection function, and its internal digital sequential and logic circuits are not accessible. Therefore, we can both implement the digital circuits function and protect intelligence property of IP designer by generating and reusing IP cores.

REFERENCES

[1] Julez. 1959. A Method of Coding TV Signals Based on Edge Detection. *Bell Syatem Tech* 4(38): 1001–1020.

[2] Han, Y. Qiao, M. & Wan, X. et al. 2014. Application of image analysis based on canny operator edge detection algorithm in measuring railway out-of-gauge goods. *Advanced Materials Research* 912–914:1172–1176.

[3] Zhang, Y. Sun, L. & Shen, X. et al. 2013. Cartoon avatar processing based on image edge detection. *Advanced Materials Research* 605–607:2232–2235.

[4] Akinsowon, O.A. & Alese, B.K. 2013. Edge detection methods in palm-print identification. *2013 8th International Conference for Internet Technology and Secured Transactions. 23–25 March 2013*:422–426.

[5] Ahmad, A. Amira, A. & Nicholl, P. et al. 2013. FPGA-based IP cores implementation for face recognition using dynamic partial reconfiguration. *Journal of Real-Time Image Processing* 8(3):327–340.

[6] Chandrakanth, V. & Tripathi, S. 2013. Customized architecture for implementing configurable FFT on FPGA. *Proceedings of the 2013 3rd IEEE International Advance Computing Conference. 22–23 February 2013*:1279–1282.

[7] Bendaoudi, H. & Khouas, A. 2013. Stereo vision IP design for FPGA implementation of obstacle detection system. *2013 8th International Workshop on Systems, Signal Processing and Their Applications. 12–15 May 2013*:145–150.

[8] Di, C.S. Prinetto, Paolo. & Rolfo, D. et al. 2013. AIDI: An adaptive image denoising FPGA-based IP-core for real-time. *Proceedings of the 2013 NASA/ESA Conference on Adaptive Hardware and Systems. 25–27 June 2013*:99–106.

[9] L.G. Roberts. 1965. Machine Perception of Three-Dimensional Solids. *Optical and Electro-Optical Information Processing*:159–197.

[10] L.S. Davis. 1975. A Survey of Edge Detection Techniques. *CGIP* 14: 248–270.

[11] J. Prewitt. 1970. Object Enhancement and Extraction. *Picture Processing and Psychopictorics.*

Engineering Management and Industrial Engineering – Leung (Ed.)

ISBN 978-1-138-02773-2

Algorithm SESP of Wireless Sensor Network node

X.H. Huang
School of Information Science and Engineering, Fujian University of Technology, Fujian, China

ABSTRACT: In order to extend lifespan of the Wireless Sensor Node, the reseach propose a Self-Adaptive Algorithm (SESP) based on energy assessment of WSN nodes with identical hardware model and same parameters. The algorithm SESP is able to choose the cluster head without sending message to learn the surplus energy of other nodes, thus reducing in-cluster communication and simplifying the calculation of energy assessment, and eventually reducing energy consumption of WSN nodes and extending their lifespan.

Keywords: WSN; energy assessment of Wireless Sensor Node node; self-adaptive algorithm

1 INTRODUCTION

The node energy of Wirelss Sensor Network (WSN) is finite. And WSN is out of use with the exhaustion of the node energy. Many studies have shown that the wireless transmission consumes the largest proportion of the whole node energy [1][2], with the consumption of transmitting a bit of datum 800 times more than that of executing a command [1]. Therefore, minimizing the transmission energy consumption is the key to reduce the node energy consumption and prolong the life of WSN. Also important for the extension of the lifespan is to balance node energy consumption and make full use of the energy of each node. To tackle the issue, a self-adaptive algorithm SESP is proposed based on the assessment of the residual energy of nodes.

2 THE CONSUMPTION OF NODE ENERGY AND ENERGY-SAVING MEASURES

The major energy consumers of WSN node are sensor modulating circuit of data collection module, micro controller and memory of data processing module and the RF circuit of data transmission module.

There is a little room left to tap the sensor modulating circuit for lessening the energy consumption as it consumes quite a little. The key then lies in reducing the energy dissipation of micro-controller and memory of data processing module and the RF circuit since the energy consumed in the process of data transmission results mainly from signal processing and offsetting transmission path attenuation.

The power dissipation of microprocessor consists of dynamic power dissipation and static one [4], with the former as the main contributor to energy saving. According to Vasanth & Michael (2005)// item 5 of the references, the dynamic power of microprocessor is closely related to supply power voltage, physical capacitance and physical capacitance, which is formulated as follow:

$$P_D \propto \alpha C V^2 f \tag{1}$$

In the formula, P_D stands for dynamic power, V is supply power voltage, C physical capacitance, f physical capacitance and α active factor.

Therefore, reduction of dynamic power dissipation can be realized by lowering the clock frequency and supply power voltage. Vasanth & Michael (2005) have demonstrated that lowering both the supply power voltage and clock frequency can result in the reduction of dynamic power dissipation. And changing the working conditions of the microprocessor from 200 MHz and 1.5 V to 150 MHz and 1.2 V could reduce up to 52% of the dissipation.

The management of dynamic power dissipation could also be realized by adopting Dynamic Voltage Supply (DVS) technology. DVS will dynamically adjust the working voltage and frequency of the microprocessor with the change of the node's working load, thus reducing the unnecessary power output in the comparatively free moments.

The RF circuit, among the components of the node, consumes the largest proportion of the enery. In accordance with the technological requirements of WSN node, RF circuit usually is made from popularly utilitized parts of low energy consumption, low price with small size.

In a WSN with nodes of N number, apart from reducing the power dissipation of the hardware of each node, energy saving can also be achieved by reducing the workload of nodes through the energy management of in-between nodes communication.

Suppose the set of nodes is U, and $U = \{u_i = (x_i, y_i), i = 1, 2, ..., n\}$, with (x_i, y_i) indicating the location of a node. With the completion of the distribution of a WSN of general purpose, its nodes locations are fixed.

The model for adjusting energy is proposed by item 2 of the references. Given that the needed energy for transmitting a unit of data from node i to node j is E_{ijs}, the energy for signal processing E_{proc}, and the energy for offsetting transmission path attenuation E_{was}, then

$$E_{ijs} = E_{proc} + E_{was} \tag{2}$$

E_{proc} here is relevant to the energy for encoding, modulation and wave filtering, and E_{was}, to the distance between node i and node j. And if the distance is designated as d_{ij}, then

$$E_{was} \propto dij \tag{3}$$

Node j also needs energy to receive information. Suppose the energy for node j to receive a unit of data is E_{jr}. E_{jr} then is relevant to the energy consumed by decoding, demodulation and wave filtering.

As the nodes locations are fixated when completing the distribution of a WSN of general purpose, the value of d_{ij} becomes fixed. And when the node hardware of WSN is determined, E_{proc} and E_{jr} remain almost unchanged too. Accordingly, reducing the volume of transmitted data is the key to save energy.

The major tasks done by WSN are collecting data from sensors and transmitting them to base station. And since its transmitting power is rather low, a WSN node has to communicate directly with its neighboring nodes. To ensure that those comparatively distant nodes could transmit data to base station, the nodes could be distributed in the form of a multiple hops self-organizing network. In the self-organizing network, clustering networking algorithm is capable of effectively reducing the volume of data transmission [6]. Every cluster elects a node as its own head which then collects data transmitted from other nodes and merges them. However, the cluster head would consume more energy than other nodes. In order to balance the energy consumption and prolong the life of the network, every node in a cluster should be chosen to serve as the head in turn periodically. Thus, assessing the node energy consumption and electing a suitable one as the head would be of great help to extend the lifespan of the network.

3 ENERGY ASSESSMENT OF NEIGHBORING NODES

Within any of the clusters of WSN, if the neighboring nodes' residule energy could be learnt, the node with much residule energy could be chosen as new cluster head so as to balance energy consumption and maximize the lifespan of each cluster.

Different nodes in a WSN typically are made from the same model of hardware or parts. Therefore, the energy E_{proc} needed by different nodes to transmit a unit of data is approximately equivalent. So the energy E_{jr} is needed for receiving a unit of data. With the completion of the distribution of WSN of general purpose, d_{ij} is fixated, and $E^{ij}{}_{was}$(from node i to node j) and $E^{ji}{}_{was}$ (from node j to node ij) are approximately identical.

Given that node t is the old cluster head, node i is one of the other nodes in the same cluster, E_{tis} represents the needed energy for transmitting a unit of data from node t to node i, E_{proc} the energy needed for signal processing, E^{si}_{was} the energy offsetting the transmission parth attenuation, according to (2), then

$$E_{tis} = E_{proc} + E^{si}_{was} \tag{4}$$

Let E_{its} stand for the energy needed for transmitting a unit of data from node i to node t, and E^{it}_{was} for the energy offsetting the transmission parth attenuation, then

$$E_{its} = E_{proc} + E^{it}_{was} \tag{5}$$

Because

$$E^{it}_{was} = E^{it}_{was} \tag{6}$$

Substitute formula (6) for its counterpart in formula (5), and then compare the resulted with (4), then

$$E_{tis} = E_{its}$$

That is, the energy demanded for transmitting a unit of data from node i to node t is equivalent to that of from node t to node i.

So the energy E_{tr} is needed for node t to receive a unit of data and E_{ir} is needed for node i to receive a unit of data.

If in a time cycle node i receives m units of data and dispatches n units of data, then the total energy consumption of node i $e_i(k)$ is

$$e_i(k) = mE_{tr} + nE_{tis} \tag{7}$$

In other words, so long as the data volume transmitted and received by node i with its neighboring nodes in a time cycle can be obtained, the total energy consumption of node i in the time cycle can be calculated and thus its residule energy can be figured out.

With new cluster head being chosen, the old heads will be transmitted with minimal transmitting power within them the messages from the new head. And the new head, with its own confirmation of the identity, will broadcast among its neighboring heads and make them know its identity message with rather big transmitting power.

4 SESP SELF-ADAPTIVE IN-CLUSTER ROUTE ALGORITHM

SESP assumes that every node has different scales of transmission power. The lowest scale is used for in-cluster transmission, with a transmission radius of R_1. All nodes of the cluster are located within the circular area with a radius of $R_1/2$, which means all cluster nodes can communicate directly with each other. The higher scales of transimission power are used for inter-cluster transmission, with a transimission radius of R_2.

If the network lifespan is divided into several life cycles based on a fixed time step length, and the length of a life cycle is l, then the time step length is $1/l$. And the whole network life cycle L is

$$L = Nl \tag{8}$$

Before the end of every cycle, the current cluster head assesses each node's residule energy with the above-mentioned SESP, and chooses the next head for head shift.

Apart from the head, other nodes within the cycle (suppose it is the k-th one) communicate only with the head, thus forming a head-centered star-network structure as indicated in Figure 1.

The data volume received is n and the data volume transmitted i is m. At the cluster head, the data volume m and n of node i can be calculate. At the end of the k-th cycle, the data volume n received by the head are respectively equivalent to that transmitted by node i and the data volume m transmitted by the head are respectively equivalent to that received by node i. Substituting m and n for their counterparts in the formula (7), the energy consumption $e_i(k)$ in the cycle of node i can be obtained. Combined with the energy consumed in the previous cycles, the total energy consumption at present of node i $E_i(k)$ is

$$E_i(k) = e_i(k) + \sum_{r=1}^{k-1} e_i(r) \tag{9}$$

Let the initial energy of each node be E_i^0, and the present residule energy of node i be $E_i^r(k)$, then

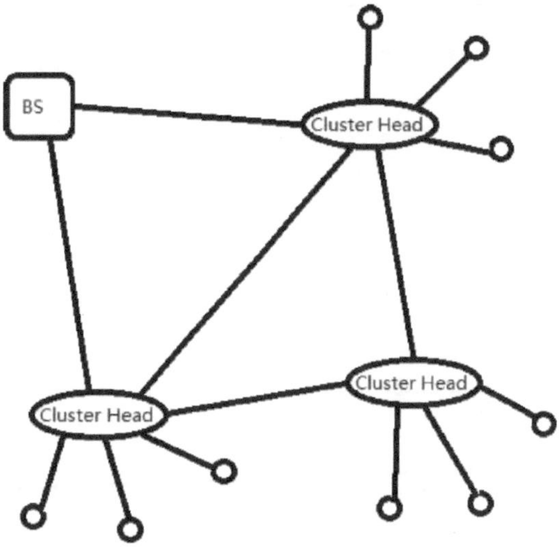

Figure 1. Network structure.

$$E_i^r(k) = E_i^0 - E_i(k) \tag{10}$$

Choose the node with maximal residule energy $E_i^r(k)$ as the next cluster head.

Every node saves a message of its energy consumption in the cluster. Meanwhile, the head builds a list of total energy consumption, recording each in-cluster node's $\sum_{r=1}^{k-1} e_i(r)$, with the head's own message included, and the locations of the nodes. In the head shift, the list will be transferred to the new head. At the end of next cycle, the new head can use the list to assess the energy consumption of the other nodes and choose the subsequent head. Such a mode will be repeated till the end of the life of the network.

5 SIMULATION

Simulation of the SESP algorithm is compared with WAF [7] algorithm. WAF algorithm calculates small CDS (Connected Dominating Set, CDS) and connects them to form a cluster.

The number of nodes in the simulation were set to 50, 100, 150, 200, 250 and 300, and these nodes randomly distributed in the area of 200 m by 200 m; each node has the initial energy of 1 j; R1 is set to 30 m. Nodes in sending, receiving, dormancy and idle mode respectively consumption power of 24 mW, 14.4 mW, 14.4 mW and 0.015 mW. These values are from literature [8].

Figure 2 shows energy consumption of the clustering process. It can be seen from the figure that the performance decreased with the increase of the node density. Because there are more nodes

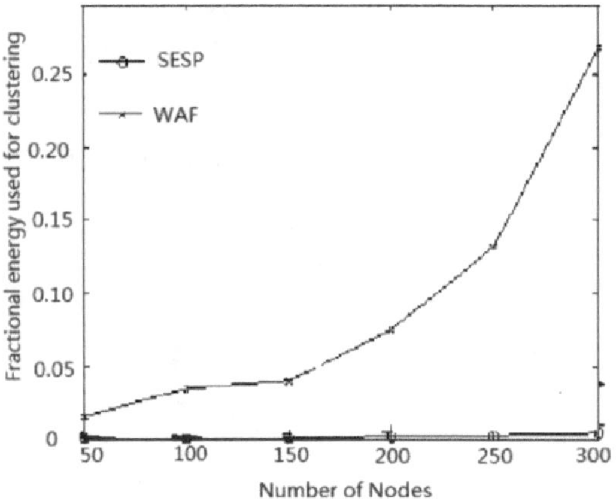

Figure 2. Energy consumption of the clustering process.

to participate in local information exchange and competition of wireless channel with the increase of nodes density. SESP algorithm consumes less energy, because it does not need to initiate communication to learn the cluster surplus energy of other node in the process of choosing cluster head. So it reduces the amount of communication within the cluster, also simplifies the estimate the computational complexity of node energy and prolongs the life of network. WAF algorithm consumes most energy, mainly because the sequence selection of cluster head and some small sets merged into a great one need to consume a lot of energy.

6 CONCLUSION

The self-adaptive algorithm SESP is based on energy assessment in a WSN network in which there are a lot of nodes with same parameters. By the algorithm, no communication is needed to obtain the residule energy of other nodes when choosing cluster head, thus reducing both the in-cluster communication volume and the complexity of energy assessment calculation. Simulation of the SESP algorithm compared with others illustrate that SESP would save the node energy and effectively extend the lifespan of the network.

ACKNOWLEDGMENT

This research is partially supported by Natural Science Foundation of Fujian Province Of China (2008 J0178).

REFERENCES

[1] Samuel Madden, Michael J. Franklin, Joseph M. Hellerstein, Wei Hong. Tag: a tiny aggregation service for ad-hoc sensor networks [C]. Proc. of OSDI, 2002.
[2] W. Heinzelman, A. Chandrakasan, H. Balakrishnan. An Application-Specific Protocol Architecture for Wireless Sensor Networks [J]. IEEE Trans. Wireless Comm., Oct.2002, 1(4), pp. 660–670.
[3] S. Amit, C. Anantha. Dynamic Power Managementin Wireless Sensor Networks [J]. Selected & test of computer, 2001, 18(2): 62–74.
[4] Ju Zhipei, Hua Yanqi. Wireless sensor network node hardware modular design [J]. Journal of University of Shanghai For Science and Technilogy, 2009, (6):20–22.
[5] V. Vasanth, F. Michael. Power reduction techniques for micropro cessor system [J]. ACM computing survey, 2005, 37(3): 195–237.
[6] A.D. Amis, R. Prakash. Load Balancing Clusters in Wireless Ad Hoc Networks [A]. Proceedings of ASSET, March 2000, pp. 25–32.
[7] P.J. Wan, K.M. Alzoubi, O. Frieder. Distributed Construction of Connected Dominating Sets in Wireless Ad Hoc Networks [C]. ACM/Kluwer Mobile Networks and Applications, no. 2, Apr. 2004, pp. 141–149
[8] S. Basagni, M. Mastrogiovanni, A. Panconesi, C. Petrioli. Localized Protocols for Ad Hoc Clustering and Backbone Formation: A Performance Comparison [J]. IEEE Trans. Parallel and Distributed Systems, April 2006, 17(4): pp. 292–306.

Engineering Management and Industrial Engineering – Leung (Ed.)
© 2015 Taylor & Francis Group, London, ISBN 978-1-138-02773-2

Assembly system process complexity analysis from different dimensions

F. He & D.B. Li
Nanjing University of Science and Technology, Nanjing, Jiangsu, China

H.P. Zhu
Huazhong University of Science and Technology, Wuhan, Hubei, China

ABSTRACT: Information-theory has been used to research the complexity of the manufacturing system for decade years, and the process complexity is one of the important complexities in the assembly system. In order to recognize the complexity more accurately, the process complexity is analyzed from three dimensions which are operation dimension, hierarchy dimension and time dimension. And under these dimensions, several kinds of process complexities are classified to demonstrate the different aspects of the assembly production process. Then an integrated process model is proposed based on the classification.

1 INTRODUCTION

When coping with today's mass customization and globally competitive market environment, companies strive to increase the range of products and reserve the system capability for unexpected market changes. The configuration of manufacturing systems should has sufficient capability with a large number and variety of resources such as facilities, tools, materials and employees to cope with the environment variation, so recognizing and controlling the system complexity are very important for today's manufacturing system.

According to information and entropy, the complexity is defined as an analytical form for manufacturing systems as a measure of how resources variety complicates the production configuration and process. From the researches of Frizelle [1,2], the manufacturing complexity is classified into static complexity and dynamic complexity. Static complexity is arising from the impact the product structure has on the resources, and the dynamic complexity is determined by the operational behavior from direct observations of the process, in particular by how queues behave. Related work by Deshmukh [3] and Efstathiou [4–6] also developed other entropy-based analytical framework for assessing the complexity in manufacturing system.

In recent years, more detailed and reified complexity characteristics are classified and measured to describe the complexity under actual manufacturing environment. Xiaowei Zhu [7] defined an operator choice complexity that integrates both product variety and assembly process information for evaluating complexity in multistage mixed-model assembly systems. S.N. Samy et al. [8] focused on the product structure, and defined the product assembly complexity as a degree to which the individual part/subassemblies contain physical attributes that cause difficulties during the handling and insertion processes in manual or automatic assembly. H. wang et al. [9,10] researched the complexity for product variety and used the complexity to analyze its impact on assembly system and supply chains. However, the process complexity which is defined as a measure of how the production variety complicates the production process, and is one of the most important expression of the manufacturing complexity that haven't been pay enough attention to in this research field.

In order to recognize the process complexity better, the assembly system and its operation process are analyzed from different views, and the classification of an integrated process complexity is proposed in this paper. The paper is organized as follows. In section 1, we have given a brief introduction to the process complexity. In section 2, the process complexity in assembly system is first defined, and different dimensions to realize the process complexity is proposed. In section 3, 4 and 5, the process complexity is discussed from operation dimension, hierarchy dimension and time dimension respectively. In section 6, the integrate model is proposed. In section 7, we draw the conclusions and point out future works.

2 PROCESS COMPLEXITY IN ASSEMBLY SYSTEM

According to the researches by Frizelle and Deshmukh, the general manufacturing system complexity can be separated as static and dynamic complexity. As one of the typical manufacturing system, the assembly manufacturing system may also contain the characteristics of the corresponding complexity. Generally, the assembly system can be separated as three primary layers to describe the whole system operation. The basic layer is system configuration, which always points to all of the manufacturing resources and their relationship. The middle layer is production process, which points to the execution process of every production tasks arranged by the planning. The top layer is control strategy, which contains all the management and control methods. And these three layers may also point to relative complexity which are structure complexity, process complexity and control complexity.

Among the complexities, the assembly process complexity is defined to describe the complexity of the detailed process of one or batches of products that is or are assembled. The factors such as operation times, setup times, equipment maintenance task, products switch time, failure rate, and product qualified rate must be taken into full consideration to measure the process complexity. By the measure of assembly process complexity, the uncertainty and instability of the assembly process can be controlled and optimized.

3 PROCESS COMPLEXITY ANALYSIS FROM OPERATION DIMENSION

For assembling an entire product, lots of work should be executed and all the activities should influence the value of the process complexity. The process complexity should always focus on all the activities and events in the assembly system.

Generally, the production process can be separated as a basic production process, a subsidiary production process and a service production process. Due to the assembly system, the basic production process usually points to the execution of manual or automatic assembly operation. The subsidiary production process points to the product switch, equipment maintenance, tools prepare and so on. The service production process always points to material distribution, technical checking, quality inspection etc. All these activities constitute the entire product process, and every activity contains its own independent information.

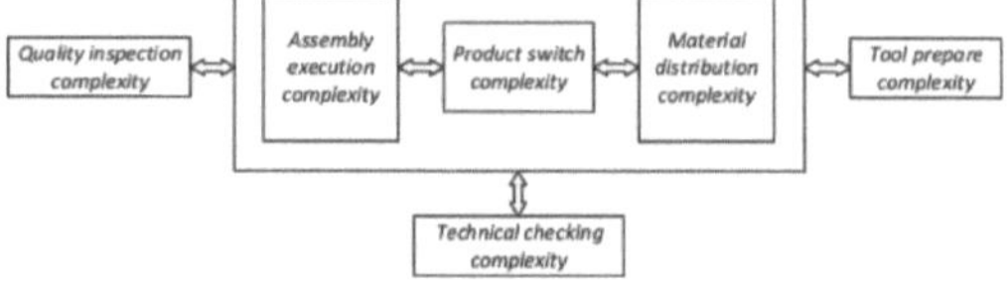

Figure 1. The complexity based on operation dimension.

From the view of the operation dimension, all these activities can be separated and extracted relative process complexity, such as assembly execution complexity, product switch complexity, equipment maintenance complexity, material distribution complexity, tools prepare complexity, etc. But among them, the assembly execution complexity, product switch complexity and material distribution complexity are more important for the system of the assembly line, because they directly determine the fluency of the production. And other process complexities are services for them.

1. **Assembly execution complexity** is defined as the function of the detailed process of one or batches of products to be assembled. By the measure of assembly process complexity, the uncertainty and instability of the assembly process and the system can be controlled and optimized.
2. **Product switch complexity** describes the complexity about the activities when the assembly system changes the different products. The factors, such as preparation for technical data, frocks changes, and equipment adjustment, must be considered in the complexity. It can be used to quantify the difference between the switched products.
3. **Material distribution complexity** describes the complex degree of the distribution of parts and components when batches of products are assembled, and it is used to divide the material flow in the work-in-process between the assembly processes.

4 PROCESS COMPLEXITY ANALYSIS FROM HIERARCHY DIMENSION

Under an actual assembly manufacturing environment, several kinds of planning are performed to guide different layers of process. For example, the short-term planning contains several batches of one day planning, but the detail scheduling is established to guide every product assembly process even the specific station operation. So the assembly process complexity can be more decomposed from the hierarchy dimension.

Layer 1: Production cycle

From the view of short-term planning, an entire production process always contains some necessary works such as production arrangements, batches of production, equipment maintenance, line switching, adjusting equipments. By implementing these works continuously, a whole production period can be formed, and we can call this process as production cycle.

Layer 2: Production sequence
The production cycle is utilized to describe the entire production process, but each of these works can also be decomposed further, such as production arrangements, product inspection, finished products transfer, equipment repair, products repair. All of these decomposed sub-processes are the same as the whole production period. So the process contains these activities that can be called as production sequence.

Layer 3: Assembly flow
The assembly process of every product is selected here to make a further decomposition, and the process can be called as product assembly flow. In addition to implement operation in every work station, the process also contains the operations such as material up line, semi-finished products transmission, process inspection and finished product down line.

Layer 4: Station operation
Based on the standard work method, a simple work station operation task can further be decomposed by several actions to rationalize the operation procedure, decreasing the labor intensity and increasing the efficiency. Therefore, the action step can be considered as the sub-process of the operation process, and this process can be called as station operation.

Figure 2 shows the decomposition of the whole production period based on the above four layers analysis of the different production processes. Beside these four dimensions, the production processes can still be expanded for much more coarse or fine grit activities, such as long or medium planning with more coarse grit activities, or the detail programming of automatic robots. And these activities will not be discussed in this paper.

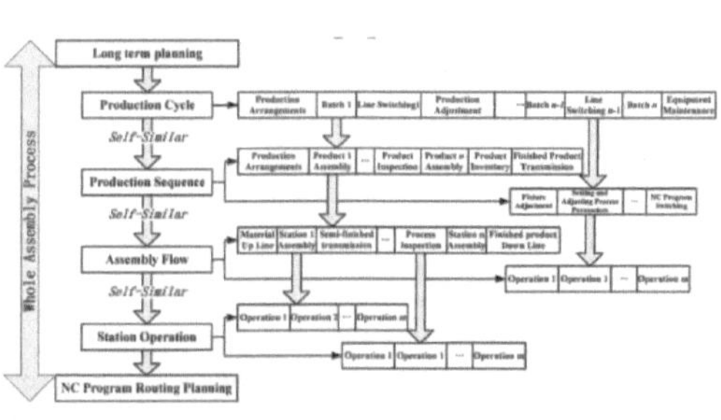

Figure 2. Decomposition of the whole production period.

From the hierarchy dimension, the different assembly activities can be distinguished better. In an aspect of the assembly execution complexity, the processes complexity can be classified as follows.

1. Production cycle complexity
Production cycle complexity can describe as the process of complexity within a whole short-term planning which mainly contains the batches of unlike products, and its value is decided by the execution time of unit batch production, products switching, and equipment maintenances and so on.

2. Production sequence complexity
Production sequence complexity is defined to describe the complexity of the process with one batch of same products, and its value is decided by the single product assembly time, production arrangement time, equipments repair tasks and so on. The production sequence complexity can be seemed as the complexity of the schedule programming, and it could provide the parameters to optimize the production scheduling.

3. Assembly flow complexity
Assembly flow complexity is used to describe the complexity of the process with one product, and its value is decided by the process routing, assembly facilities, operation times, semi-finished product transmission time and so on. By the research on the assembly process of the single product, the assembly flow complexity can be utilized to specify the assembly process diversity between the different products.

4. Station operation complexity
Station operation complexity is used to describe the complexity of the assembly operation process within every station, and it always focuses on the execution of every standard operations and their execution time.

Beside the assembly execution complexity, other process complexity such as product switch complexity and material distribution complexity can also be decomposed by the similar hierarchy.

5 PROCESS COMPLEXITY ANALYSIS FROM TIME DIMENSION

Under an actual manufacturing environment, lots of stochastic events such as equipment failure, operation delay and temporary tasks may also influence the execution process of the planning, and the value of process complexity must be changed. By coping with the diversification of the production process from the time dimension, the process complexity must be separated as static process complexity and dynamic process complexity to face the different environments.

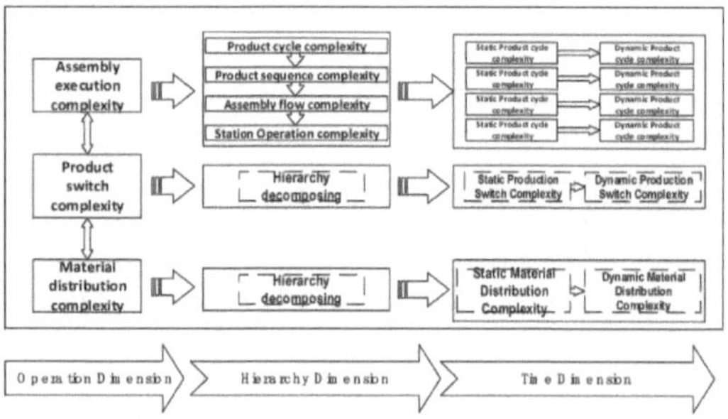

Figure 3. Assembly process complexity classification model.

Static process complexity is primarily used to measure the information required to describe the expected state of the pre-decided manufacturing process when the system is under a quiescent state. For a given manufacturing system, the expected state of production process always depends on the production planning, so the static process complexity can be measured by the information contained in the planning. It should be noted that the static process complexity is independent of the behavior of the system.

Dynamic process complexity concerns about the actual state of every resource during the operation, and it is used to describe the information requirement when the system is in operation. This complexity can be measured either as the predicted future of the system state, or the unpredictability of the system performance measures. Therefore, when the dynamic process complexity is measured, besides the normal processing procedure of every task, the amount changes of the planning such as equipment failure, operation delay, temporary task, and the state changes of every resource also must be observed and measured.

6 INTEGRATED CLASSIFICATION MODEL FOR PROCESS COMPLEXITY

According to the analysis of the process complexity, the process assembly system complexity can overall be recognized. In order to understand the effect of every process complexity and the relationship among them, the conceptual model of the process complexity is given in Figure 3.

As Figure 3 shows, the model can be separated as three dimensions, and there is a progressive relationship between the dimensions. More and more delicate factors are decomposed to recognize the process complexity. Therefore, by the model of assembly system complexity with these three dimensions, the process complexity which is an abstract and fuzzy concept can be comprehended more concretely and roundly.

7 CONCLUSIONS

In this paper, we present the result of our work aiming at the process complexity of the assembly manufacturing system. In order to understand the complexity of the assembly manufacturing system, three dimensions are proposed to analyses the process complexity. From the operation dimension, several process complexity especially assembly execution complexity, production switch complexity and material distribution complexity are classified. From the hierarchy dimension, every kind of these process complexities can be more decomposed by layers to face the coarseness in the different activities. And From the time dimension, regardless of rough or fine process complexity, all can be separated as static and dynamic process complexity to distinguish the complexity between ideal and factual production process. At last, an integrate classification model is established, and more delicate factors are proposed to understand the whole process complexity.

ACKNOWLEDGEMENTS

This research work is supported by the National Natural Science Foundation for Young Scholars of China under Grant No. 51105157.

REFERENCES

[1] Frizelle G., Wookcock E., 1994. Measuring complexity as an aid to developing operational strategy. *International Journal of Operations & Production Management,* 15(5), pp. 26–39.

[2] Frizelle G., Suhov Y.M., 2001. An entropic measurement of queueing behaviour in a class of manufacturing operations. *Proceedings of The Royal Society*, 457, pp. 1579–1601.

[3] Deshmukh, A.V., Talavage, J.J., Barash, M.M., 1998. Complexity in manufacturing systems. Part 1: Analysis of static complexity. *IIE Transactions*, 30(4), pp. 35–44.

[4] Efstathiou J., Tassano F., Sivadasan S., Alves J., 1999. Information complexity as a driver of emergent phenomena in the business community, *Proceeding of the International Workshop on Emergent Synthesis*. Kobe University, Japan, pp. 1–6.

[5] Sivadasan S., Efstathiou J., Frizelle G., Schrin J., Calinescu A., 2002. An information-theoretic methodology for measuring the operational complexity of supplier-customer systems. *International Journal of Operations & Production Management*, 22(1), pp. 80–102.

[6] Zhang T., Efstathiou J. 2006. The complexity of mass customization systems under different inventory strategies. *International Journal of Computer Integrated Manufacturing*, 19(5), pp. 423–433.
[7] Xiaowei Zhu, Hu S.J., Koren Y., Marin S.P. 2008. Modeling of manufacturing complexity in mixed-model assembly lines. *Journal of Manufacturing Science and Engineering*, 130/051013: 1–10.
[8] S.N. Samy, H. ElMaraghy. 2010. A model for measuring products assembly complexity. *International Journal of Computer Integrated Manufacturing* 23 (11): 1015–1027.
[9] H. Wang, S.J. Hu. 2010. Manufacturing complexity in assembly systems with hybrid configurations and its impact on throughput. *CIPR Annals—Manufacturing Technology* 59:53–56.
[10] Hui Wang. 2010. Product variety induced complexity and its impact on mixed-model assembly systems and supply chains, *PhD thesis*, University of Michigan.

Engineering Management and Industrial Engineering – Leung (Ed.)
© 2015 Taylor & Francis Group, London, ISBN 978-1-138-02773-2

The research application about large deformation FEM in the pile-soil interaction mechanism of jack-up platform

Z.D. Zhang & X. Zhang
School of Naval Architecture and Ocean Engineering, Zhejiang Ocean University, Zhoushan, Zhejiang, China

ABSTRACT: The large deformation Finite Element Method (FEM) is widely used in the study of pile-soil interaction mechanical problems to solve the high mesh distortion, especially for Arbitrary Lagrangian-Eulerian (ALE) and Coupling Eulerian-Lagrangian (CEL) algorithms. The advantages and disadvantages of ALE and CEL algorithms in the pile-soil interaction mechanism of jack-up platform are decided by their research and development strategies. The initial disturbance of geostatic stress, the ultimate bearing capacity, the soil surface failure mechanism and the development processing of cavity are compared by the two algorithms. Then the numerical simulation results are compared with the experimental data in published documents. The comparison shows that the ALE algorithm is more convenient and accurate in the pile-soil interaction mechanism in geotechnical engineering.

Keywords: geostatic stress; bearing capacity; cavity; numerical simulation

1 INTRODUCTION

Jack-up rigs, as an important mobile platform, due to their proven mobility, low cost and high efficiency (Dean ETR, 2010), are widely used in water depth less than 150 m offshore industry. Owing to the short of operating cycle and frequent shift, as well as the different bearing capacity of the operating sea area, the catastrophic accidents often occurred. Scholars conducted extensive researches on both the bearing capacity of foundation and the pile-soil interaction mechanism (Yi et al., 2012) studied pore pressure generated during spudcan installation in undrained kaolin clay and its impact on the bearing capacity with CEL algorithm in Abaqus/Explicit. Implicit/Explicit/CEL in FEM has been explored in the penetration depth related to the pile reactive force and the soil flow rule at pile shaft (G. Qiu, S. Henke & J. Grabe, 2005). Hossain et al. (2005) demonstrated the formation of cavity and the associated soil back flow around the spudcanin similar kaolin clay with both centrifuge tests and finite element analyses with ALE algorithm. Cassidy et al. (2010) compared the experimental pushover loads and displacements on the hull and spudcans in dense sand to numerical simulations using different assumptions of spudcan stiffness.

In this study, the spudcan penetration process under a given soil in centrifuge tests (Hossain et al., 2006) and Yi document was analyzed numerically using a generic finite element program. The main purposes of the study are to test the suitability of the ABAQUS/Explicit ALE and CEL models for solving large-deformation jack-up preloading process, to understand possible soil movements in such a stratified soil structure in the process of spudcan penetration, and to determine the impact of the soil structure to the ultimate bearing capacity of the spudcan foundation. In the end, the stress of the soil, the bearing capacity of the foundation and the risk of punch-through failure relevant to the specific stratified soil structure were investigated. Finally, the conclusion shows that ALE algorithm is more easily accessible and stable to handle the initial state of geostatic and bearing capacity than CEL, and the back-flow of soil is more visual and actual. The ALE method is better in geotechnical engineering, the penetration problems is especial.

2 NUMERICAL METHOD

2.1 *ALE theory*

The ALE method was introduced to soil mechanics (Donea, Fasoli-Stella & Gilulíani, 1977) as a generalization of the classical Lagrangian and Eulerian formulations to minimize the commonly known limitations of both original forms while solving for large-deformation soil mechanics problems (Donea, Huerta, & Rodriguez-Ferran, 2004).

In the ALE description, a so-called arbitrary referential coordinate system is introduced in addition to the Lagrangian and Eulerian coordinates. Thus substituting the relationship between material time derivative and the referential-domain time derivative leads to the ALE equations,

$$\frac{\partial f(X_i,t)}{\partial t} = \frac{\partial f(x_i,t)}{\partial t} + w_i \frac{\partial f(x_i,t)}{\partial x_i} \quad (1)$$

where X_i is the Lagrangian coordinate; x_i is the Eulerian coordinate; w_i is the relative velocity. Letting v being the velocity of the material, u the velocity of the ALE mesh, the relative velocity becomes $w = u - v$. Thus the governing equations for the ALE formulation can then be written as follows.

Mass equation:

$$\frac{\partial \rho}{\partial t} = -\rho \frac{\partial v_i}{\partial x_i} - w_i \frac{\partial \rho}{\partial x_i} \quad (2)$$

Momentum equation:

$$\rho \frac{\partial v_i}{\partial t} = \sigma_{ij,j} + \rho b_i - \rho w_i \frac{\partial v_i}{\partial x_j} \quad (3)$$

Energy equation:

$$\rho \frac{\partial E}{\partial t} = \sigma_{ij} v_{i,j} + \rho b_i v_i - \rho w_i \frac{\partial E}{\partial x_j} \quad (4)$$

where σ_{ij} is the stress tensor defined by $\sigma = -p + \tau$, τ is the shear stress from the constitutive model and p the pressure; b is the body force vector per unit mass.

2.2 *CEL theory*

CEL method is the key technology of fluid-structure coupling calculation by using Abaqus software. This method comprehends the advantages of Lagrangian and Eulerian grids and adopts the type that the material can flow in free in the fixed grid to solve the malpractice of grid distortion in large displacement.

The governing equations for the CEL formulation can be written as follows,

Continuity equation:

$$\frac{D\rho}{Dt} + \rho \nabla \cdot v = \frac{\partial \rho}{\partial t} + v \cdot \nabla \rho + \rho \nabla \cdot v \quad (5)$$

Momentum equation:

$$\rho \cdot \frac{Dv}{Dt} = \nabla \cdot \sigma + \rho b$$

$$\frac{D(v)}{Dt} = \frac{\partial (v)}{\partial t} + v \cdot \nabla (v) \quad (6)$$

Energy equation:

$$\rho \frac{DE}{Dt} = \sigma \cdot \dot{\varepsilon} + \rho \dot{Q} \quad (7)$$

where $\dot{\varepsilon} = 1/2(\nabla v + \nabla v^T)$ is strain tensor ratio, while the heat transfer factor is ignored.

3 TEST CASE AND APPLICATION

3.1 *Soil and pipe model*

The most commonly used in modeling of soil deformation and bearing capacity is the Mohr-Coulomb model, which is applicable to both drained and undrained soil strengths. The soil structure and the associated properties used in the model were based on documents (Yi et al., 2012). The mesh models of ALE and CEL as shown in Figure 1 and Figure 2. It should be marked off districts, so the pile is convenient to penetrate into the soil and also it increases the efficiency of the FEM. The rigid pile (Hossain et al., 2005) model is shown in Figure 3.

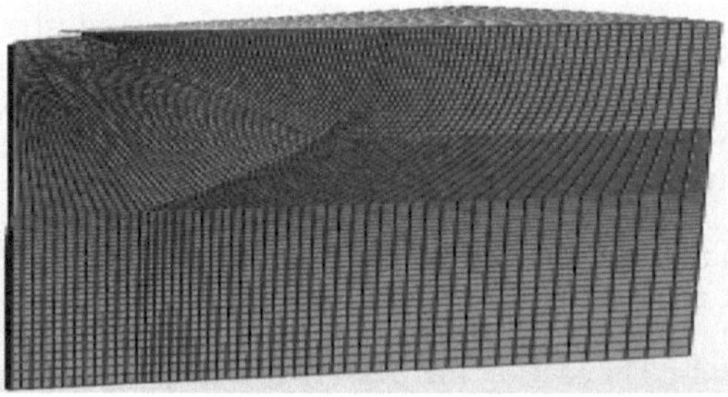

Figure 1. ALE mesh model.

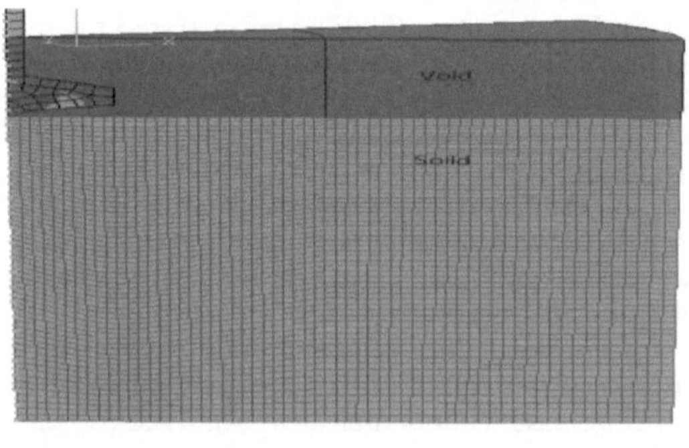

Figure 2. CEL mesh model.

3.2 *The initial disturbance of geostatic stress*

The geostatic stress distribution is calculated by ALE and CEL algorithms, at different widths of the foundation. The results are shown in Figure 4.

At the near boundary field (L/D = 0, L/D = 2), the CEL initial disturbance of geostatic stress are more severe than ALE, but at the far field boundary (L/D = 5, L/D = 10), the boundary conditions have little or no effect on the distribution of stress. In the CEL algorithm, the model of the soil displacement is volatile and leaping at the boundary field, it needs to recalculate and balance the next displacement node. Whereas, the displacement nodes on the boundary field are constant in ALE algorithm.

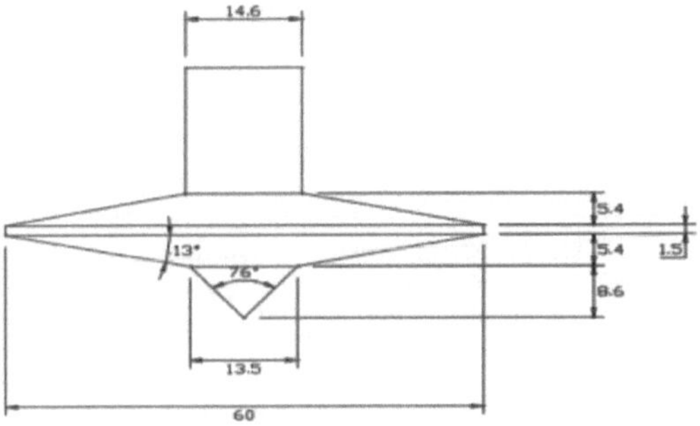

Figure 3. Rigid pipe model.

3.3 *The ultimate bearing capacity*

In a recent numerical study by Yi et al. (2012) and Hossain et al. (2010) the bearing resistance of soft Kaolin clay to the spudcan penetration was studied using the ABAQUS/Explicit Eulerian model. In order to verify the effectiveness of the ABAQUS ALE model adopted in this paper, the same study case is first analyzed and the results of penetration resistance are compared to the results of study.

The short-term, or undrained, bearing capacity of a shallow foundation at a specific depth, d, under the action of purely vertical loading, can be determined as

$$q_u = \frac{\gamma' V_b}{A} + \gamma' d + NcoSuo \tag{8}$$

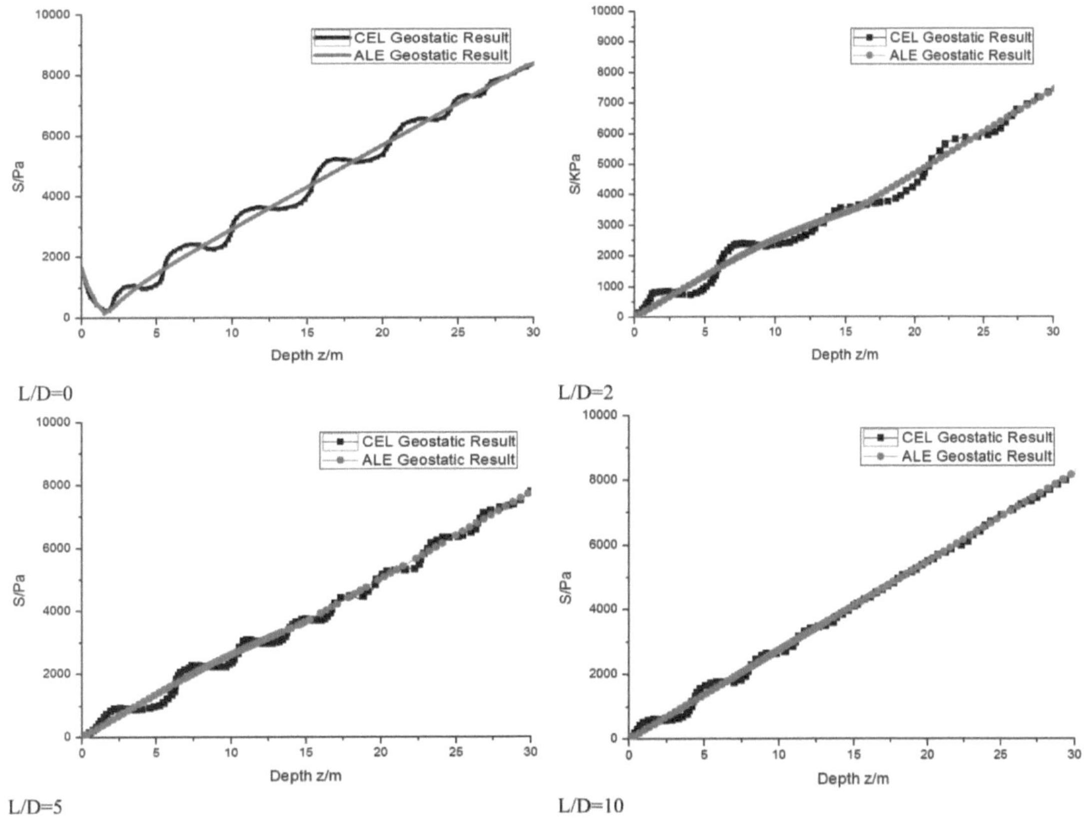

Figure 4. The geostatic stress distribution at different width of the foundation.

where V_b and A are the embedded volume (below the maximum diameter) and cross-sectional area of the spudcan respectively, *Suo* is the shear strength at the lowest maximum spudcan diameter; *Nco* is taken from Figure 5 as Hossain test needed. Figure 6 shows the spudcan penetration resistance, in its infancy, the two algorithms are consistent. Deeply, because the material point flows freely in the grid, but the boundary keep still in Euler algorithm (Goh TL, 2003), so the actual resistance is smaller than ALE algorithm. It is consistent with the actual situation, the ALE algorithm is better to deal with the relationship of the grid, boundary and material.

3.4 *The soil surface failure and the development processing of cavity mechanisms*

The soil surface failure mechanism and the development processing of cavity are calculated by ALE and CEL algorithms at different depth. The results are shown in Figure 7 and Figure 8.

The failure mechanism for both soil surface and cavity are coincident. At the initial penetration, the spudcan pushes the soil near the spudcan edge outward and upward. This step leads to the surface heave just outside the spudcan rim. At deeper penetration, there is a considerable rotation of the velocity vectors near the spudcan edge, resulting in the soil elements along the cavity face displacing radically inward, the effect of which is close to the cavity. Backflow of soil occurs until the spudcan is fully covered, leading to a reduction in surface heaving. At further penetration, soil back-flow reaches a steady pattern, characterized by localized rotational movement of soil around the spudcan edge, while the soil near the ground surface is largely unaffected. It is more visualized for ALE algorithm to allow the grid, boundary and material

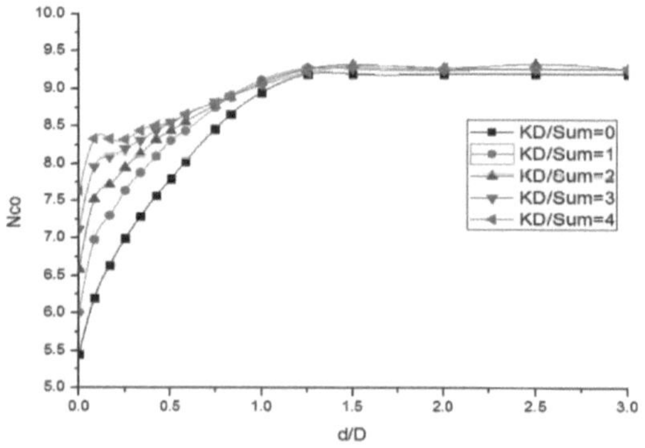

Figure 5. Bearing capacity factors for spudcan in weightless soil.

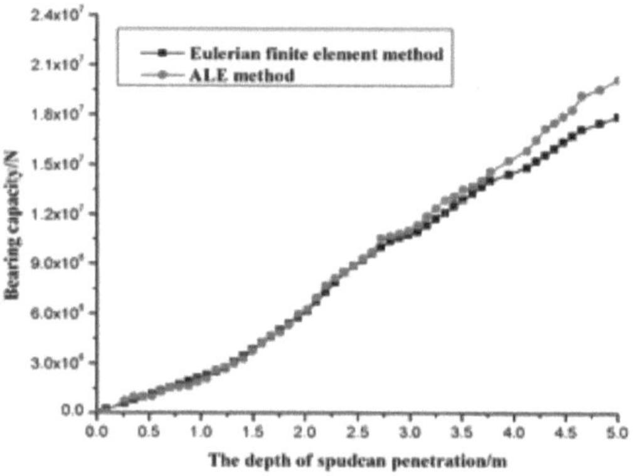

Figure 6. Spudcan penetration resistance profiles for Smooth base.

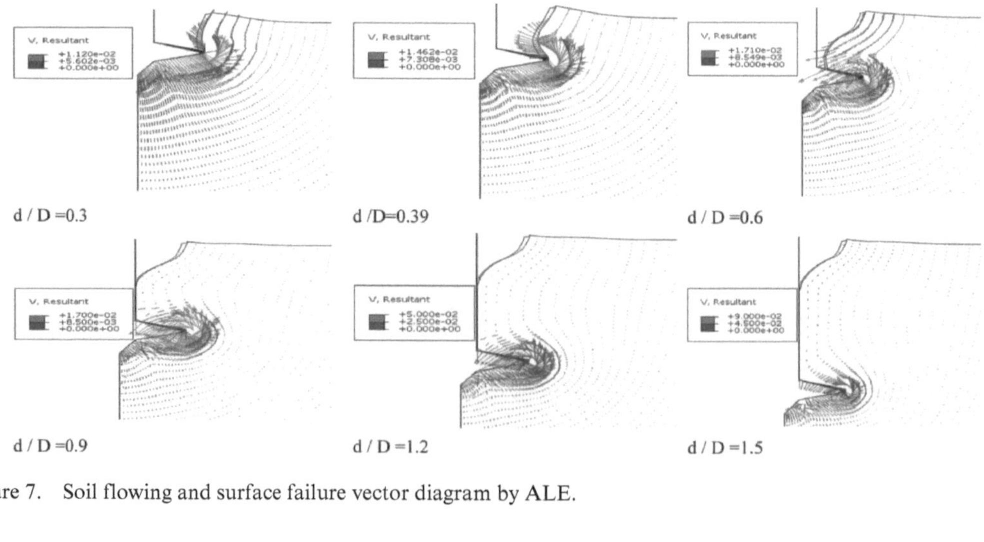

Figure 7. Soil flowing and surface failure vector diagram by ALE.

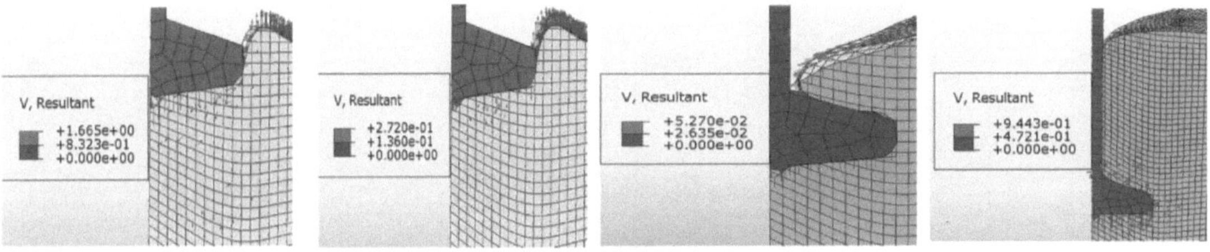

d /D=0.16 d /D=0.32 d /D=0.48 d /D=1.2

Figure 8. Soil flowing and surface failure vector diagram by CEL.

movements to get together. However, once the material flows out of the Euler mesh, this part loses efficacy.

4 CONCLUSIONS

The numerical results of the ultimate bearing capacity and the penetration depth during preloading of jack-up is presented by using ALE and CEL algorithm, the following are some significant conclusions deduced from the simulation results.

1. The soil movement mechanisms have been given. During the initial penetration, the soil of surrounding spudcan moved to lateral and the surface soil heaved latterly. With further penetration, soil began to flow back and a cavity formed above the spudcan. When penetration became deeper, soil back-flow continued to occur, while the initial cavity was still unchanged. Similar flow mechanisms and cavity formation are also found in Hossain centrifuge test.
2. From the previous discussion in Abaqus software, the feasibility by using of ALE algorithm in large soil deformation problems is verified. The results indicate the applicability of ALE method for the existing soil constitutive model, and can be widely applied to investigate large soil deformation problems.
3. For the penetration problems in geotechnical engineering, the geostatic stress is the initial and important step, so the initial disturbance of geostatic stress should be taken seriously. For further research about the pile-soil interaction, the ALE algorithm is prior to CEL algorithm.

ACKNOWLEDGMENTS

This research work is financially supported by The National Natural Science Foundation of China (51179173 and 51379189).

REFERENCES

Cassidy et al. compared the experimental pushover loads and displacements on the hull and spudcans in dense sand to numerical simulations using different assumptions of spudcan stiffness (2010).

Dean ETR. Offshore geotechnical engineering: principles and practice. London: Thomas Telford;2010.

Donea. J., Fasoli-Stella, P., and Giluliani S. 1977. Lagrangian and eulerian finite element techniques for transient fluid-structure interaction problems. In Trans. 4th SMIRT Conf.

Donea J., Huerta A., Ponthot J.P., and Rodriguez-ferran, A. 2004. The encyclopedia of computational mechanics. John Eiley & Sons Inc., New York.

Goh T.L. Stabilisation of an excavation by an embedded improved soillayer. Ph.D. Thesis, National University of Singapore; 2003.

Hossain M.S., Hu Y., Randolph M.F., White D.J. Limiting cavity depth for spudcan foundations penetrating clay. Geotechnique 2005;55(9):679–690.

Hossain M.S., Randolph M.F., Hu Y., White D.J. Cavity stability and bearing capacity of spudcan foundations on clay. In: Proceedings of the offshore technology conference (OTC'06);2006.

Hossain M.S., Randolph M.F. Deep-penetrating spudcan foundations on layered clays: numerical analysis. Geotechnique 2010;60(3):171–184.

Qiu G., S. Henke, and J. Grabe. Applications of Coupled Eulerian-Lagrangian Method to Geotechnical Problems with Large Deformations, 2009 SIMULIA Customer Conference.

Yi, Jiang Tao, Fook Hou Lee, Siang Huat Goh, Xi Ying Zhang, Jer-Fang Wu. Eulerian finite element analysisof excess pore pressure generated by spudcan installation into soft clay. Computers and Geotechnics; 42(2012):157–170.

Information processing and engineering

Engineering Management and Industrial Engineering – Leung (Ed.)
© 2015 Taylor & Francis Group, London, ISBN 978-1-138-02773-2

Research on the integration solution for information platform in power enterprises

Y.J. Jiang, X.S. Yu & M. Yu
China Three Gorges University, Yichang, China

ABSTRACT: This paper puts forward a kind of enterprise-class solution to building information integration platform in power enterprises based on Service-Oriented Architecture (SOA) by analyzing the problems in the construction process of Management Information System (MIS). In SOA, the enterprise's applications will be encapsulated as Web Services that provides a uniform interface and data transmission services. For its expansibility and reusability, this information integration architecture based on SOA can solve the key problems like the enterprise data concentration, applications integration, business cooperation and so on.

Keywords: information integration; power enterprises; Service-Oriented Architecture; web services

1 INTRODUCTION

As a key sector of the national economy China's electric power industry has separated power plants from grid and is undergoing bidding toin generation side. In the new reform plan, the electric power will become a true commodity. Therefore, it's very necessary for the power enterprises to enhance theirs information construction level in which I think that the construction of MIS is a key problem. By strengthening the information feedback to enhance accuracy of decision-making, the power enterprise's management level can be improved to meet the requirements of market economy and the characteristics of electric power production. On this basis, the existing application systems and data can be integrated efficiently into a unified information management platform by implementing the technology of Enterprise Application Integration (EAI) and Enterprise Information Portal (EIP), and then are used to solve the problem of isolated information, to realize the data centralized management, and to unify the interface to the user. Finally, we can build a rapid and smooth information exchange channel for the enterprise. In addition, we can enhance the collaboration between companies through business processes.

2 ANALYSIS OF MIS CONSTRUCTION

2.1 *The situation of MIS construction*

In order to adapt to the development requirements of the power system reform, power enterprises have carried out different levels of information construction in the aspects of production, operation and management. A batch of business application systems have been developed and put into use, such as Power Marketing Management System, Energy Management System, Property Management System, Production Management System, and Distribution Management System. It should be said that each subsystem or application or module has played an active and effective role, meeting the power enterprise's business needs. However, due to the lack of scientific planning, the information generated in these systems cannot be shared very well and collaborative business cannot be carried out. Current construction situation of power enterprise's MIS is blind, dispersive and farraginous.

2.2 *The causes of the problem*

The problems are mainly manifested in:

Lack of business process integration. There was not Work-Flow Engine in software implementation by reason of technical, so every business process is relatively independent, and its openness is poor[1].

Lack of the unified information technical standards and technical specification. For example, the production system had been developed by C/C++ language in general for its high performance requirements, in which the core programs were developed independently by manufacturers including more and more private and proprietary technology. Besides, the database, file format and access interface of these systems are unavailable [1].

The information systems are isolated, which cannot enjoy the overall benefit. Because of a lack of overall planning, redundant and repetitive data can not be effectively integrated into uniform information management platform, which can not be fully used for management, decision and data mining to serve power enterprises[1].

3 NEW REQUIREMENTS OF POWER ENTERPRISE INFORMATION

With the further deepening of the power system reform, electrical power monopoly will gradually be broken. In future, the organization structure of power industry will conform to a typical supply chain structure, that is: Power Plant—Power Grid Company—Power Sale Enterprise, with different focuses on information construction.

Power Plant focuses on ensuring the power information at all levels of the enterprise information system, quick and safe interaction;

Power Grid Corp is more concerned with automation of electric power transmission and distribution, the real-time safety monitoring of power grid and treatment of accident disaster.

Power Sale Enterprise places more emphasis on power market analysis and customer relationship management.

Power enterprises present the development trend of collectivization, industrialization, standardization and intelligent since the reform. The group operation needs collaborative operation among member enterprises in each link of procurement, production, sales, and management. With the industry linkage and collaboration, the difficulty in management is increasing and the demand for data analysis, decision support, and intelligence management is gradually highlighted. These factors make the work focus shift from production operations to the management and strategic decision, and the need for 'Integration' is particularly urgent. In the software application, the decentralized procurement from different vendors in the traditional mode is gradually being replaced by the whole industry solution.

4 THE CONSTRUCTION OF INFORMATION INTEGRATION PLATFORM

4.1 *The leader of information integration platform*

After the reform, The Power Grid Corp should become a public enterprise in charge of power supply management and optimization, and its primary mission is to maintain the safety of transmission network, supervise the electricity market, integrate supply chain information, and make a power supply plan to make sure that the power supply is running well. At the same time, it will provide valuable information service for the production and sales of power enterprises in different sectors, to achieve maximum profit of power supply with completing the tasks above. Power Grid Corp should also have to participate in the planning and be responsible for the construction of electric power industry. Therefore, we should take the Power Grid Corp as the core to form the system control center, organize and build the electric power information platform, and complete the information resources integration, application integration and other global integration.

4.2 *Technical architecture of the information integration platform*

In the heterogeneous network environment, building a universal technology layer which is independent of the operating system and programming language to realize the application integration between enterprises, is an urgent need for the power enterprises.

1. The SOA-based Enterprise Architecture
   SOA is a design pattern and design method but not a specific technique, which essentially implies an application architecture made up of loosely coupled 'services'. Based on SOA, the loosely coupled coarse-grained application components are dispersedly deployed, composed and used in the network. The service layer is the foundation of SOA, which can be directly applied to callso as to control the artificial dependence between the system and software agents[2]. The service is designed to solve the interaction problems between heterogeneous system such as data integration, data mining etc.

   In the scheme of SOA, by considering the situation of enterprise's information systems availability and integration requirements, we will have not only to retain the original information system of power enterprise data storage scheme, but to share information with other enterprises. So the enterprise information integration platform not only covers all application areas within the enterprise, but also relates to other enterprises and customers, in which the business process, supply chain and customer relationship will be integrated, and the firewall will be set to ensure the safety of them[4]. (see Fig. 1).
   Implementation: (see Fig. 2)

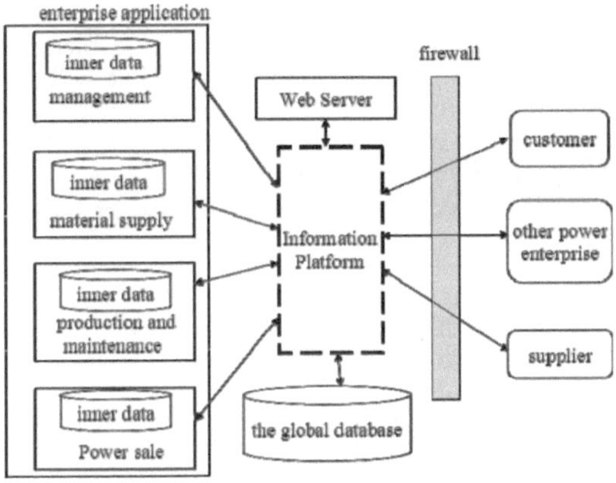

Figure 1. Information integration framework of power enterprise based on SOA.

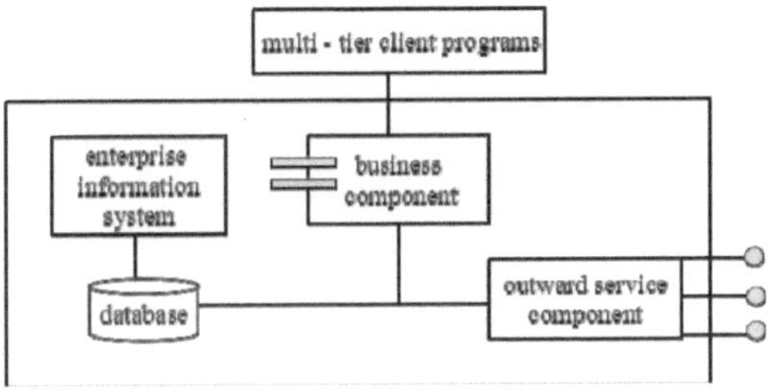

Figure 2. Departmental system package.

For each information system within a sector, applying the local central database scheme;

For the interaction of date or business process among departments, applying the architecture based on Web service.

Shared information resources in each department's will be encapsulated as Web Services that provides a uniform interface and data transmission services. We will use the word 'software component' to describe Web Services, the granular pieces of functionality an SOA relies on, to equip with various enterprise applications. The components may be a distributed application, a internal MIS encapsulated with Web Service, which can shield heterogeneous between them, like different hardware and software, diverse development and deployment platform. On a functional level, components allow for the benefits of a service-oriented architecture because they deliver a level of abstraction, hiding the details of precisely how data is manipulated, but at the same time offering the manipulated data to other components in a standardized fashion. By defining a standard service interface, not the real implementation details, all kinds of function programs can be called conveniently. In the Architecture, XML of self describing properties, as the standard format of message passing, and XML Schema as the standard structure of information exchange, will be used to realize information exchange and interconnection between heterogeneous systems.

Enterprise Service Bus (ESB) is the core component of enterprise information integration platform in SOA architecture, providing a unified service registration and query mechanism, and supporting multi-bus service mode. ESB has the ability of services arrangement to support the construction of the synthesis service and adaptation to files, databases and message queues and other resources. It can provide transaction controlling and XML data transfer function, and realize the real-time monitoring of all registration services[5].

2. The establishment of enterprise-level data standard

We will have to establish a enterprise-level data standard, including Common Information Model (CIM) and data coding standard, provides the standard for the data exchange between application systems within a enterprise. The CIM will provide a unified data definition standard, code specifications and exchange format. In addition, it will support mainstream style and format in data exchange and business processing synchronously or asynchronously. We have to straighten out the channels for information collection, adhering to the principle of a standard & single data source to avoid more-and-long report, to ensure the data consistency and integrity.

In this, I think that it's necessary to introduce the IEC61970 series standards under the International Electrotechnical Commission. The standard includes the Common Information Model (CIM) and Component Interface Specification (CIS) two aspects. Its purpose and meaning lies in: 1) Easy integration for Enterprise Management System (EMS) developed; 2) Interconnection between EMS and other systems in scheduling center; 3) Model transformation of EMS in dispatching center[6].

Getting these international standards into our standards and implementing 'plug and play' of software products as soon, to make EMS and other systems interconnect and interaction, is significant to improve automation level of China power grid. CIM is a major part of IEC61970, providing a standardized definition for power system. According to this standard, a variety of applications in control center or distribution applications at different sites can share power system data for applications integration.

3. Carrying out technical transformation for the existing systems.
   At first, it is necessary to complete data integration of the related system involved in main management activities covering enterprise core value chain. The systems comprise of the Grid Control System, Enterprise Resource Plan (ERP), Peripheral Production management System, Marketing System, Customer Service System. However, the conversion from the existing system data model to CIM is quite complex because of the large difference between the two data model. In many cases, the direct mapping can not solve the problem, and often need to clean up and reorganize data to meet CIM.
4. Interface Integration.
   We can use the enterprise portal technology to provide unified application system entry to company staff for messaging and collaborative work, which can display the image of company very well. With all levels of the directory service, unified authentication for all kinds of application system will be implemented. The Portal can also provide content management function and personalized service according to the user's permissions, personalized data, application and content requirement.

5 CONCLUSIONS

In the integration platform, the interoperation between different business systems can be achieved, and mass data will be provided by these systems for mining to discover more valuable and useful information and knowledge. That will be a set of monitoring, management and decision-making in one integrated system, which provides a strong guarantee for the smart grid building.

REFERENCES

[1] Chen, Z.Y. 2007. Application of EAI/EIP Technology in Electric Power Enterprise. Power System Technology 31(2).

[2] Barai M. et al. 2008. Service Oriented Architecture with Java. Packt Pulishing: 33–38.

[3] IBM. IBM's SOA Foundation:An Architectural Introduction and Overview[EB/OL]. Available from:. http://download.boulder.ibm.com/ibmdl/pub/software/dw/webservices/ws-soa-whitepaper.pdf.

[4] Liu, J.F. 2008. Application of Web Services in Power Enterprises Information Integrating. Computer Engineering Design 29(14).

[5] Xu, B.Y. 2008. Research and practice of information integration platform for power grid enterprises based on SOA. Electric Power Information Technology 6(1).

[6] Zhang, N.H. 2006. Research of Informationalization for Power Enterprise Based on Information Resource Integration. Master's degree thesis of North China Electric Power University: 17–19.

Engineering Management and Industrial Engineering – Leung (Ed.)
 ISBN 978-1-138-02773-2

Adaptive pushover analysis for the seismic response prediction of structures

S.F. Qin
Research Center for Numerical Tests on Material Failure, Dalian University, Dalian, China

H.L. Wang
Institute of Bridge Engineering, Dalian University of Technology, Dalian, China

ABSTRACT: The principal objective of this investigation is to propose a new practical pushover procedure because the classic pushover analysis procedure based on force can't account for the structural stiffness degradation caused by structural yielding. By studying the change of the displacement model and the participating coefficient of vibration mode under different damage states during the push over process of the structure, an adaptive Displacement-Spectra-based Pushover Analysis procedure (DASPA) is proposed in this paper. The DASPA procedure is based on inelastic displacement response spectra. It provides more superior accuracy and reasonability because the effect of higher mode is considered and the behavior of the structure in the earthquake is controlled by displacement in the whole processing. While the damage of the structure reaches a level, pushover analysis is performed by using invariant displacement pattern in DASPA method. It has the conceptual simplicity and computational attractiveness for routine application in structure engineering practice.

Keywords: seismic performance; pushover; displacement response spectra

1 INTRODUCTION

The results of the classic pushover analysis procedure based on force for estimating the seismic performance of the bridge depend on the selection of the lateral load pattern. And the classic pushover analysis procedure based on force can't account for the structural stiffness degradation caused by structural yielding because that structural internal force and loading history is not the single value correspondence relationship. Earthquake damage, experiment and theoretical analysis all indicate that the deficiency of transfiguration capacity and capacity of energy dissipation are the main reasons of structure collapse. The degree of damage of the structures in large earthquake relates to their displacement responses and deformability. So it is more reasonable to control structural behavior by displacement[1,2,3].

Recently, some researchers begin to explore and develop the displacement-based pushover analysis methods. Chinese researcher Wang Meng-fu has also studied the displacement-based non-linear static analysis method, and proved its effectiveness by analyzing the typical high-rise building with shear wall[2]. But in his method the displacement model is determined only by considering the effect of participating coefficient of vibration mode, not by combination of the response spectrum of the input ground motion. Chopra's investigation indicates that the displacement model is related to both participating coefficient of vibration mode and response spectrum. So in this paper the structural displacement pattern is determined directly by inelastic displacement response spectra. And an adaptive Displacement-Spectra-based Pushover Analysis procedure (DASPA) is developed by studying the change of the displacement model and the participating coefficient of vibration mode under different damage states during the push over process of the structure.

2 ADAPTIVE DISPLACEMENT-SPECTRA-BASED PUSHOVER ANALYSIS

2.1 *Determination of the displacement pattern and the target displacement*

Determination of the displacement pattern and the target displacement are the key problems to establish the adaptive displacement-spectra-based pushover analysis procedure. The research shows that the structural displacement pattern is related not only to participating coefficient of vibration mode but also to displacement response spectra[4]. So the structural displacement pattern and the target displacement can be described by the following equations:

$$\Delta_{ij} = \Gamma_j \phi_{ij} S_d(j),\ \Delta_i = \sqrt{\sum_{j=1}^{N} \Delta_{ij}^2} \tag{1}$$

$$u_{rjo} = \Gamma_j \phi_{rj} S_d(j),\ u_{ro} = \sqrt{\sum_{j=1}^{N} u_{rjo}^2} \tag{2}$$

$S_d(j)$ is spectral displacement corresponding to jth mode; Γ_j is mode participation factor for jth mode; ϕ_{ij} is mode shape value at ith node corresponding to jth mode; Δ_{ij} is lateral node displacement at ith node for jth mode; Δ_i is combined node displacement at ith node for jth mode; u_{rjo} is structural target displacement corresponding to jth mode; u_{ro} is combined structural target displacement.

2.2 *Structural model for the study*

A 6-storey reinforced concrete frame with height of 22 meters is selected as the numerical example. The elevation of the frame is shown in Figure 1 and the section calculation parameters of the beams and columns are shown in Table 1. The concrete strength grade is C30; the intensity grade of the main reinforcement is II grade and the stirrup is grade reinforcement. And assembled monolithic reinforced concrete floors and roofs are used[5].

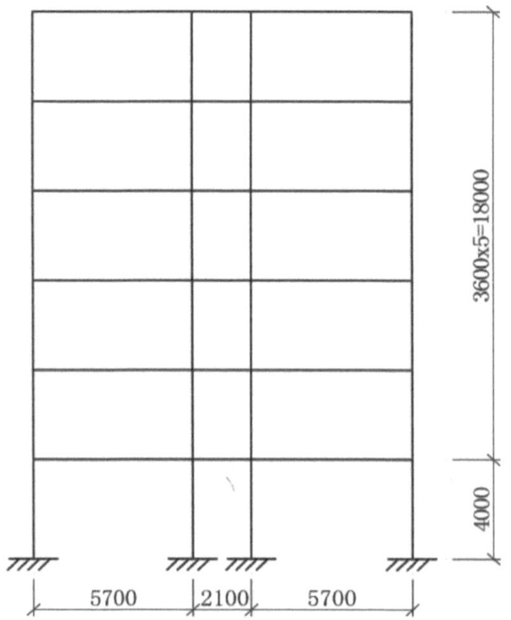

Figure 1. Elevation of the frame (unit: mm).

2.3 *The selected typical ground motion record*

El Centro ground motion (1940) is selected in the paper. The acceleration time histories curve (normalization the maximum) is shown in Figure 2. The displacement ductility ratio of the frame is got by using iterative method[6] that $\mu = 4$. And the inelastic displacement spectra (5% damped, $\mu = 4.0$) is shown in Figure 3.

2.4 *The vibration characteristic analysis*

Carry out modal analysis, and calculate the initial displacement model of the structure. Pushover the structure by using the calculated initial displacement pattern, and carry out modal analysis again after the structure yields. Recalculate the structural displacement model and pushover the structure by using the new displacement pattern. Therefore, the mode participation factor of each mode under different damage states (Fig. 4) and the structural displacement pattern under different damage states (Fig. 5) can be obtained. Case1 represents elastic state, and the damage state become more and more severe from case 2 to case 13. Figures 4 and 5 show that, the participation coefficients of high modes are large and the structural displacement pattern varies significantly under elastic or slight damage states. But with the degree of damage deepening, the effect of high mode is becoming smaller and the structural displacement pattern tends to be uniform.

2.5 *Basic steps of adaptive displacement-spectra-based pushover analysis*

The conclusions can be drawn according to the above analysis of the frame calculation example: 1) The participation coefficients of higher modes are large and the structural displacement patterns varies significantly under elastic or slight damage states. Under this situation, the structural displacement pattern should be adjusted as the change of the structural dynamical characteristic. 2) While the damage of structure reaches a level (the ratio of the effective mass of the main mode and the total effective mass is more than 90%[7]), the effect of high mode can be neglected, and the structural displacement pattern tends to be uniform. Under this situation, the next pushover analysis can be carried out with invariant displacement distribution. So the detailed steps of the adaptive Displacement-Spectra-based Pushover Analysis procedure (DASPA) can be described as the follows:

1. Build the calculation model, and carry out modal analysis to compute initial period and mode of the structure;

Table 1. The section calculation parameters of the beams and columns.

	Beams		Columns		
Floor	Section size (mm × mm)	Sectional reinforcement	Section size (mm × mm)	Sectional reinforcement	Floor quality (×10^5 kg)
1	650 × 250	4Φ25	400 × 400	4Φ16	1.16
2~5	650 × 250	4Φ25	400 × 400	4Φ16	1.03
6	650 × 250	4Φ25	400 × 400	4Φ16	0.78

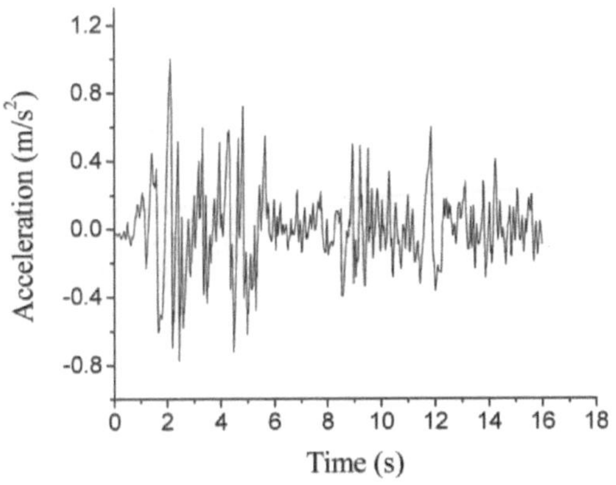

Figure 2. Acceleration time history.

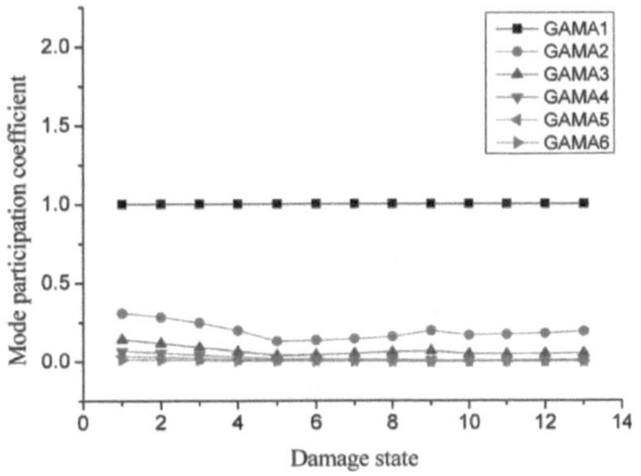

Figure 4. Mode participation coefficient.

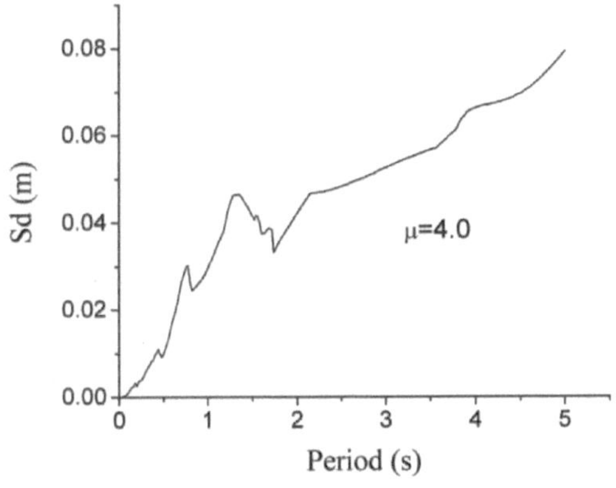

Figure 3. Inelastic displacement response spectra.

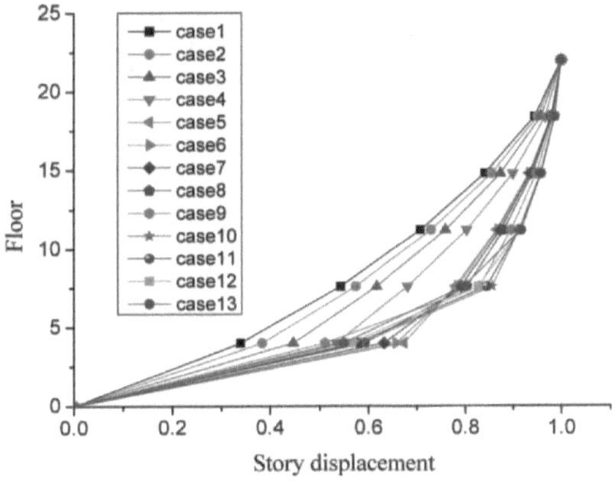

Figure 5. Mode participation coefficient.

2. Calculate the displacement ductility ratio μ and the inelastic displacement response spectra S_d of the structure;
3. Calculate the structural initial displacement pattern and target displacement according to the equation (1) and (2);
4. Carry out pushover analysis for the structure with the initial structural displacement pattern obtained from the above step till some component of the structure yield;
5. Re-compute the member and global stiffness matrices, and carry out modal analysis again, and calculate the new displacement pattern of the structure according to the equation (1);
6. Continue to carry out pushover analysis for the structure with the new structural displacement pattern till the new component of the structure yield;
7. Repeat the steps (5)~(7), till the main mode and the total effective mass is more than 90%;

8. Push the structure with the invariant displacement pattern to the target displacement.

3 CONCLUSIONS

In this paper, an adaptive Displacement-Spectra-based Pushover Analysis procedure (DASPA) is put forward by studying the change of mode participation coefficient and displacement pattern of a six-story-frame in pushover process. A pushover analysis is performed for a beam bridge by using DASPA method. And the result obtained by DASPA is compared with non-linear RHA procedure and other two classical pushover methods. Main conclusions are drawn as follows.

1. When the structure suffer different earthquake shaking the displacement pattern of the structure is variant. The structural displacement pattern under elastic or slight damage states varies significantly. While the damage of structure reaches a level, the structural displacement pattern tends to be uniform.
2. The loading displacement pattern of DASPA method is based directly on inelastic displacement response spectrum. Under elastic or slight damage states, the effect of higher mode is considered, and the structural displacement pattern is adjusted as the change of the structural dynamical characteristic. So DASPA method is applicable to the complicated multi-degree-of-freedom structures. While the damage of the structure reaches a level, pushover analysis is performed by using invariant displacement pattern in DASPA method. It can simplify calculation.

CONFLICT OF INTERESTS

The authors declare that there is no conflict of interests regarding the publication of this article.

ACKNOWLEDGEMENTS

This work is supported by National Natural Science Foundation of China (51108052).

REFERENCES

[1] Mao Jian-meng, Xie Li-li, Zhai Chang-hai. Studies on and improvements in modal pushover analysis [J]. *Earthquake Engineering and Engineering Vibration*, 2006, 26(6):50–55.

[2] Wang Meng-fu, Wang Rui. Study on displacement-based pushover procedure [J]. *Earthquake Engineering and Engineering Vibration*, 2006, 26(5):73–80.

[3] Balram Gupta, M. Eeri, & Sashi K. Kunnath. Adaptive Spectra-Based Pushover Procedure for Seismic Evaluation of Structures [J]. *Earthquake Spectra*, 2000,16(2):367–391.

[4] Anil K. Chopra, & Rakesh K. Goel. A Modal Pushover Analysis Procedure for Estimating Seismic Demands for Buildings [J]. *Earthquake Engineering and Structural Dynamics*, 2002, 31:561–582.

[5] Wei Wei. The research on comparison of several Push-over analysis methods [D]. Harbin: Institute of Engineering Mechanics, China Seismological Bureau, 2001.

[6] Anil K. Chopra. Dynamics of Structures: Theory and Applications to Earthquake Engineering [M]. Beijing: Tsinghua University Press, 2005, 264–265.

[7] Ministry of Construction of P.R.C. Seismic design code [S]. China building industry press, 2001.

Engineering Management and Industrial Engineering – Leung (Ed.)
© 2015 Taylor & Francis Group, London, ISBN 978-1-138-02773-2

Strategies to promote corporate technological diversification ability under the network environment

L.Q. Zhang
Department of Industry Engineering, Changchun University, Changchun, China

ABSTRACT: Under the present network environment, enterprises have extensive technical capabilities. By combining them with the creative use of these skills, enterprises can effectively reduce the risk of product market in the future and effectively enhance the enterprise competitive advantage. Based on the literature review and practical research, organizational learning, innovation culture, internal resources, external relations, corporate technological diversification were put forward. Theoretical assumptions were promoted. Based on the theory of open innovation, multiple regression method study hypothesis was used.

1 RESEARCH STATUS OF CORPORATE TECHNOLOGICAL DIVERSIFICATION

In the 80s, scholars began to do research on how enterprises in the third world countries improve their own technological level through the introduction of technology. Technological reserve and management ability have been emphasized by many enterprises. Technical ability is regarded as one of the key factors in the development of enterprises and is widely concerned as a complete concept (Cohen & Levinthal, 1990; Kim, 1997; Leonard-Barton, 1995; XieWei, 2001; Piscitello, 2004; WeiJiang, 2006). Among them, Cohen & Levinthal (1990) put forward the famous "absorption capacity concept", emphasizing the meaning of enterprise innovation activities on the absorptive capacity, they thought that absorption capacity is a function of enterprise. Leonard-Barton (1995) thought the technical ability should include four aspects, which are staff knowledge and skills system, material and technical system, management system, enterprise value and ideological system and so on. The core technical ability to knowledge, values and other "soft" elements both extends the idea of core competence of Prahalad & Hamel (1990), but it is also a good convergence of the enterprise knowledge theory.

In theory, the strategy to improve the existing literature on corporate technological diversification ability is still very immature. Many scholars are still describing the qualitative, quantitative effectiveness verification strategy rarely. Since China has joined the WTO, the traditional development strategy of low cost advantage of China's enterprises has been hitherto unknown challenge. Implementation of the transformation and technological innovation is based on strategy. Enterprises of our country are in the main direction in the future for the international competitive advantage. Therefore, this paper tries to take Chinese enterprise as the case. Through analysis of typical cases of product innovation and the theory of hypothesis, promotion strategy proposed enterprise diversity of technology capability. Statistical hypothesis testing was used to verify the effectiveness of the proposed method in the hope of corporate technological diversification, and provides new ideas and useful supplement.

2 THE IMPACT FACTORS OF ENHANCING CORPORATE TECHNOLOGICAL DIVERSIFICATION ABILITY UNDER THE NETWORK ENVIRONMENT

In fact, technological diversification is the accumulation of enterprise technology knowledge and ability, which is a variable as time-dependent, and has the properties of evolution (Fai, 2003). Various internal factors are likely to influence the expansion of enterprise technology ability (Malerba & Orsenigo, 1990; Bas & Patel, 2005). Among them, the impact of external factors (such as the aggravation of the market competition, industry technology cycle is shortened, the supply chain becomes more and more complex, and so on) are most expressed for enterprises to carry out technological diversification strategic motive. Among internal factors, some were the performance for motivation and some were the promoting factors.

Motivation analysis helps us to understand the formation of technological diversification enterprises. Promoting the factor analysis can provide guidance for enterprises to carry out technological diversification strategy. Based on the theory of technology strategy and the existing related research factors of corporate technological diversification effect, this paper constructs the conceptual model of corporate technological diversification ability promotion way. The various factors of the mechanism are described one by one, and the theory hypothesis is put forward.

Based on the above analysis, this paper put forward the following hypothesis:

H1: organizational learning has a significant positive effect on corporate technological diversification, the higher the level of organizational learning, the higher the level of technology diversification.

H2: innovation of organizational culture has a significant positive effect on corporate technological diversification, the more open innovation organizational culture is, the higher technological diversification level is.

H3: the internal resources of corporate technological diversification has a signification positive effect, the more fully the enterprise's internal innovation resources is, the higher the level of technological diversification.

H4: technology source external enterprises have a significant positive effect on corporate technological diversification, technology source enterprise and external enterprises more closely linked, the higher the level of technology diversification.

H5: external technology sources have a significant positive effect on corporate technological diversification, corporate and external technology sources more closely linked, the higher the level of technology diversification.

3 ANALYSIS METHODS OF EMPIRICAL UPGRADE CORPORATE TECHNOLOGICAL DIVERSIFICATION ABILITY

The empirical analysis of the hypothesis is conducted by using multiple regression. The first of five promoting factors measurement items are described, and then the reliability and validity analysis are analyzed. Correlation analysis and regression analysis use SPSS19.0 software to verify the hypothesis.

3.1 *Control variables*

The industry and enterprise scale as control variable. First, larger scale enterprises may behave more money and demand accumulated technology knowledge and ability in a larger scope, so the technology diversification level may be higher; second, requirements of technical ability of different industries will also have certain difference. High technology industry technology changes very fast, the product structure is more complex, and enterprise need to grasp the diversity of technical knowledge and ability level of technology. Diversification of these enterprises may be higher; therefore, the control of these two variables may influence corporate technological diversification.

3.2 *The validity test*

On organizational learning culture, innovation, internal resources, external technical, contact measurement table 23 items for the principal component analysis, validity verification scale before the factor analysis. This paper uses KMO sample adequacy measurement and Bartlett test to see if they are suitable for factor analysis. The results showed that the correlation data, suitable for factor analysis.

The factor analysis results and five factors were identified, which were named organizational learning culture, innovation, internal resources, technology source, external technological sources of enterprise. Each item has good load to the expected measurement factor and the factor analysis results. The factor loading coefficient is greater than 0.5 (the maximum value is 0.844 and the minimum value is 0.566). Cumulative characteristic root of five factors explained variance of 73.046% ideal effect, factor analysis. Therefore, these are the organizational learning of 4 items, the innovation culture of 7 items, 4 items of internal resources, external technological sources of enterprise 4 items, external professional technology source 4 items respectively. Then we use the simple average of the average value into the regression model for subsequent analysis as the variables of the sample value.

4 MULTIPLE REGRESSION ANALYSIS

As the diversification hypothesis verification technology influence antecedent (hypothesis H1~H5), it is based on the technology of diversification for the explanatory variables in organizational learning, innovation culture, internal resources, external technologic sources of enterprise, external professional sources of enterprise, external professional technology source as the explanatory variable, with firm size and industry attributes as the control variable, the regression model, the results of the regression analysis are shown in Table 1.

Table 1. The technological diversification strategy promoting factors regression analysis results.

	Standardized regression coefficient	T value	Significant probability	Variance inflation factor
Variables				
Industry type	0.068	1.379	0.170	1.107
The scale of the enterprise	0.032	0.640	0.523	1.159
Organizational learning	0.182	2.357	0.019	2.686
Innovation culture	0.173	2.033	0.043	3.270
Internal resources	0.267	2.913	0.000	2.101
External enterprises technology source	0.153	2.444	0.015	1.761
External professional technology source	0.149	2.299	0.023	1.910
Model statistics				
DW value	1.916			
R^2	0.595			
After the adjustment of the R^2	0.579			
F	38.386			

Note: the regression coefficients reported in the table is the final step in the regress the standard regression coefficient (N = 191), the explanatory variable, technology diversification, $p^*<0.05$; $p^{**}<0.01$; $p^{***}<0.001$.

5 CONCLUSION

In the regression model, the results show that the every hypothesis has a significant positive effect on technological diversification. That is to say, the more full every aspect is, the higher the level of technology diversification is. Therefore, assuming H1~H5 is verified in this sample data.

At the same time, the model is significant in statistics (F = 38.386, P < 0.001), the statistical result has certain stability. In addition, the influence on control of regression coefficient, industry type and scale of enterprises is positive (respectively 0.068 and 0.032), but not significantly different from zero (respectively significant probability for 0.170 > 0.05 and 0.523 > 0.05). In the Chinese current background high and new technology, enterprises and traditional manufacturing enterprises have no significant difference. The size of the enterprise also has no significant influence on technological diversification level.

Comprehensive analysis of the above regression five promoting factors hypotheses is confirmed through research, in line with expectations. The internal resources of the standardized regression co-efficient were the highest, followed by organizational learning.

REFERENCES

Hamel, G. & Prahalad, C. Competing for the future[M]. Harvard Business School Press, USA, 1994.

Hannan, M.T. & Freeman, J. The population ecology of organizations [J]. American Journal of Sociology, 1977(82)5:929–964.

Helat, C.E & Raubitschek, R.S. Product sequencing, coevoltion of knowledge, capabilities and products[J]. Strategica Management Journal, 2000, 21(10–11): 961–979.

Mendonca, S. The revolution within ICT and the shifting knowledge base of the world's largest companies[J], Economics of Innovation and New Technology, 2006(15);777–799.

Meyer, M. &Utterback, J.M. The product family and the dynamics of core capability[J]. Sloan Management Review, Spring, 1993:29–47.

Miles, R.E. & Snow, C. Organizations; new concepts for new forms[J]. California Management Review, 1986, 28(3);62–73.

Engineering Management and Industrial Engineering – Leung (Ed.)
© 2015 Taylor & Francis Group, London, ISBN 978-1-138-02773-2

The Android system localization of Tibetan studies

Y.G. Dai, X.Z. He, Y.M. Shan & F.C. Wan
National Language Information Technology, Northwest University for Nationalities, Lanzhou, Gansu, China

ABSTRACT: This paper researched Android system source code and Tibetan text features, and elaborated Tibetan localized operating on Android system. The main contents include downloaded and compiled Android 4.1 source, adding the Tibetan language settings options, modifying the ICU4C module, adding Tibetan font, translating into Tibetan language, solving the problem of Tibetan superposition display, and adding Tibetan input method. The implementation of the Android platform Tibetan localization is the foundation for Tibetan localization on Android devices.

1 INTRODUCTION

Android is the open mobile phone operating system based on the Linux kernel. It is characterized by its openness promotes technological innovation, and reduction of development costs, which has become the global smartphone market leader. By the end of 2013, with a global market share of 79%, the Android platform become one of the most popular mobile phone and tablet system. Faced with such a high market share, in order to solve global areas' language differences, Android system has been supported by more than 57 kinds of languages. But the system still does not support Tibetan Android localization, and there is no superimposed deformation Tibetan problem correctly.

Tibetan is a complex text, which is the main language of the Tibetan people's daily communication. Although the application of the Android platform has emerged in numerous, the Tibetan applications are very scarce. In order to implementing Tibetan localization of the Android platform, to improve the level of Tibetan area mobile information and the society as a whole information level, Tibetan has important implications for the long-term development.

2 THE ANDROID SYSTEM OVERVIEW

Android is an open platform for the mobile phone operating system, Android is consisted of four layers, from senior to junior, including application layer, application framework, system libraries and kernel operation.

Application layer contains the necessary text messages and contact management the core of the Android system application in addition, and also includes numerous third party applications. Simple, clear, open are for the application framework layer's quick agile development in the Android. The runtime system including the Android library and the Android runtime library. Library is an application framework supporting Android applications, connected to the application framework and the Linux kernel layer important link. The Android library including system C library, media library, FreeType, SQLite, ICU4C, and Skia. Android runtime provides the Java programming language core library in function. Most of the Android core system services depend on their kernel, including memory management, process management, security, and network protocol stack.

ICU4C was developed by IBM and its open source organizations can support the internationalization of open source software projects for version of C/C++ platform. It's C/C++ platform that provides powerful internationalization development capabilities, which can help developers to make it according to customs and language habits around the implementation of digital, money, time, date, message formatting and parsing, the string case conversion, such as sorting, searching, and sorting internationalization operation.

Skia is the underlying core graphics engine of Android. As a very good vector display engine, it can be a high quality of the 2nd graphics appeared on the low-end equipment, and can realize animation processing, word processing, image processing, and other functions.

3 THE TIBETAN FONT CHARACTERISTICS

As a kind of alphabet writing, Tibetan contains thirty consonants and four vowels. Hidden text usually has a letter as the core, the rest of the letters

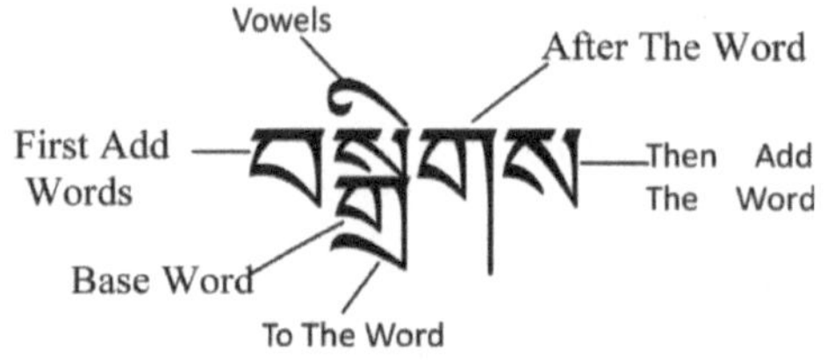

Figure 1. Tibetan glyph structure.

Figure 2. Tibetan glyphs superimposed deformation examples.

are attached before and after this letter and stack up and down.

Core letters are called base word, the word that is in front of the base word called first add words. A word behind the word is named after the word, the base word at the top of is named add words, the words below is named to the word, vowels is added to the top or bottom of the consonants. Here is the superposition of a typical Tibetan writing sample as shown in Figure 1.

In addition, Tibetan in writing is usually not only a simple superposition of letters, but need for some letters to do deformation, etc. As shown in Figure 2, when the surface of the current and the following letter alphabet vertically superimposed become the third character. Unicode plane 0F00-0FFF identified 194 Tibetan characters, collectively known as the Tibetan basic set, which includes all kinds of basic letters and symbols, and there is no word on the entire coding.

Tibetan language is different from the Chinese characters which written from left to right order. When it needs to write down overlay, superimposed on the Android platform to the Tibetan display makes it difficult.

4 THE REALIZATION OF THE TIBETAN LANGUAGE LOCALIZATION

The Android platform has no Tibetan language. Realize the Tibetan shows common method is to /system/fonts/Droid-SansFallback.ttf replaced by Himalayan font. But after transformation of the Himalayan font specification handwriting, Tibetan and English typesetting specification, Android applications have no Tibetan language pack, and this method has some problems, such as the need of root authority, replaced Chinese cannot display normally, Tibetan overlay display is also abnormal. The Android system and Android applications development should follow international standards, to facilitate internationalization and localization, therefore the realization of the Tibetan localization is to use Unicode standard code, to add Unicode Himalayan font. In order to realize localization of the Tibetan Android platform, there is a need to modify the Android source code and recompiling. The basic idea is as follows: download and compile Android 4.1 source code. Tibetan option is added in the language settings, modify ICU4C module, add a Tibetan fonts, Tibetan language pack, solve the problem of Tibetan overlap, and add the Tibetan input method.

4.1 *Download the compiled Android source code*

This paper also talks about implementation of the Android platform Tibetan localization, compile environment for Google recommends Ubuntu12.04 LTS system, the Android source code version for Android 4.4.1.

First, configure the Android source code compiler environment, download the Java JDK1.6, install additional software needed to compile.

Second, add ~ / bin into the path, download the repo tool, add the executable permissions.

```
$ mkdir ~/bin
$ PATH = ~/bin: $PATH
$curl http: //commondatastorage.googleapis.com/git
-repo-downloads/repo >~/bin/repo
$ chmoda+x ~/bin/repo
```

And then establish a new source directory, initialize the repo to the client.

```
$ mkdir WORKING_DIRECTORY
$ cd WORKING_DIRECTORY
$repo init –u https://android.googlesource.com/
platform/manifest -b android-4.1.1_r1
```

Next, download Android 4.1 source code.

```
$ repo sync -f
```

Finally, initialization compiler environment, specify the compile target, compile all Android 4.1 source.

```
$ sourcebuild/envsetup.sh
$ lunchfull -eng
$ make -j4
```

Now, downloading the Android source code to compile work has been finished, the next step is to modify the source code.

4.2 *Language settings and add Tibetan options*

View in WORKING_DIRECTORY root file Makefile.Makefile just jump execution /build/core/main.mk. Followed by analysis makefile, build/target/product/full_base.mk calls the same directory locale_full.mk for the list of languages, as follows:

```
#Get a list of language
$(callinherit-product,$(SRC_TARGET_DIR/product/
locale_full.mk))
```

Annotation locale_full. Mk add PRODUCT_LOCALES, the file called language_small.Mk, laguage_small.Mk to add bo_CN PRODUCT_LOCALES, retain en_US zh_CN zh_TW, modified content is as follows:

```
PRODUCT_LOCALES: = en_US bo_CN zh_CN
zh_TW
```

In the frameworks/base/core/res/res/ New to add a values-bo-rCN folder (the bo said Tibetan, the CN said China, Copy the values-zh-rCN folder strings.xml to values-bo-rCN folder, the contents of the strings.xml file translated into Tibetan. Recompile the source code, the language setting option appears in English in Figure 3.

To make displayed in Tibetan option in Tibetan, it needs to modify the ICU4C module, ICU4C under external/ICU4,folders.external/icu4c/stubdata/readme.txt is the method of adding language module.

According to this description modified icu-data-default.txt.Add curr/bo.res,lang/bo.res, region/bo.res, zone/bores, bo_res, bo_CN.res, Execute.Icu_dat_generator.py command. Tibetan language modules will be added to the icudt50l-default.dat file. Recompile, Tibetan option shows in the form of Tibetan, but at the moment, there is no Tibetan fonts, Tibetan displayed as Spaces. Adding Tibetan fonts are needed by the following operation.

4.3 *Add the Tibetan font*

The location of the system font library in frameworks/base/data/fonts folder, the Himalayan font renamed DriodSansTibetan.ttf, and added to the folder, modify Android.mk and add a new font files.

Modify the fonts.mk and add new fonts.

According to the /external/skia source content Skia will call system_fonts.xml and fallback_fonts.xml, forming font list, font drawing operations. So in order to make the Android system displays Tibetan, you need to modify the fallback_fonts.xml file, and add the following content.

Recompile and run the simulator, Tibetan option shown in Figure 4.

```
font_src_files :=\
  DroidSansTibetan.ttf\
```

```
PRODUCT_PACKAGES :=\
  DroidSansTibetan.ttf\
```

```
<family>
  <fileset>
    <file>DroidSansTibetan.ttf</file>
  </fileset>
</family>
```

4.4 *Add the Tibetan language resource bundles*

Internationalization and localization of Android in applications have a good solution. In the res directory of the Android project, it can implement multi-language support by defining a special folder name. Establish values-bo-rCN folder, put a strings.xml file in folder, strings.xml is Tibetan language string, when the Android system set up the Tibetan language, internal system is called values-bo-rCN Tibetan language string to display the folder.

System resources not include the Tibetan language pack, it needs translated into Tibetan

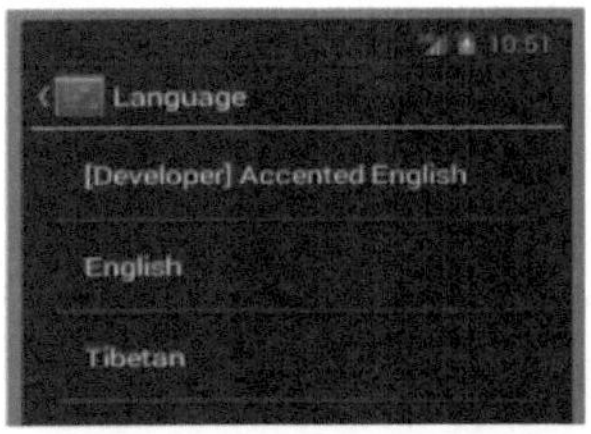

Figure 3. Tibetan options displayed in English.

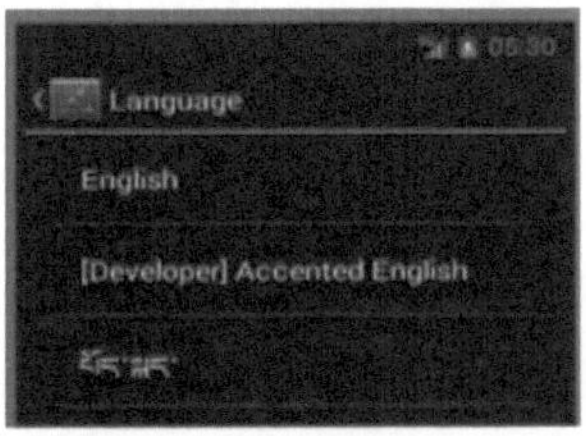

Figure 4. Tibetan options displayed in Tibetan.

strings.xml. The translation strings.xml file into the values—bo—rCN folder according to the values-zh-rCN Chinese language pack.

strings.xml, the part of the string values-zh-rCN, folder as follows:

```
<string name = "settings_label">"Setting"</string>
<string name = "header_category_wireless_networks">" Wireless and Internet "</string>
<string name = "radio_controls_title" > "More..."</string>
<string name = "header_category_device">" equipment "</string>
```

Strings.xml, the part of the string values-zh-rCN, with Chinese corresponding string as follows:

```
<string name = "settings_label">"སྒྲིག་བཀོད།"</string>
<string name = "header_category_wireless_networks">"སྐུད་མེད་ད་དང་དྲ་རྒྱ།"</string>
<string name = "radio_controls_title" >" དེ་ལས་མང་བའི།..."</string>
<string name = "header_category_device" >"སྒྲིག་ཆས།"</string>
```

4.5 *The Tibetan overlay display*

The Tibetan overlay display is always a difficult problem to Tibetan shows under the Android system. Although the Android system has the font layout engine, Tibetan still cannot be superimposed deformation. The realization of the Android Tibetan localization is the most important superposing of Tibetan display.

There is a need to add the Tibetan OpenType font processing module in the system, and to read Tibetan OpenType font file, as well as the Tibetan coding of start-stop, Tibetan consonants vowels characters, Tibetan font properties, such as modular and Tibetan OpenType fonts.

After modular, you need to use the Skia font rendering module design, to realize the display of Tibetan characters.

Complete the above operation, the Android system can correct superposition of drawing deformation of Tibetan. Android 4.2 system under the Tibetan display effect as shown in Figure 5.

Figure 5. Tibetan display system settings.

5 CONCLUSION

To get through the Android system framework, the Android system source code, Android internationalization and localization strategy and Tibetan font characteristics of research, then to the Android source code for the Tibetan localization is easy, but there is so many Android devices, Tibetan localization method and implement specific equipment. Therefore, Tibetan on realizing the Android source localization on the basis of realizing the Tibetan localization on the Android device and implementation of the Android platform other language localization, remains to be further research.

ACKNOWLEDGEMENTS

This work is supported by Science and technology major special project in Gansu province (1203FKDA033).

REFERENCES

Behdad Esfahbod. Harf Buzz. [EB/OL]. [2013-11]. http://www.freedesktop.org/wiki/Software/HarfBuzz/.

Conder S. Darcey L. Android mobile development from entry to the master [M]. Translator: Wei Zhang, Hui Li. Beijing: People's posts and telecommunications publishing house. 2010. 72–82.

Google How to check out and build Skiaon Android.[EB/OL].[2013-11]. https://sites.google.com/site/skiadocs/user-documentation/quick-start-guides/android.

Google. Downloading and Building Android Resources [EB/OL]. [2013-11]. http://source.Android.com/source/building.html.

Google Project Hosting. Soft keyboard. [EB/OL]. [2013-11]. http://code.google.com/p/softkeyboard.

ISO. ISO639-2Code. Codes for the Representation of Names of Languages [S/OL]. [2013-11]. http://www.loc.gov/standards/iso639-2/php/code_list.php.

Engineering Management and Industrial Engineering – Leung (Ed.)
© 2015 Taylor & Francis Group, London, ISBN 978-1-138-02773-2

Realistic space rendering

T. Ma
College of Information Engineering, Qingdao University, Qingdao, China

J.R. Wang
College of Mechanical and Electronical Engineering, Qingdao University, Qingdao, China

Y. Li & J. Li
College of Information Engineering, Qingdao University, Qingdao, China

ABSTRACT: Real-time realistic space rendering is significant to geography, astronomy and other strategic fields. A solution to the key points of real-time realistic space rendering, atmosphere, surface details, light beams and lens flares is introduced. Physically believable atmosphere is simulated using an improved algorithm of Rayleigh and Mie scattering effects; photographic star rendering is implemented by an enhanced post radial blurring process and a HDR procedure; lens flare is implemented in an additional functional layer that supports more image processing methods. The solution is simple to implement on current consumer hardware and it can provide a much more realistic result for minimal hardware consumption.

1 INTRODUCTION

Realistic space environment rendering can do great contribution to many territories in academic researches such as geography, meteorology, cosmology, disaster forecasts, and so on, and even to many space projects and national defense. Along with these facets, it also can be applied in a lot of entertainments such as movies and video games.

In this paper, realistic space rendering is divided into 2 representative rendering processes: realistic Earth rendering and more visually alike star effect. That's because an implementation method of the earth with atmosphere can comprise almost all features most planets would have, and most stars in current visible range of human can appear in a similar form, just differing in the facets of some trivial parameters such as volume, color, and so on.

The method of rendering the earth is divided into 2 parts, planet rendering and atmosphere rendering. Planet rendering has applied some classic algorithms in computer graphics, and atmosphere is rendered according to some scattering theories physically.

For the rendering of a star, this paper has rendered a typical star, and the sun. Visual effects, such as the volumetric light beams of the sun and light flooding effect due to the reaction of retina are implemented.

2 EARTH RENDERING

2.1 *Rendering the earth*

Rendering the earth itself can be a classic implementation version of basic techniques and algorithms in the field of computer graphics. The real earth is an ellipsoid but not visually palpable, so it can be generated in the form of a sphere. In the implementation of this paper, surface details are applied via multi-texture mapping and bump mapping[1]. One noteworthy aspect is that artificial lights can be normally on the surface of the dark night earth hemisphere, and this effect could be processed by an application version of procedural mapping by a determination coefficient of the diffusion factor calculated in Lambertian illumination model; reflection of surfaces covered by water such as oceans can be processed with an additional gloss map; cloud can be processed via another texture channel, with a blending factor and respective channels of RGB[2]. The rendering result of the earth itself is shown in Figure 1.

Figure 1. Earth rendering.

2.2 *Rendering the atmosphere*

Atmosphere rendering can be quite important to a planet. The reason why the result in Figure 1 is not so realistic can be attributed to the absence of atmosphere. In most cases, a planet has an atmosphere around it, except some very particular planets such as the Mercury and Pluto (now a dwarf planet). Even many planets themselves are clusters of gas. Planets without atmosphere can be very easy to render and methods can be quite similar to the method in 1.1, the differences may just be something such as radius, textures, bumps, and so on.

Several methods of atmosphere rendering have been introduced. Nishita 1993[3] provided some useful solution theories to the calculation of atmospheric scattering. Hoffman et al. 2003[4] developed a method of rendering the outdoor atmospheric scene taking into account effects by other scene objects. O'Neil 2005[5] had demonstrated a very practical rendering method of atmospheric scattering, without transformation process, this paper modified this method with transformation support added in so it can be applied more flexibly in various situations.

2.2.1 *Scattering theory*

There are 2 main scattering effects occurring in atmosphere: Rayleigh scattering and Mie scattering[3]. Rayleigh scattering is caused by small molecules in the air. The shorter wavelength the light is, the greater it will be scattered. So it contributes to the blue appearance of the sky due to the scattering of short wavelength blue light all over the atmosphere which then arrives at one observer. Mie scattering is caused by some particles much larger, such as dust. They are called aerosols. It scatters all wavelength lights equally so it contributes to a gray appearance of the sky in a cloudy weather due to its attenuation effect as a gray scaler[5].

Three equations are critical to the rendering algorithm of atmosphere[3],[5]. They are the Phase Function, the Out-Scattering Equation and the In-Scattering Equation.

Assuming a scattering situation is illustrated in Figure 2.

The Phase Function (1) describes the possibility of whether one ray is going to be scattered to the direction of the observer by the parameter of θ

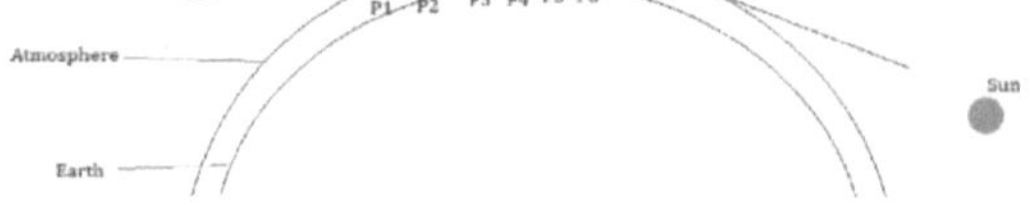

Figure 2. Atmosphere scattering illustration.

(angle between the two green rays) and a constant g that determines the symmetry of scattering[6]. If $g = 0$, the phase function can be considered as the phase function of Rayleigh scattering; $g < 0$ represents most light is scattered forward; $g > 0$ then backward. For Mie scattering, g is a value somewhere between −0.999 and −0.75. Notice that g cannot be 0, otherwise the phase function will permanently be zero.

$$F(\theta,g)=\frac{3\times(1-g^2)}{2\times(2+g^2)}\times\frac{1+\cos^2\theta}{(1+g^2-2\times g\times\cos\theta)^{3/2}} \tag{1}$$

The Out-Scattering (2) describes how many light particles are scattered away from the ray into atmosphere.

$$t(P_aP_b,\lambda)=4\pi\times K(\lambda)\times\int_{P_a}^{P_b}\exp\left(\frac{-h}{H_0}\right)ds \tag{2}$$

Here λ is the wavelength of light and $K(\lambda)$ is the scattering constant which is relative to the density of the atmosphere at the sea level. Rayleigh scattering constant is usually divided by $\lambda 4$, and Mie scattering by $\lambda 0.84$[5]. The integral determines the optical depth, the average atmospheric density across the ray from P_a to P_b, and it can be calculated by some discrete segments. h is the height of the sample point from 0 (sea level) to 1 (top of the sky) and H_0 is the scale height, the height of the point average atmosphere density is, and this paper adopts 0.25.

The In-Scattering Equation (3) describes how much light has reached to the observer from the total light beam from the sun.

$$I_v(\lambda)=I_s(\lambda)\times K(\lambda)\times F(\theta,g)\times\int_{P_a}^{P_b}\exp\left[\left(\frac{-h}{H_0}\right)\times\exp(-t(PP_c,\lambda)-t(PP_a,\lambda))\right] \tag{3}$$

For each point P from P_a to P_b, PP_c is the ray from the sample point to the star and PP_a is the ray from the point to the observer. $I_s(\lambda)$ is the intensity of the sunlight.

And one more point must be highlighted. Observer position must be subtracted by the translation information vector of the model transformation matrix as the following GLSL codes so that weird visual errors can be avoided.

```
vec3v3NewCamPos
  = v3CamPos-vec3(m_Model_Matrix[1])
```

Atmosphere result only and atmospheric earth are shown in Figure 3 and Figure 4.

Figure 3. Atmosphere.

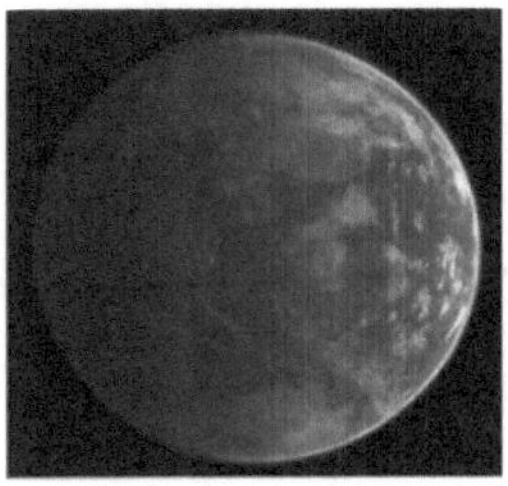

Figure 4. Earth with atmosphere.

3 STAR RENDERING

3.1 *Rendering the volumetric light beams*

Many means for rendering a valid volumetric light beam have been introduced. A multiple light sources crepuscular rays rendering method in non-real-time was introduced by Nishita et al. 1987[7] using a modified shadow volume algorithm. Dobashi et al. 2002[8] introduced a method by using a volume rendering way based on slices and an improved version is provided by Mitchell in 2004[9]. James 2003[10] provided a method by using accumulation of overlapping volumes by using frame buffer blending with depth peeling. In this paper, an improved image processing method based on radial blurring method developed in post process volumetric light rendering algorithm (Mitchell 2007)[11], with an extra HDR (High Dynamic Range) pass is applied.

A light scattering model is introduced by Hoffman and Preetham 2002[12] and it works well in most situations, but in this paper, this model is simplified as (4).

$$L(s,\theta,\phi) = intensity \times \sum_{i=0}^{n} decay^{i} \times \frac{L(si,\theta_i)}{n} \tag{4}$$

Light intensity at view position ϕ can be obtained by (4), *intensity* represents intensity of light shining in the screen space, and the accumulation of $decay^i$ enables smooth attenuation of volumetric rays.

Figure 5. Post-process volumetric light and HDR applied rendering result.

3.2 *HDR image processing*

HDR (High Dynamic Range)[1] is a term used in the field of Computer Graphics and Digital Image Processing. HDR image processing preserves more color details that might be lost due to limited contrast range by allowing multiple passes of getting bright and dark information of a certain scene under different exposure by sampling. Also, restriction in traditional color format is no longer a problem due to the flexible support to floating-point color format provided by current consumer hardware. In this paper, brightness information will be extracted from the original scene, down sampled 4 times, processed with respective Gaussian smooth passes, and bloom back to the original scene. The result of post-process volumetric light and HDR processed scene is shown in Figure 5. Light beams and the bloomy effect can be easily recognized.

4 IMPLEMENTATION TECHNIQUS AND DETAILS

The demo for this paper is developed in C++ and Open GL&GLSL, comprising 3 main steps to render a simple space scene including a glowing sun and a fully textured earth with abundant surface details and atmosphere, and it is being tested on the platform powered by an AMD IIN970 CPU, 2GB DDR3 RAM and an AMD HD6470M graphics car with 512MB graphics memory.

4.1 *Posterior pixel processing*

The posterior pixel processing is applied in screen space. Before applying, the scene has been rendered to one properly generated frame buffer object. After applying a model-view-projection transformation, the position of light source where rays are coming from in screen space has also been calculated in the scene rendering procedure according to the transform sequence of the Open GL rendering pipeline. Then the posterior pixel processing is applied to a full screen quad mapped with the texture which is bound to the previous scene FBO. The result after this pass is shown in Figure 6.

4.2 *Light occlusion*

In Figure 6 some uncomfortable streaks of the earth appeared due to the depth information loss in 2D image space which is indispensable to correct light occlusion. This paper provides a solution with an additional pass. First passes renders the scene as normal but light source will be rendered with pure black color, second passes renders light source normally but occluding objects black. Accumulation of the 2 above passes as a solution implements a better result as in Figure 7, and the final result with all processes is shown in Figure 8.

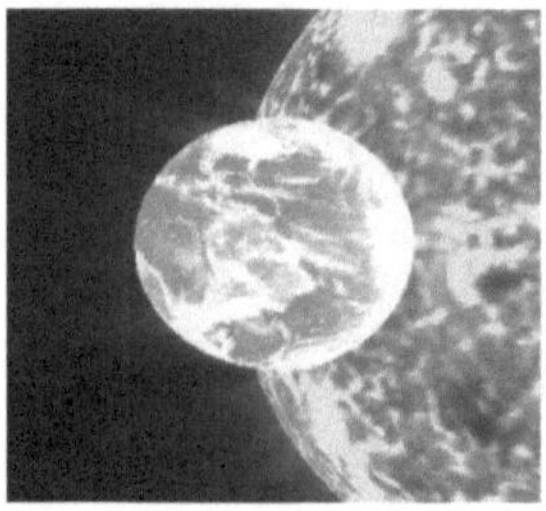

Figure 6. Result of pixel processing.

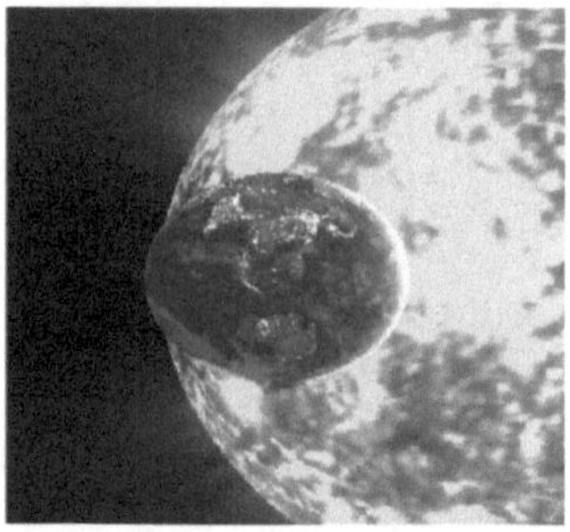

Figure 7. Modified with an additional pass.

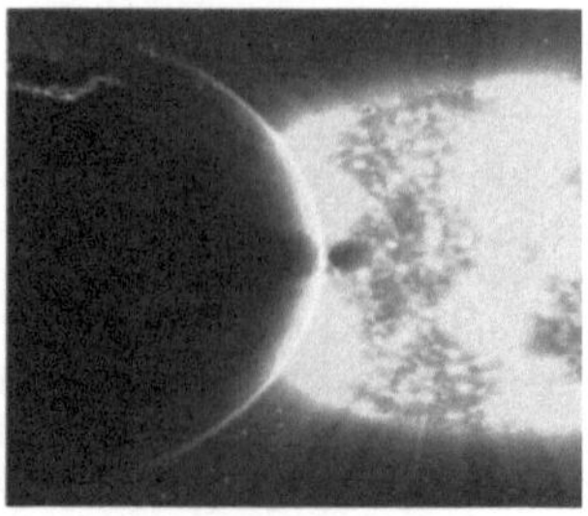

Figure 8. Final results with sky box.

5 ISSUES AND FURTHER WORK

Though a convincing result has been generated, some undesirable problems are still existing, mainly in the volumetric light generation part. The shape of radiating rays is constrained by the color of the texture mapped to the light source polygon mesh. And full screen light flooding takes place after the HDR process. Volumetric light will also simply vanish if the light source is out of the projection viewing frustum. Solution to the 3 above problems is the main work in future.

Also, star clusters are important to a more realistic scene of space. Now a basic star cluster rendering method via Open GL point sprite has been developed. More details of stars in a cluster can be redesigned by instanced imposter technic with a procedural processing and rendering to extend a local space to a vast but much more vivid cosmos. The final solar system rendering result with lens flare added is shown in Figure 9 and Figure 10.

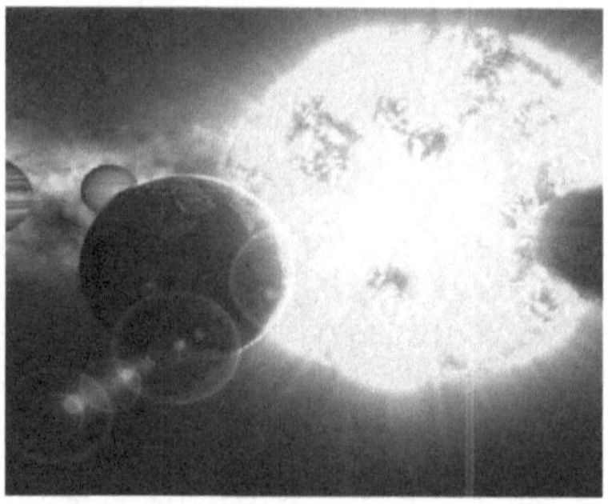

Figure 9. Lens flare.

Figure 10. Final solar systems.

ACKNOWLEDGEMENTS

The research work is supported by Natural Science Foundation of Shandong Province under Grant No. ZR2011GM008.

REFERENCES

[1] Nishita, T., T. Sirai, K. Tadamura & E. Nakamae. Display of the Earth Taking into Account Atmospheric Scattering. In Proceedings of SIGGRAPH 93, pp. 175–182, 1993.
[2] Sean O'Neil. Accurate Atmospheric Scattering. GPU Gems 2[M]. 2nd. Nvidia, 2005.
[3] Wright R.S., Haemel J., Sellers G., et al. OpenGL SuperBible 5th Edition[M]. 5th. Pearson Education, 2010.
[4] Nguyen H. GPU Gems 3[M]. 3rd. Nvidia, 2007.
[5] W.M. Cornette, J.G. Shanks. Physical reasonable analytic expression for the signle—scattering phase function [J]. Applied Optics, 31(16), pp. 3152–3160, 1992.
[6] Dobashi, Y., Yamamoto T., etal. Interactive Rendering of Atmospheric Scattering Effects Using Graphics Hardware[C]. //Graphics Hardware, 2002.
[7] James, R. True Volumetric Shadows[C]. //Jeff Lander. Graphics Programming Methods, Charles River Media, pp. 353–366, 2003.
[8] Hoffman, N., Mitchell A.K. Methods for Dynamic, Photorealistic Terrain Lighting[C]. //D. Treglia. Game Programming Gems 3, Charles River Media, pp. 433–443, 2002.
[9] Hoffman, N., Preetham A.A. Real-Time Light-Atmosphere Interactions for Outdoor Scenes[C]. //Jeff Lander. Graphics Programming Methods, Charles River Media, pp. 337–352, 2003.
[10] Mitchell, J. Light Shafts: Rendering Shadows in Participating Media[R]. Game Developers Conference, 2004.
[11] Nishita, T., Miyawaki Y., etal. Atmospheric Scattering Considering Luminous Intensity Distribution of Light Sources[R]. Proceedings of SIGGRAPH 87, 1987.
[12] Rost R.J. OpenGL Shading Language[M]. 3rd. Addison-Wesley, pp. 304–309, 2009.

Engineering Management and Industrial Engineering – Leung (Ed.)
© 2015 Taylor & Francis Group, London, ISBN 978-1-138-02773-2

Study on the regularity of perceived usefulness and user's satisfaction affecting continued usage intention of information system: An example of mobile banking

Z.Y. Liu & C.H. Chen
School of Management, Xiamen University, China

ABSTRACT: Although former researchers have found perceived usefulness and user satisfaction influencing users' intention to continue IS usage, they hadn't analyzed the relationship of them. This paper reviewed the theories about information system continued usage intention, and took an investigation, with 373 effective questionnaires collected to explore the law of the relation of them. It is found that continued usage intention improves quickly while perceived usefulness and user satisfaction are lower, and improves slowly while perceived usefulness and user satisfaction reach a higher level. Finally, some implications on management practice are suggested.

1 INTRODUCTION

With the rapid growth of the study on information system, continued IT/IS usage has recently gained increasing attention of researchers. The overall goal of this study is to understand users' intention to continue using an IT/IS-based technology. There are plentiful studies using the Technology Acceptance Model (TAM) and Expectation Disconfirmation Theory (EDT) to explain IT/IS usage. Some of the studies showed that user satisfaction and perceived usefulness cause an individual's intention of continued usage of a system.

However, there is no study which is attempted to explain how user satisfaction and perceived usefulness influence intention of continued usage of a system. The purpose of this research is trying to find the effect regularity of user satisfaction and perceived usefulness on the continued usage of information system. The results will guide the service company to improve intention of continued usage in different phase. And the company will gain more profits as a result of users' continued usage of information system.

The layout of this paper is as follows. First, it explains user satisfaction and perceived usefulness influence intention of continued usage of a system. Second, it discusses the regularity of perceived usefulness and user satisfaction affecting continued usage intention of information system. Finally, it concludes the results and suggests implications on management practice.

2 THEORETICAL BACKGROUND

2.1 *Perceived usefulness affecting continued usage intention*

Among the theoretical perspectives advanced to address IT adoption and usage, the Technology Acceptance Model (TAM) advocated by Davis[1] is widely accepted as a framework to understand users' IT acceptance processes. TAM is preferred over alternative models such as the Theory of Reasoned Action (TRA) in diverse user contexts. In its formulation, users' intention to adopt an IT is explained by two major perceptual factors: perceived ease of use and perceived usefulness. Perceived usefulness is, in turn, influenced by perceived ease of use.

Many researchers think TAM framework is appropriate to be applied to the study of users' acceptance in an e-commerce, software, and network context. From these studies, while both perceived ease of use and perceived usefulness were found to be dominant constructs that determine the initial acceptance of a technology, perceived usefulness was found to be a predominant predictor of user intention of continued IT/IS usage. (Venkatesh et al., 2003[2]; Karahanna et al., 1999[3]; Berends et al., 2007[4]). Perceived usefulness is about the degree to which a person believes that using a particular system will make his/her life easier (Davis et al., 1989)[1]. They believe that the system is useful to their lives and they will continue to use it. Although the usefulness-intention association was originally derived in an acceptance context, it is likely to hold true in continuance contexts, because human tendencies for subconsciously pursuing instrumental behaviors or striving for rewards are independent of the timing or stage of such behaviors (Bhattacherjee, 2001)[5].

Previous empirical evidence has shown that perceived usefulness is one of the major cognitive beliefs in determining users' continuous usage of IT. For instance, Bhattacherjee (2001) investigated online banking usage by considering perceived usefulness, confirmation, satisfaction and IT continuance intention. His findings highlighted that the significance of perceived usefulness and satisfaction was influencing IT/IS continuance intention. Young Sik Kang and Heeseok Lee (2009)[6] investigated website service and found that the significance of perceived usefulness was influencing user satisfaction and web service continuance intention. Vincent Cho et al. (2009)[7] put forward a continuous usage integrate model based on TAM, and examine users' usage of self-paced e-learning tools. The results demonstrate that perceived usefulness and user satisfaction are the major determinants of continued usage, while perceived usefulness is influencing user satisfaction. Moez Limayem and Christy M.K. Cheung (2008)[8] used the Expectation-Confirmation Model (ECM) in an internet-based learning context and found that the significance of perceived usefulness and satisfaction was influencing IT/IS continuance intention. Hong et al. (2006)[9] integrated TAM and ECM model and investigated continuous system usage of mobile internet. They found that perceived usefulness plays an important role on user satisfaction and continuance intention. R. Naidoo et al. (2007)[10] extended the TAM model by adding perceived service quality and loyalty. They found not only the quality and loyalty of perceived service, but also the directly positive effects of perceived usefulness on continuance intention.

2.2 *User satisfaction affecting continued usage intention*

Customer satisfaction refers to "the summary psychological state resulting when the emotion surrounding disconfirmed expectations is coupled with the consumer's prior feelings about the consumption experience" (Oliver, 1981). If the customer has good experiences of using IT/IS over time, then he will have cumulative customer satisfaction and continue to use it. If the customer has poor experiences of using it, then he will complain and switch to other IT/IS. Since customer satisfaction reflects the degree of a customer's positive feeling for a service provider in a mobile commerce context, it is important for service providers to understand the customer's vision of their services. Only when the service provider offers satisfactory service to the customer, will the customer continue to use the IT/IS system, after that, the provider could make a success.

Empirical evidence on IS continuance supports that user satisfaction is a major factor of IS continuance intention (Bhattacherjee, 2001[5]; Hong, Thong & Tam, 2006[9]; Thong et al., 2006[9,11]). Bhattacherjee (2001)[12] found that the significance of user satisfaction was influencing online banking usage continuance intention. Bolton and Lemon (1999)[13] showed the strong relationship between service usage and satisfaction by using the customer confirmation disconfirmation paradigm. According to Bolton and Lemon, if a customer uses a service more than he or she expected, the customer rates the exchange as more satisfied because the outcomes, such as increase in the levels of usage, are in accordance with his or her normative expectation. Szymanski and Henard (2001)[14] found that repeat purchasing is one of the most frequently assumed consequences of customer satisfaction. Ming-Chi Lee (2009)[15] clarified the important role of user satisfaction in users' continuous intention to use e-learning by using a new paradigm integrating ECM, TAM, TPB and flow theory. Jeoungkun Kim et al. (2011)[16] presented a conceptual model including three quality measures—System Quality, Information Quality, and Service Quality. Data analysis results revealed that satisfaction and trust have significant effects on users' continuance intention. Young Sik Kang and Heeseok Lee (2010)[17] proposed a model by extending the user satisfaction perspective into research on online service continuance. The analysis results found that website information satisfaction and system satisfaction play key roles in forming continuance intention through perceived usefulness and perceived enjoyment.

Several studies have been conducted to understand the influence of perceived usefulness and user satisfaction on users' continuance. However, there are not the researches to descript how perceived usefulness and user satisfaction affect users' continuous usage intention. The paper tries to find out whether perceived usefulness and user satisfaction have a stable continuous influence on users' continuous usage intention.

3 EMPIRICAL BACKGROUND AND MEASURES

Mobile banking, a typical IT system, consists of mobile terminals, wireless networks of communication carriers, and mobile banking system of banks. We have carried out a questionnaire survey to collect the analysis data.

Table 1. List of model constructs and items.

Factor	Item	Source
Perceived usefulness	PU1: I find mobile banking to be useful to me. PU2: Using mobile banking can improve my job performance. PU3: Using mobile banking can increase my job convenient. PU4: Using mobile banking can increase my job effective. PU5: Using mobile banking can increase my live effective.	Davis (1989)[1]
User satisfaction	SAT1: I am satisfied with the performance of mobile banking. SAT2: I am pleased with the experience of using mobile banking. SAT3: My decision to use mobile banking was a wise one. SAT4: In general, using mobile banking would give me a sense of satisfaction.	Spreng et al. (1996)[18] Tsai and Huang (2007)[19]
Continuous usage intension	CU1: I will use mobile banking in the future. CU2: I intend to use mobile banking more than other banking channels. CU3: I intend to try new function of mobile banking in the future.	Bhattacherjee (2001)[5] Vincent Cho et al. (2009)[7]

This questionnaire survey was designed to collect data on mobile banking users' perceptions of mobile banking. Empirical data was collected via a paper questionnaire or telephone survey with 7 branches of China Construction Bank and 2 branches of Bank of Communications. A total of 403 responses were gathered, with 30 responses discarded due to incomplete or invalid answers; thus, the final number of valid responses was 373.

Detailed information about the constructs and the sources of the questionnaires are shown in Table 1. All the items are measured on seven-point Likert scales, with anchors ranging from "strongly disagree" to "strongly agree".

4 DATA ANALYZE

After we collect the data, the reliability and validity of research variables were checked. Then we calculated the factor scores through Component Score Coefficient Matris in SPSS, and standardized the data and data analysis by using the Matlab Curve fitting tools.

First, while every variable has three observations, we need to count the factor core from the observation value of items. In SPSS we can consider the factor scores as linear combination of the observation.

$$F_j = \beta_1 X_{1j} + \beta_2 X_{2j} + \ldots + \beta_i X_{ij} + \ldots + \beta_p X_{pj} \quad (1)$$

Formula (1) is the factor score function. Let F_j, X_{ij}, β_i, p denote, respectively, the variable score of sample j, observation value of sample j of item i of the variable, Coefficient Matrix of item i, the number of the items of the variable. Coefficient Matrix of item comes from the Component Score Coefficient Matris in SPSS. Then the results of F_CU (Continuous Usage Intention), F_PU (Perceived Usefulness), and F_SAT (User Satisfaction) are the following.

$$F\_CU = 0.380XCU1 + 0.398XCU2 + 0.397XCU3$$
$$F\_PU = 0.214XPU1 + 0.239XPU2 + 0.240XPU3 + 0.232XPU4 + 0.232XPU5$$
$$F\_SAT = 0.288XSAT1 + 0.291XSAT2 + 0.291XSAT3 + 0.275XSAT4$$

Second, we should standardize the data to transform the data to number between 0 and 1 through the Min-Max data standardization process.

$$X' = \frac{X_1 - \min A_1}{\max A_1 - \min A_1} \quad (2)$$

Third, we analyzed the standardized data using the Matlab Curve fitting tools.

4.1 *The relation between perceived usefulness and users' continuous usage intention*

Robust curve fitting is an extension of standard regression (standard non-linear Least Squares Fitting (LSF)) that can even out individual outliers in a data set and neutralize their effect on the ultimate result. We used Matlab Curve Fitting with Robust (LAR) option to analyze the influence of perceived usefulness on users' continuous usage intention by using the standardized data, as shown in Figure 1.

Correspondingly, the function of the curve is

$$f(x) = -0.736x^4 + 2.731x^3 - 3.454x^2 + 2.294x + 0.108 \quad (3)$$

The points in Figure 1 are the standardized data of variables of perceived usefulness and users' continuous usage intention. Table 2 shows

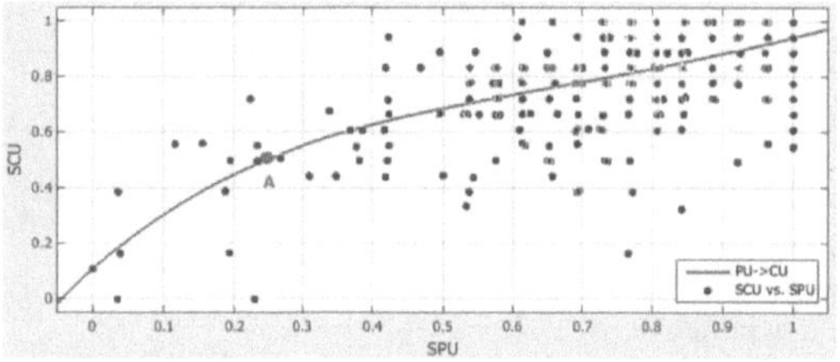

Figure 1. Relation between perceived usefulness and users' continuous usage intention.

Table 2. Goodness of fit.

Fit name	SSE	R-square	Adjusted R-square	RMSE
Results	0.442	0.965	0.964	0.035

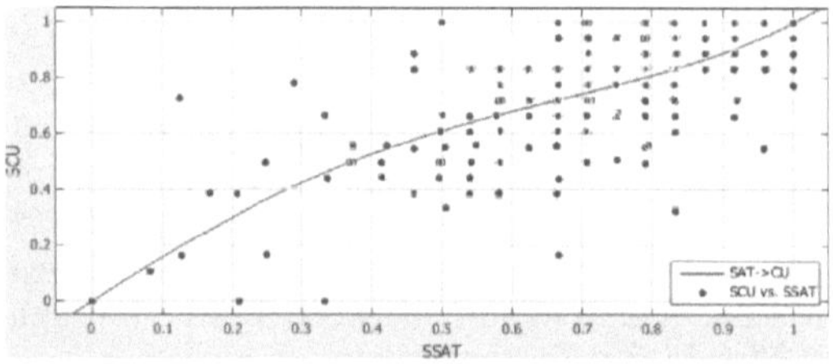

Figure 2. Relation between user satisfaction and users' continuous usage intention.

the goodness of fit. From Table 2, the curve line showed an adequate fit.

The analysis results of the data are strongly supportive of the ECM propositions that the perceived usefulness is the key determinants of the user's satisfaction. More perceived usefulness cause more user satisfaction.

4.2 *The relation between user satisfaction and users' continuous usage intention*

Then we use Matlab Curve Fitting with Robust (LAR) option to analyze the influence of user satisfaction on users' continuous usage intention using the standardized data, as shown in Figure 2.

Correspondingly, the function of the curve is

$$f(x) = 1.089x^4 - 1.206x^3 - 0.542x^2 + 1.659x - 2.051e - 017 \quad (4)$$

The points in Figure 2 are the standardized data of variables of user satisfaction and users' continuous usage intention. Table 3 shows the goodness of fit. From Table 3, the curve line showed an adequate fit.

Table 3. Goodness of fit.

Fit name	SSE	R-square	Adjusted R-square	RMSE
Results	0.373	0.970	0.970	0.032

The analysis results of the data are strongly supportive of the ECM propositions that the user satisfaction is the key determinants of the user's satisfaction. More user satisfaction brings about more user satisfaction.

5 DISCUSS

5.1 *Regulation of perceived usefulness impact on continued IT usage*

One interesting finding is that, among the process of the impact of perceived usefulness on continued IT usage intention, continued IT usage intention increases more quickly than the higher level when perceive usefulness is in the lower level. When perceive usefulness arrives a higher level (e.g. point A), continued IT usage intention increases more gently. We conjecture that the stronger influence of perceived usefulness on continued IT usage intention may be due to the fact that perceive usefulness stimulate the usage intent in the beginning when perceive usefulness is in the lower level. Our results imply that the banks should take an effort on rising perceive usefulness as quickly as possible when perceived usefulness is low. Therefore, continued IT usage intention can increase quickly, and more and more users will find the performance of IT system and tend to continue using the IT system.

Figure 1 also shows that even in the highest level of perceived usefulness, users' continued mobile banking usage intention could not increase to 1. Probably there are some other factors influencing the continued intension. The result implies that while the perceive usefulness reaches a higher level, the bank should not only continue promoting the perceive usefulness, but also consider other factors (e.g. user satisfaction). As a result, the users' continued IT usage intention will reach the highest level and the users will then become loyal customers.

5.2 *Regulation of user satisfaction impact on continued IT usage*

It is also interesting to find that, among the process of user satisfaction affecting continued IT usage intention, continued IT usage intention increases more quickly than higher level when user satisfaction is in the lower level. When user satisfaction arrives more about 0.5, continued IT usage intention increases more gently. We conjecture that the stronger influence of user satisfaction on continued IT usage intention may be due to the fact that user satisfaction stimulates the usage intent in the beginning when user satisfaction is in the lower level. Our results imply that when user satisfaction is lower, the banks take an effort to raise user satisfaction quickly by improving IT function, adding service channel, and raising the quality of service. Accordingly users' continued IT usage intention will increase quickly and more and more users will tend to continue using mobile banking. Even user satisfaction is higher and higher, the bank should not only continue promoting user satisfaction, but also consider the other factor (e.g. perceived usefulness). As a result, the users' continued IT usage intention will achieve the highest level and the users will become loyal customers.

6 CONCLUSION

This study explores the law of perceived usefulness and user satisfaction influencing on users' IT/IS continued usage intention. Empirical testing strongly supports ECM. In view of the findings on mobile banking, perceived usefulness or user satisfaction has a nonlinear influence on users' IT/IS continued usage intention during the usage phase. The opinion that users' IT/IS continued usage intention will increase 100 percent if perceived usefulness or user satisfaction improves 100 percent is not supported. This study has enriched the IT continuance researches in China. The banks should take different methods to promote the continued usage intention in different phase. Further empirical validations on the continued usage intention in different IT contexts are encouraged.

LIMITS

Our study has some limitations. One limitation originates in the biases inherent in most survey-based research. The current study only includes users from two commercial banks in China. The findings in this study may be specific to the culture in China. Another limitation is that our findings regarding regulation of the continued use of mobile banking might not be generalizable to continued use of other IT systems.

REFERENCES

[1] Davis, Fred D. Usefulness, Perceived Ease of Use, and User Acceptance of Information Technology[J]. MIS Quatterly. September 1989: 319–340.

[2] Venkatesh, Viswanath, Morris, Michael G., Davis, Gordon B. and Davis, Fred D. User Acceptance of Information Technology: Toward a Unified View[J]. MIS Quarterly. 2003, 27(3): 425–478.

[3] Karahanna, Elena, Straub, Detmar W. and Chervan, Norman L. Information Technology Adoption Across Time: A Cross Sectional Comparison of Pre-Adoption and Post-Adoption Beliefs[J]. MIS Quarterly. 1999, 23(2): 183–213.

[4] Berends, H., Vanhaverbeke, W. and Kirschbaum, R. Knowledge management challenges in new business development: case study observations[J]. Journal of Engineering and Technology Management (JET-M). 2007, 24(4): 293–400.

[5] Bhattacherjee, Anol. Understanding Information SystemsContinuance: AnExpectation-Confirmation Model[J]. MIS Quarterly. September 2001, 25(3): 351–370.

[6] Kang, Young Sik, Hong, Soongeun and Lee, Heeseok. Exploring continued online service usage behavior: The roles of self-image congruity and regret[J]. Computers in Human Behavior. 2009, 25(1): 111–122.

[7] Cho, Vincent, Cheng, T.C. Edwin and Lai, W.M. Jennifer. The role of perceived user-interface design in continued usage intention of self-paced e-learning tools[J]. Computers & Education. 2009, 53(2): 216–227.

[8] Cheung, Moez Limayem, Christy. Explaining Information Systems Adoption and Post-Adoption: Toward an Integrative Model[J]. International Conference on Information Systems (ICIS) 2003 Proceedings.

[9] Hong, Se-Joon, Thong, James Y.L. and Tam, Kar Yan. Understanding Continued Information Technology Usage Behavior: A Comparison of Three Models in the Context of Mobile Internet[J]. Decision Support Systems. 2006, 42: 1819–1834.

[10] Naidoo, R. and Leonard, A. Perceived usefulness, service quality and loyalty incentives: Effects on electronic service continuance[J]. South African Journal of Business Management. 2007, 38(3): 39–48.

[11] Thong, James Y.L., Hong, Se-Joon and Tam, Kar Yan. The effects of post-adoption beliefs on the expectation-confirmation model for information technology continuance[J]. International Journal of Human-Computer Studies. 2006, 64(9): 799–810.

[12] Bhattacherjee, Anol. An empirical analysis of the antecedents of electronic commerce service continuance[J]. Decision Support Systems. 2001, 32: 201–214.

[13] Bolton, R.N. and Lemon, K.N. A dynamic model of customers' usage of services: Usage as an antecedent and consequence of satisfaction[J]. Journal of MarketingResearch. 1999, 36(5): 171–186.

[14] Szymanski, D.M. and Henard, D.H. Customer satisfaction: A meta analysis of empirical evidence[J]. Journal of the Academy of Marketing Science. 2001, 29(1): 16–35.

[15] Lee, Ming-Chi. Factors influencing the adoption of internet banking: An integration of TAM and TPB with perceived risk and perceived benefit[J]. Electronic Commerce Research and Applications. 2009, 8(3): 130–141.

[16] Kim, Jeoungkun, Hong, Soongeun, Min, Jinyoung and Lee, Heeseok. Antecedents of application service continuance: A synthesis of satisfaction and trust[J]. Expert Systems with Applications. 2011, 38: 9530–9542.

[17] Kang, Young Sik and Lee, Heeseok. Understanding the role of an IT artifact in online service continuance: An extended perspective of user satisfaction[J]. Computers in Human Behavior. 2010, 26(3): 353–364.

[18] Spreng, R.A., MacKenzie, S.B. and Olshavsky, R.W. A Reexamination of the Determinants of Consumer Satisfaction[J]. Journal of Marketing. July 1996, 60: 15–32.

[19] Tsai, Hsien-Tung and Huang, Heng-Chiang. Determinants of e-repurchase intentions: An integrative model of quadruple retention drivers[J]. Information & Management. 2007, 44(3): 231–239.

Engineering Management and Industrial Engineering – Leung (Ed.)
© 2015 Taylor & Francis Group, London, ISBN 978-1-138-02773-2

Research of financial crisis mechanism information system based on middleware

W.N. Zhang
Tianjin Maritime College, Tianjin, China

ABSTRACT: Middleware is emerging as an interconnection bridge of heterogeneous platforms on distributed network. Design and development of large-scale application system with middleware have become a trend. Design and development features of financial information system are analyzed, aiming at financial information system—the special field. A financial system solution instance is combined for reaching transaction requirement of financial system through implementation of middleware; thereby information system can be applied for analysis of the financial crisis. Duffing-Holmes equation is utilized for discussing formation causes and internal mechanism of financial crisis according to chaotic characteristics of the financial crisis. Chaotic control method is proposed to aim at chaotic characteristics of financial crisis. Finally, countermeasures of controlling financial crisis are discussed through combining with practical condition of Asian financial crisis.

1 INTRODUCTION

Global financial crisis has constantly erupted since 1990s. Japanese bubble economy was shattered in August 1990. Italy and the United Kingdom suffered from currency crisis in 1992[1–2]. Mexico suffered from financial crisis in 1995. Southeast Asia underwent financial crisis in 1997, Russia financial crisis was caused in 1998, Brazil also suffered from financial crisis in 1999, and Argentine financial market was collapsed in 2001. Author believes that financial market is a complex economic system, chaos theory can be adopted for handling complex system effectively, and it can provide new method and thought for solving financial crisis and related issues. Since financial crisis has huge influence on, important position and role in financial system and national economy, special requirement of financial business are studied[3]. Financial system not only has characteristics of general system development, but also has the following features, namely stability, high performance, robustness and security. Techniques of test key and digital signature can be adopted for ensuring data integrity and non-repudiation in aspect of financial transaction safety. System confidentiality can be ensured through encrypted transmission of data among nodes. Security of systems management can be realized through system connection control and tight operation authority management. Network and hardware equipment can be managed mainly through network management and system safety features in the paper, middleware technology is introduced into financial crisis analysis. A set of more complete analysis mechanism and analysis methods is designed, and advantages of the method are verified through examples.

2 INTRODUCTION OF MIDDLEWARE

A very complex process is required for system design and development according to requirements and characteristics of financial system. If independently developed traditional methods are adopted for the whole development process, stability, high performance, robustness and security implementation. It cannot be guaranteed even without consideration of huge workload. In addition, it is quite difficult to update and expand system in the future. Therefore, middleware can be widely and deeply applied in financial system.

Middleware belongs to a major category in basic software, which is regarded as a scope of reusable software. Middleware is above operating system, network and database, but below application software. It is generally used for providing operating and development environment for application software, and assisting users to develop and integrate complex software applications flexibly and efficiently[4–5].

3 CHAOS ECONOMICS CHARACTERISTICS

3.1 *Three states of financial market*

There are three states in financial market according to the views of chaos economics: first it is unscrambled state of financial market, namely, the financial market is in a relatively stable state; second, chaotic state after the financial market is disrupted, financial market state under the condition is called financial chaos or financial crisis; third, the original system of financial market is collapsed and located in a completely different state[6].

3.2 *Duffing model*

Duffing model is a model that is frequently adopted for studying chaos in economics. Its concrete forms are shown as follows: $x1+ax+bx3 = -\varepsilon\delta x2+\varepsilon f\cos(\omega t)$ wherein: x represents state of financial market; x1 represents state change speed of financial market; x2 represents state change acceleration of financial market; a and b respectively represent linear and nonlinear rigid coefficients in financial market; f represents external force strength affected by the financial market; ω represents frequency of external force; δ represents defensive force of the government on financial risks; ε represents control parameter of government.

4 IMPLEMENTATION PRINCIPLE OF TRANSACTION MIDDLEWARE

4.1 *Reference model of transaction middleware*

X/OPEN organization developed Distributed Transaction Processing (DTP) model as shown in Figure 1. The model divides DTP system into four major functional modules: ① application system program AP: operational functions applied by the user are described; a series of operations on resources can be defined by each application, including definition of boundary of the whole transaction, withdrawal and deposit of resources in the transaction boundary scope, decision for submitting or rolling transaction. If one transaction is related to many applications and exists in many different computer nodes, DTP model also provides interface modes which are used among applications on different computer nodes for information exchange. ② Transaction manager TM: global transaction operations are managed,

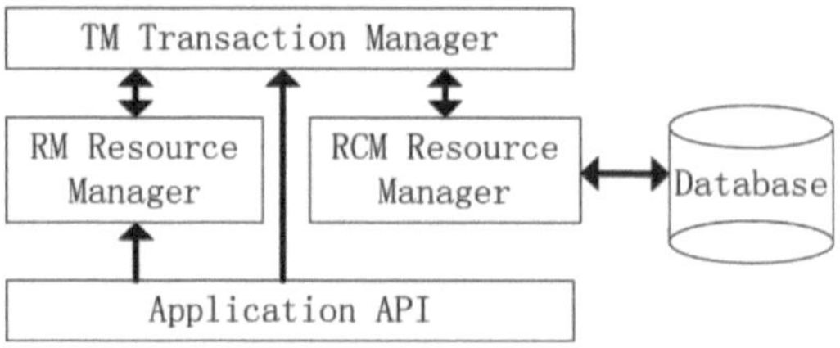

Figure 1. DTP model diagram.

transaction submission or rollback actions are coordinated to ensure integrity of transaction, in addition, the module can be used for coordinating recovery of failed transaction during failure of system; ③ Resource manager RM: shared resources in the system can be managed, these shared resources can serve application programs through interfaces provided by resource manager, typical resource managers include database management system, file access system, printing shared server, and so on. Changes of all transactions on resources can be restored by resource manager in DTP system with integrity assurance. Resource manager RM allows that transaction manager TM can coordinate and manage completed transaction operations and operations instructed by other resource managers, thereby ensuring their action coordination and transaction integrity; ④ Communication resource manager CRM: communication among distributed application is controlled, communication can be achieved within control scope of a transaction manager or within the control scope of different transaction managers, CRM can realize communication with CRM in other TM management cope through TCP/PI interface[7]. CRM allows a complete transaction in different TM management scopes, and allows the transaction manager TM to coordinate submission or rolling of the whole transaction.

4.2 *System structure of transaction middleware system*

DTP model functional positioning and interface definition for transaction middleware are followed. General system structure of transaction middleware is composed of the following parts: ① General management core: it is used for managing node information, maintaining communication connection, and completing transfer for application data and management control information on network; ② Transaction management core: it is used for managing transaction status, coordinating transaction submission/rollback, managing, scheduling and monitoring operation of application process; ③ Safety management core: it is used for exchanging encryption key, managing and maintaining safety information; ④ Local system management tool: it is used for providing user with means in configuration, monitoring and maintenance of system; ⑤ Remote system management tool: it can provide users with means for remote configuration, monitoring and maintenance of systems, which is convenient for users to uniformly manage nodes; ⑥ Shared memory and message queue: It is used for registering operation control information of core and application, and transferring application data between application and core as communication mechanism among cores and between cores and application process.

5 ANALYSIS EXAMPLES OF FINANCIAL SYSTEM

5.1 *Introduction of demand analysis*

The financial system covers several provinces and cities for a large number of cross-city transactions. Transaction host machine central control and multi-layer audit modes are adopted. The system is physically a three-layer structure, which is composed of central transaction host machine, city sub-center and city transaction terminal. Transaction front end 1 issue one transaction, which is handled by central transaction host machine, data flowchart of the result to transaction front end 2 is shown in Figure 2.

The following requirements are proposed on the system in the entire business process: ① every transaction must be integral, namely indivisible. Success or failure for whatever reason should be suitably handled by the system; ② Submission and processing of critical business can be ensured under the condition of crowded and busy business; ③ Timeliness of communication transmission and business processing should be ensured, system synchronous and asynchronous transaction coexistence is allowed for the system, synchronous transactions are applicable to transaction requests with demand of immediate response, asynchronous transactions are generally targeted at general transaction requests; ④ Security of transactions should be ensured, and access permissions should be strictly required. ⑤ the system can realize failure tracing and diagnostic remote monitoring and other management functions, thereby controlling system resources and business processing. One transaction middleware is used in the financial system as system platform, and the whole system is shown in Figure 3 after addition of middleware.

When a transaction is initiated from the transaction front end, client of middleware is called first,

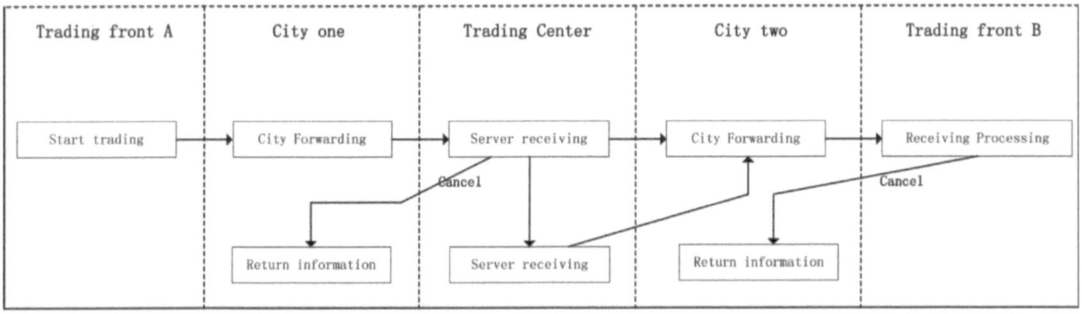

Figure 2. Financial business data flowchart.

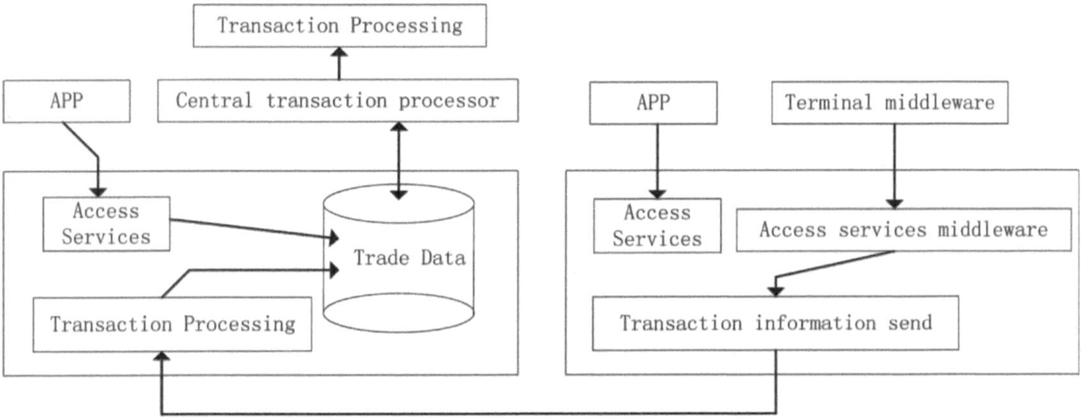

Figure 3. System overall framework figure.

middleware can transfer the transaction instructions from the transaction front end to city sub-center, transmission reliability and security can be guaranteed by the middleware itself. City sub-center is responsible for a series of business processing, such as application calling, storage, and so on. Then transaction instructions can be sent to the central transaction host machine through middleware, central transaction host machine can call different calculation modules for handling according to contents of transaction instructions. The results can be returned back to transaction front ends in the original way. Synchronous or asynchronous middleware transaction mode can be automatically selected according to different business.

5.2 *Realization of transaction safety by middleware*

Only lower access application set is opened for lower node. Techniques can be deployed through middleware. Central transaction host machine can be visited by lower access service application set of central transaction host machine. Application of lower node can not directly visit two application sets of host machine application services and terminal access services. Lower access service application set of central transaction host machine can forcibly check access password when the lower node submits visit requirement. Similarly, the node also can be adopted for service division of urban sub-center. However, one transaction message submission service should be increased. Different safety strategies can be implemented on different application sets through such deployment, thereby preventing hackers from intruding into host machine transaction module, and realizing mutual independence of all service groups, and improving business processing performance. Meanwhile, service relation among nodes can be simplified, and system maintainability can be improved by such deployment.

5.3 *Realization of monitoring management stability through configuration of local and remote system management tools of middleware*

Middleware provides monitoring tools so that users can understand the system operating conditions timely. Monitored contents include the following: use condition of system resource, node connection status operation condition, process operation condition, and so on. System resource monitoring provides use condition information of memory and queue. Various key memory tables in report system, current use quantity, use peak and limit quantity of message queue can assist the system administrator to determine rationality of configuration parameters, the administrator can adjust accordingly, thereby optimizing resources.

6 CONCLUSION

Various existing applications of financial system are constantly expanded, and new application forms are continuously increased with network development.

More and more demand should be solved in financial system development process. Traditional system software development can not meet such requirement. Financial middleware platform provides me with a perfect solution; it can be introduced to analysis mechanism of financial crisis mechanism for studying. The following conclusions can be obtained: first, after financial crisis is generated, large-amount fund assistance should be rapidly provided for controlling financial crisis as basic keys. Tiny tuning on financial policies should be frequently conducted, and we should do well in mental preparation on long-term control of financial crisis; second, when the financial market is relatively stable without disturbance, the government should intensify financial system reform; third, when the financial markets are in an unstable or chaotic state, the government should not make major change on financial system, international financial institutions should not delay large-amount fund assistance due to financial restructuring; fourth, when the financial crisis is eased, financial assistance amount from international financial organizations such as IMF, etc. may be gradually reduced, but government's effort on financial system reform must be gradually strengthened.

REFERENCES

[1] E. Ott. Physical Review Letter Vol. 64–67(2007), p. 1196–1199.

[2] Z. Wan, Y. Zhang, X. Wang. Journal of Transport Geography Vol. 34–36(2014), p. 1–6.

[3] S.R. IMF. Nber Paper Vol. 10–14(2012), p. 7–13.

[4] B. Jared. European Journal of Work and Organizational Psychology Vol. 23–26(2004), p. 45–47.

[5] G. David. International Economics Vol. 54–58 (2010), p. 13–31.

[6] C.R. Nelson, A.F. Siegel. Journal of Business Vol. 24–27(2007), p. 473–489.

[7] Diebold, X. Francis. Journal of Econometrics Vol. 10–15(2008), p. 351–363.

Engineering Management and Industrial Engineering – Leung (Ed.)
© 2015 Taylor & Francis Group, London, ISBN 978-1-138-02773-2

A literature review on cloud computing by using social network analysis

D.L. Hu, N.X. Wang, Y. Jia & S.L. Ge
Jiangsu University of Science and Technology, Zhenjiang, Jiangsu, China

ABSTRACT: In order to understand the intellectual structure of cloud computing research in Information System (IS) discipline, social network analysis was used to analyze 168 cloud computing related articles which were retrieved from 18 IS prestigious journals and 2 top IS conferences. 19 important articles were identified based on degree centrality and betweenness centrality, and the results of main path analysis indicate that cloud computing studies in IS can be divided into initial and growing stages. This study helps to understand the state-of-art and guide the future direction of cloud computing research in IS domain.

1 INTRODUCTION

Under the uncertain environment of global market competition, enterprises must continuously improve the agility of strategy and technique; at the same time make the business and information technology operation less complex to obtain and maintain competitive advantage (Petter, et al., 2012; Zissis and Lekkas, 2012). Cloud computing is a kind of information technology service mode which can meet the needs of many users' at the same time over the internet (Armbrust, et al., 2010; Plummer, et al., 2008; shanlin, et al., 2012). Cloud computing is characterized by dynamic resources configuration, on-demand self-service, broad network access, measurable service and virtual resources (Mell and Grance, 2011b). It provides technology support to the two goals of enterprises' stated above at the same time. Cloud computing has become one of the most rapidly growing emerging information technologies at present. According to the prediction of Gartner, the revenue of global cloud computing services market will exceed \$150 billion in 2013 and \$180 billion in 2015. Cloud computing will account for most of the new information technology investment in 2016.

The rapid development of cloud computing also causes wide attention of scholars in IS domain (Kappelman, et al., 2013). In recent years, some scholars in IS domain have analyzed and summarized some research results on cloud computing. For example, Will Venters (Will Venters, 2012) has divided characteristics of cloud which are important for users into technology and service dimension, and then made comments on the existing research of cloud computing; Yang and Tate (Yang and Tate, 2012) have summarized existing research of cloud computing on the topics including technology, business, concepts and applications; Hoberg's and Wollersheim's (Hoberg and Wollersheim, 2012) review synthesized the existing research on cloud computing from a business perspective; the results are structured according to the four dimensions following: cloud computing characteristics, adoption determinants, governance mechanisms, and business impact; Ermakova (Ermakova, et al., 2013) summarized research topics including developing cloud-based applications, platforms or brokers, security and privacy mechanisms, and benefit assessments for the use of cloud computing in healthcare; Shanlin (shanlin, et al., 2012) analyzed the research status of multi-source information service system based on cloud computing; Hongchang (hongchang, et al., 2013) summarized the research progress and question of cloud computing in the field through the analysis of technology characteristics, usage pattern, the potential commercial value and application risk of cloud computing.

The reviews of cloud computing above help to understand the state of the research, define the research deficiencies and determine the potential areas of future research on cloud computing. Moreover, the reviews are of great reference significance for the scholars of IS to conduct cloud computing research. But it is difficult to objectively and comprehensively reveal the history, focus and tendency of the research on cloud computing due to the following limitations: the scholar's knowledge background and research perspective, and they retrieve the literatures only in the Web Of Science (WOS) for convenience. Unlike the existing review of cloud computing research, we retrieved papers from 18 IS prestigious journals and 2 top IS conferences (ICIS and AMCIS), using social network analysis method to analyze the citation network of cloud computing and display the overall view, structure feature and evolution path in visible form. This study aims to help us understand the state-of-art and guide the future direction of cloud computing research in IS domain.

2 RESEARCH METHODOLOGY

2.1 *Data collection*

Research on cloud computing is interdisciplinary which involves computer science and management science (Birman, et al., 2009). The existing researches on cloud computing are mainly in computer science domain, focusing on technical or operational problems; in management science domain, the researches on cloud computing are concentrating on IS discipline. In order to fully define the research status of cloud computing in IS domain, we choose 13 journals in IS domain including the MISQ, ISR, JMIS, ISJ, JAIS, JSIS, JIT, EJIS, I&M, DSS, CAIS, CACM, ACM TMIS, 5 top journals in management domain including Service Science, Management Science, Decision Sciences, Harvard Business Review, Sloan Management Review, and 2 top IS conferences including ICIS and AMCIS as sources. We got 159 literatures after searching with cloud computing, cloud service, on demand service, IaaS, PaaS, SaaS as keywords and the retrieval time ranges from 2004 to 2013.

Nine literatures are found frequently cited and are not included in the 159 articles after analyzing the references of the 159 articles. In order to make cloud computing research review more comprehensive, the 9 literatures are also added to the analysis sample. The 168 literatures are carried out in accordance with the publishing year, and the literature quantity distribution of each year is as shown in Figure 1. The literatures related cloud computing in the field of IS are less from 2004 to 2008. With the rapid development of cloud computing technology and market, cloud computing literatures begin to increase rapidly in 2009. After 2011, literatures about cloud computing in ICIS and AMCIS are rapidly increasing, more than the number of cloud computing literatures in journals. These suggest that cloud computing research has attracted wide attention in the field of IS and has become one of the hot topics.

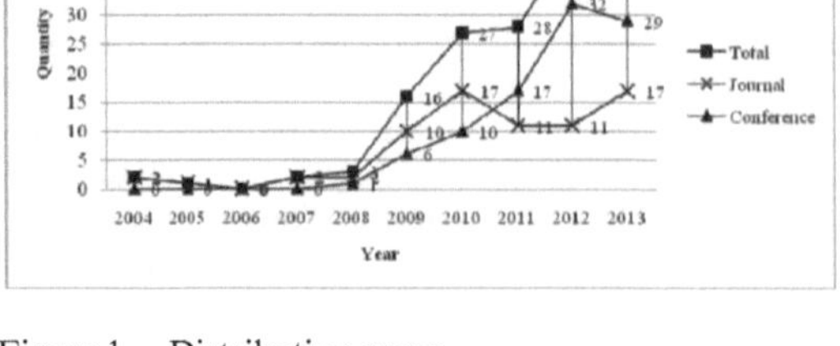

Figure 1. Distribution curve.

2.2 *Data processing*

Social network analysis is an interdisciplinary method integrating applied mathematics, graph theory and computer science. It is able to make relationship and relationship model measurable, appreciable and visible. Management scholars abroad have applied this method in operation management (Pilkington and Meredith, 2009), strategic management (Nerur, et al., 2008), supply chain management (Giannakis, 2012) and IS (Polites and Watson, 2009) to analyze the relation mode of partner network, journal network and citation network, making the overall view, structure feature and evolution path of certain field visible. Their data used to analyze the structure knowledge of the cloud computing is obtained from citation database of WOS. The research on cloud computing in IS domain has a short history; and many research results are often published in conferences of IS, but conferences and some journals are not included in the WOS.

In order to show the research state of cloud computing in IS domain more comprehensively, we choose the 159 literatures retrieved from 19 mainstream journals and 2 top conferences as well as 9 frequently cited articles as sample. After analyzing the 168 literatures, we got the mutual reference relationship which is made into a 168*168 sparse matrix. Then the matrix is imported into UCINET and PAJEK respectively to make important literature analysis and main path analysis.

3 RESULTS

3.1 *Citation network of literature*

The citation network of 168 literatures is drawn on the basis of the data analysis and is shown as in Figure 2. The citation relationship of each literature is sparse. The density of the sociogram is 0.0136, the density would be just 0.0211 even if 33 isolated points are delete, still far less than that of the general social networks. All these suggest that the research on cloud computing in IS discipline is a new field which is still in its primary stage and researches are not related closely.

The degree centralization inflects the centralized tendency of a network. The closer the degree centralization is to 1, the more centralized the

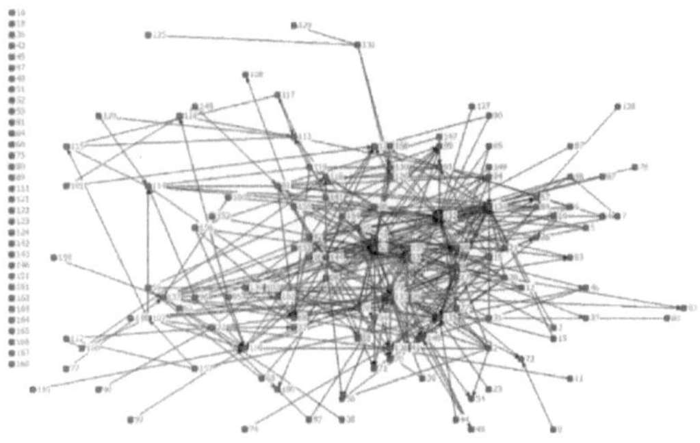

Figure 2. Citation network.

network is. The out-degree centralization of the network is 6.244%, the in-degree centralization is 14.377% and the degree centralization is 13.78%. All these show that the citation is asymmetric. The in-degree centralization is big while the degree centralization is small, which suggests that the researches on cloud computing in the field of IS are dispersive and still have not formed a paper group.

The betweenness centralization is used to analyze the overall structure of a network. If the betweenness centralization is big, it means that one or several parts of the network controls the circulation of the network. The betweenness centralization of the citation network is 0.19% and is not big, which means the diffusion of the cloud computing knowledge does not rely on one or several literatures.

3.2 *Important literature*

The importance of a literature is reflected by its role in the citation network, namely the degree centrality and the betweenness centrality. The degree centrality measures the number of nodes directly related to one node. The degree centrality can be divided into in-degree one and out-degree one. The betweenness centrality measures the ability of one node on the shortest path of other two nodes. If the betweenness centrality is big, it means that the literature is able to control the diffusion of knowledge at source. The paper counts the in-degree centrality, out-degree centrality and between centrality of all the nodes. The one with an in-degree centrality not less than 8 or betweenness centrality not less than 3 is an important literature as shown in Table 1.

Table 1 suggests that the degree centralities and betweenness centralities of numbers 132, 50, 135, 76, 133, 137 and 109 literatures are big, which means these literatures are the knowledge source of cloud computing research. The numbers 132, 135, 133 and 137 literatures are among the added 9 ones, which means that the research on cloud computing in the field of IS highly rely on the other field. The numbers 132, 50, 135, 133 and 137 literatures are about cloud computing concepts and cognition, which further show that the research on cloud computing in the field of IS is on the initial stage. The number 109 literature analyzes the factors affecting the soft adoption by users.

According to Table 1, the in-degree centrality of the literatures with the number of 24, 144, 73, 28 and 22 is not big while the out-degree centrality and the betweenness centrality are big, which suggests that these literatures play critical role in the diffusion of the cloud computing research though most of them are citation from other scholar's research achievements. After further study of the 4 literature contents, it is found that the literatures with the number of 24, 144, 73, and 28 are review papers and the number 22 literature studies the standard problem of user's choice of SaaS on the basis of expert panel achieved with scientific method and literature review.

3.3 *Evolution path*

The evolution path of research on cloud computing can be defined through the main path analysis which reveals the main development process of disciplinary research through searching the literature and mutual citation of the literature on the trunk of disciplinary development. The paper adopts SPLC to calculate the ergodic weights of the edges and draw the main path of cloud computing research as shown in Figure 3. The node in the figure represents literature. Every literature is marked by its number, author and published time. The arrow points to the cited literature. Figure 3 shows that there are 12 literatures on the main path of cloud computing research from 2007 to 2013 and 9 of the 12 literatures are important ones.

According to Table 1, Figure 3 and the contents of the 12 literatures, the development of cloud computing can be divided into two stages:

The first stage (before and 2010): the initial stage. In this stage, research literatures on clouding computing are mostly studying on concept and cognition, including the definition, technical features, development opportunity and challenge of cloud computing. For example, the number 140 literature (Weiss, 2007) describes different views of people's to cloud computing, including data center, distributed computing, a utility grid as well as software as a service and so on; the number 133 (Buyya, et al., 2009) literature provides the architecture for creating clouds with market-oriented and insights on market-based resource management strategies; the number 137 (Christof Weinhardt, 2009) literature makes

Table 1. Important articles.

No.	Author	Title	In-degree	Out-degree	Betweenness
132	(Mell and Grance, 2011a)	The NIST Definition of Cloud Computing	50	0	18.7
50	(Armbrust, et al., 2010)	A View of Cloud Computing	42	1	15.1
135	(Armbrust, et al., 2009)	Above the Clouds: A Berkeley View of Cloud Computing	25	0	6.9
76	(Marston, et al., 2011)	Cloud Computing—The Business Perspective	19	3	4.2
133	(Buyya, et al., 2009)	Cloud Computing and Emerging IT Platforms: Vision, Hype, and Reality for Delivering Computing as the 5th Utility	17	1	3.7
109	(Xin, 2008)	Software-as-a Service Model: Elaborating Client-Side Adoption Factors	14	0	2.9
137	(Christof Weinhardt, 2009)	Cloud Computing—A Classification, Business Models, and Research Directions	11	1	3.6
68	(Cusumano, 2010)	Cloud Computing and SaaS as new Computing Platforms	11	0	1.0
114	(Choudhary, 2007)	Comparison of Software Quality Under Perpetual Licensing and Software as a Service	9	1	1.7
69	(Durkee, 2010)	Why Cloud Computing Will Never Be Free	9	1	0.4
140	(Weiss, 2007)	Computing in the Cloud	9	0	0.8
138	(Foster, et al., 2008)	Cloud Computing and Grid Computing 360 Degree Compared	9	0	1.3
55	(Brynjolfsson, et al., 2010)	Cloud Computing and electricity: Beyond the Utility Model	9	1	0.7
136	(Catteddu and Hogben, 2009)	Cloud Computing Benefits, Risks and Recommendations for Information Security	8	0	1.3
144	(Will Venters, 2012)	A Critical Review of Cloud Computing: Researching Desires and Realities	3	18	5.4
24	(Hoberg and Wollersheim, 2012)	The Business Perspective on Cloud Computing—A Literature Review of Research on Cloud Computing	2	23	6.8
73	(Yang and Tate, 2012)	A Descriptive Literature Review and Classification of Cloud Computing Research	0	17	4.0
28	(Walther and Plank, 2012)	Success Factors and Value Propositions Of Software as a Service Providers—A Literature Review and Classification	1	9	3.4
22	(Repschlaeger and Turowski, 2012)	Selection Criteria for Software as a Service: An Explorative Analysis of Requirements	0	11	3.0

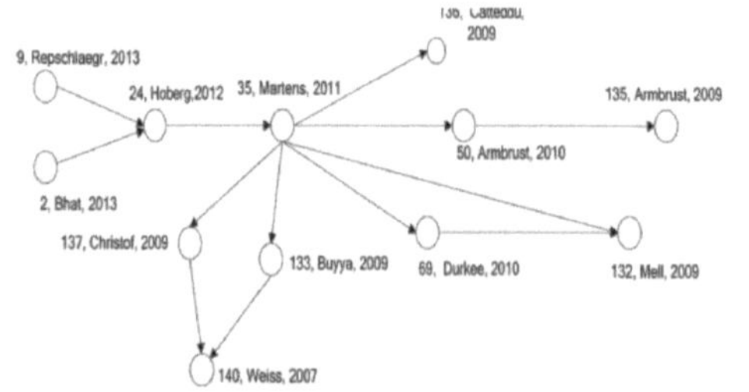

Figure 3. The main path of cloud computing research.

a comprehensive comparison between the cloud computing and grid computing, and indicates the business model and research direction of cloud computing; the number 132 (Mell and Grance, 2011a) literature gives a general definition of cloud computing and describes the characteristics of cloud computing technology; the numbers 50 (Armbrust, et al., 2010) and 135 (Armbrust, et al., 2009) literatures make comprehensive analysis of the opportunities and challenges faced by cloud computing; the number 136 (Catteddu and Hogben, 2009) literature analyzes the benefits, risks, and information security and other issues of cloud computing.

The second stage (2011–2013): the growing stage. In this stage, the research literature on cloud computing begin to emerge increasingly. The specific content, theoretical perspective and research method are defined. For example, the number 35 (Martens and Teuteberg, 2011) literature designs a reference model to help companies manage risk and compliance through the systematic literature review and requirements analysis of cloud computing service; the number 24 (Hoberg and

Wollersheim, 2012) literature discusses the characteristics of cloud computing, the determinants of cloud computing adoption, governance mechanism and business impact from the business perspective with the comprehensive literature review method and finds that there are few researches on the factors affecting cloud computing adoption and the business value of the cloud computing; the number 9 (Repschlaeger, et al., 2013) literature studies the preferences of cloud service providers with analytic hierarchy process, providing decision-making reference for the customer when choosing cloud service provider; the number 2 (Bhat, 2013) literature studies the institutional factors affecting Indian small and medium-sized enterprise's cloud service adoption on the basis of transaction cost theory. The above analysis shows that the study of cloud computing at the second stage has shifted to the adoption and application of cloud computing by enterprises from the concept and cognition of cloud computing at the first stage.

4 CONCLUSIONS

As a new information technology revolution, cloud computing has exerted huge influence on the information technology industry and information technology application in enterprise, and catching the attention of the IS scholars. 168 research literatures on cloud computing are found through retrieving 18 mainstream journals and 2 top conferences in IS domain. The 168 literatures are systematically analyzed by social network analysis, and the important literatures, development of the research on cloud computing in the field of IS are defined. The main research conclusions are as follows:

First, the citation network of the literatures shows that the density of the network of the cloud computing literature is small and the degree centrality and betweenness centrality are all relatively small. These suggest that the research of cloud computing which is a new research field is still in its primary stage and researches are not related closely; the research topics are dispersive and still have not formed a paper group; breakthrough achievement has not been made yet.

Second, most of the important research literatures on cloud computing in the field of IS are about the concepts, cognition and review of IS. This further suggests that research on cloud computing in the field of IS is at primary stage. Many important literatures are from journals and conferences in other fields, which suggests that research on cloud computing in the field of IS highly relies on related researches in other fields.

Third, research on cloud computing in the field of IS can be divided into the initial stage and the growing stage. The initial stage focuses on the definitions, technical characteristics, development opportunities and challenges of cloud computing but not the specific research method and clear research topics; the growing stage shifts to the adoption and application of cloud computing by enterprises with specific research method.

ACKNOWLEDGEMENTS

Supported by Project of National Natural Science Foundation of China (Grant numbers 71331003, 711711004, and 71101065). and Universities in Jiangsu Province plan to graduate practice and innovation, number 1254051409.

REFERENCES

Armbrust, M. Fox, A. Griffith, R. et al. 2010. A View of Cloud Computing, *Communications of the ACM*, 53(4): 50–59.

Armbrust, M. Fox, A. Griffith, R. et al. 2009. Above the Clouds: A Berkeley View of Cloud Computing Cloud Computing: An Old Idea Whose Time Has (Finally) Come, *UC Berkeley Reliable Adaptive Distributed Systems Laboratory*: 07–013.

Bhat, J.M., 2013, Adoption of Cloud Computing by SMEs in India: A study of the Institutional Factors, *Proceedings of the Proceedings of the Nineteenth Americas Conference on Information Systems*, Chicago, Illinois, 1–8.

Birman, K. Chockler, G.; and Renesse, R. v. 2009. Toward a cloud computing research agenda, *ACM SIGACT News*, 40(2): 68–80.

Brynjolfsson, E. Hofmann, P.; and Jordan, J. 2010. Cloud Computing and Elecricity: Beyond the Utility Model, *Communications of the ACM*: 3–6.

Buyya, R. Shin, C. Venugopal, S. et al. 2009. Cloud computing and emerging IT platforms: Vision, hype, and reality for delivering computing as the 5th utility, *Future Generation Computer Systems*, 25(6): 599–616.

Catteddu, D. & Hogben, G. 2009. Cloud computing: Benefits, risks and recommendations for information security, *ENISA.*: 1–125.

Choudhary, V. 2007. Comparison of Software Quality Under Perpetual Licensing and Software as a Service, *journal of MIS*, 24(2): 141–165.

Christof Weinhardt, W.A., Anandasivam, Benjamin Blau 2009. Cloud Computing—A Classification, Business Models, and Research Directions, *BISE-STATE OF THE ART*.

Cusumano, M. 2010. Cloud Computing and SaaS as new Computing Platforms, *Communications of the ACM*, 53(4): 27–29.

Durkee, D. 2010. Why cloud computing Will never Be free, *Communications of the ACM*, 53.

Ermakova, T. Erek, K.; and Huenges, J., 2013, Cloud Computing in Healthcare—a Literature Review on Current State of Research, *Proceedings of the Proceedings of the Nineteenth Americas Conference on Information Systems*, Chicago, Illinois, 1–9.

Foster, I. Zhao, Y. Raicu, I. et al. 2008. Cloud Computing and Grid Computing 360-Degree Compared, *Grid Computing Environments Workshop*.

Giannakis, M. 2012. The intellectual structure of the supply chain management discipline: a citation and social network analysis, *Journal of Enterprise Information Management*, 25(2): 136–169.

Hoberg, P. & Wollersheim, J., 2012, The Business Perspective on Cloud Computing—A Literature Review of Research on Cloud Computing, *Proceedings of the Proceedings of the Eighteenth Americas Conference on Information Systems*, Seattle, Washington, 1–10.

hongchang, W. jinghua, X.; and yongqin, X. 2013. A Review of the literature of Cloud Computing and Future Prospects: A Business Perspective, *Foreign Economics & Management*, 35(6): 13–21.

Kappelman, L. McLean, E.; and Luftman, J. 2013. Key Issues of IT Organizations and Their Leadership: The 2013 SIM IT Trends Study, *MIS Quarterly Executive*, 12(3): 227–240.

Marston, S. Li, Z. Bandyopadhyay, S. et al. 2011. Cloud computing—The business perspective ☆, *Decision Support Systems*, 51(1): 176–189.

Martens, B. & Teuteberg, F., 2011, Risk and Compliance Management for Cloud Computing Services: Designing a Reference Model, *Proceedings of the Proceedings of the Seventeenth Americas Conference on Information Systems*, Detroit, Michigan, 1–10.

Mell, P. & Grance, T. 2011a. The NIST Definition of Cloud Computing *NIST*: 1–3.

Mell, P. & Grance, T. 2011b. The NIST definition of cloud computing, *NIST special publication*, 800(145): 1–7.

Nerur, S.P. Rasheed, A.A.; and Natarajan, V. 2008. The intellectual structure of the strategic management field: an author co-citation analysis, *Strategic Management Journal*, 29(3): 319–336.

Petter, S. DeLone, W.; and McLean, E.R. 2012. The Past, Present, and Future of "IS Success", *Journal of the Association for Information Systems*, 13(5): 341–362.

Pilkington, A. & Meredith, J. 2009. The evolution of the intellectual structure of operations management—1980–2006: A citation/co-citation analysis, *Journal of Operations Management*, 27(3): 185–202.

Plummer, D.C. Bittman, T.J. Austin, T. et al. "Cloud computing: Defining and describing an emerging phenomenon,"), 2008,

Polites, G.L. & Watson, R.T. 2009. Using social network analysis to analyze relationships among IS journals, *Journal of the Association for Information Systems*, 10(8): 95–636.

Repschlaeger, J. & Turowski, K., 2012, Selection Criteria for Software as a Service: An Explorative Analysis of Provider Requirements, *Proceedings of the Proceedings of the Eighteenth Americas Conference on Information Systems*, Seattle, Washington, 1–12.

Repschlaeger, J. Wind, S. Zarnekow, R. et al., 2013, Decision Model for Selecting a Cloud Provider: A Study of Service Model Decision Priorities, *Proceedings of the Proceedings of the Nineteenth Americas Conference on Information Systems*, Chicago, Illinois, 1–11.

Shanlin, Y. he, L.; and shuai, D. 2012. Survey on multi-sources information service system based on cloud computing, *Journal of Management Sciences in China*, 15(5): 83–94.

Walther, S. & Plank, A., 2012, Success Factors and Value Propositions of Software as a Service Providers—A Literature Review and Classification, *Proceedings of the Proceedings of the Eighteenth Americas Conference on Information Systems*, Seattle, Washington, 1–14.

Weiss, A. 2007. Computing in the Clouds, *NetWorker*: 16–25.

Will Venters, E.A.W. 2012. A critical review of cloud computing: researching desires and realities, *Journal of Information Technology*, 2012(27): 179–197.

Xin, M., 2008, Software-as-a Service Model: Elaborating Client-Side Adoption Factors, *Proceedings of the Twenty Ninth International Conference on Information Systems*, Paris, 1–11.

Yang, H. & Tate, M. 2012. A Descriptive Literature Review and Classification of Cloud Computing Research, *Communications of the Association for Information Systems*, 31(1): 35–60.

Zissis, D. & Lekkas, D. 2012. Addressing cloud computing security issues, *Future Generation Computer Systems*, 28(3): 583–592.

Management technology and engineering

Engineering Management and Industrial Engineering – Leung (Ed.)
© 2015 Taylor & Francis Group, London, ISBN 978-1-138-02773-2

An experimental study of the effects of cross-cultural time orientation and corporate social responsibility on consumers' brand evaluation

Y.R. Dong
Nanjing University, Nanjing, Jiangsu, China

Y.Q. Liu
Nanjing University of Finance and Economics, Nanjing, Jiangsu, China

N. Jiang
Nanjing University, Nanjing, Jiangsu, China

ABSTRACT: From the perspective of cross-cultural psychology, this research studies the effect of cross-cultural time orientation on the relationship between corporate social responsibility and consumers' brand evaluation. Furthermore, it explores the underlying mechanism of the effect. With these efforts, this study compensates the existing literature by exploring further how cultural factors might affect the effect of corporate social responsibility. In addition, this study also contributes to the cross-cultural consumer research at large. The empirical evidence from three experiments justifies such findings as follows. First, corporate social responsibility would enhance long-term orientated consumers' brand evaluation, but not short-term oriented consumers'. Second, this effect occurs because corporate social responsibility activities reflect the company's long-term view, which is consistent with long-term oriented consumers' cultural values. Third, this effect would only occur when consumers did not have a preexisting identification with the company.

1 INTRODUCTION

In recent years, two important tendencies have exerted an influence over operations and management in most of large companies. For one thing, an increasing number of enterprises starts to operate globally or has to face competition from enterprises in other parts of the world, because of the development of economic globalization. For another thing, by means of supporting charity, preserving ecological environment and so on, more and more enterprises actively undertake social responsibilities for the sake of increasing enterprises' brand value, sales and market performance. However, few researches have taken these two important tendencies into account synthetically. From the perspective of cross-cultural psychology, this research studies the effect of corporate social responsibility on consumer behaviors in different cultures and countries.

Specifically, by considering a particular cultural difference: cross-cultural time orientation, we find that the effect of corporate social responsibility will be actuated for those who are long-term cultural orientated consumer, while for those who are short-term cultural orientated consumer, the effect will be alleviated. This is because corporate social responsibility itself reflects a kind of long-term orientation and is consistent with cultural values of long-term cultural orientated consumers, which lead to enhance consumers' brand evaluation. Examining these hypotheses, we present three experiments and recruited almost five hundred participants. Different methods are applied to measure or manipulate these variables (e.g., cross-cultural time orientation, consumer-company identification, and research situation) as well.

2 THEORETICAL FOUNDATION

2.1 *Corporate social responsibility and brand evaluation*

Corporate social responsibility is originally proposed by Bowen (1953). According to the definition of Bowen, Corporate Social Responsibility (CSR) refers to attention to the interests of the whole society that businessmen pay in the process of formulating policies, making decisions and operating activities. After summarizing the existing research of CSR, Murphy & Schlegelmilch (2013) regarded CSR as "concern about social welfare showed and spread by enterprises in the process of operations management". In marketing and other related research field of enterprise management, the most attention is paid to whether enterprises are able to gain rewards through taking CSR (Bhattacharya & Sen 2004).

In the field of consumer research, there are an abundant literature that focus on whether corporate social responsibility have an impact on consumers' brand evaluation and thus stimulate consumers' positive behavioral response (e.g., purchase intention, actual purchase behavior and oral spreading. Brown & Dacin 1997, Torelli et al. 2012). Although earlier studies generally agree that taking social responsibility can stimulate a more positive brand evaluation (Creyer & Ross 1997, Drumwright 1996), some relevant literature review and meta-analyses find that the relationship between CSR and brand evaluation is actually very weak and do not reach statistical significance (Simonson & Becker 2006, Valor 2008). Because of this, a growing body of research is focusing on the moderators. More specifically, under what circumstances or for what types of consumers CSR have a positive effect (Bhattacharya & Sen 2004, Murphy & Schlegelmilch 2013, Torelli et al. 2012). For example, according to the research at present, the effects of CSR on consumers' brand evaluation are influenced by many factors, such as enterprise competence (Brown & Dacin 1997, Biehal & Sheinin 2007), consumers' support for CSR (Sen & Bhattacharya, 2001), consumers' attributions for CSR (Yoon et al. 2006, Ellen et al. 2006, Klein & Dawar 2004) and so on.

Therefore a question unavoidably occurs: from the perspective of cross-culture, whether consumers from different cultures response differently to the same corporate social responsibility? The answer, according to existing research, is yes. For instance, it has been found that in the case of the same CSR information consumers from different countries and cultures showed different perceptions, made different brand evaluation and even led to different consumer behaviors (Gift et al. 2013). However, these current researches remain flawed and need to be enriched by further studies. First, the relationship between these researches is a mere correlation by nature, not casual. Therefore this fact does not conclude that cultural difference leads to different consumers' responses. It might result from economic development factors or even measuring error, rather than cultural difference (Steenkamp & Baumgartner 1998). Second, even if these research findings, to a certain extent, are caused by cultural difference, they also fail to answer the question, "cultural difference in which aspect will lead to different consumers' reactions to CSR?" (Ho et al. 2012). Hence, Deng (2012), Murphy & Schlegelmilch (2013) appealed for further research to explore national difference in which aspect cause different consumers' reaction to the same CSR in a deep sense. To answer this appeal, we explore how cultural differences in time orientation influence consumer to respond to CSR based on cultural dimension model (Hofstede 1980, 2001).

2.2 *Cross-cultural time orientation*

Cross-cultural value model developed by Hofstede (1980) is now widely applied to explain the consumer behavioral differences during the consumer decision making in different cultural background (Zhang & Shrum 2009). As Hofstede's model defined, cultural differences are largely measured by the following four dimensions: power distance, uncertainly avoidance, individualism/collectivism, and masculinity/femininity. Nevertheless, these cultural differences model remains an important defect. All these dimensions are from western cultural perspective. Thus, the significance of other cultures has been ignored (Hofstede & Bond 1988). So, Hofstede (1991) added a new cultural difference dimension: cross-cultural time orientation. This dimension is first developed by Michael Bond and some Chinese scholars renewably examining cross-cultural differences from Confucian perspective (Chinese Culture Connection 1987). Hofstede (1991) defined 'time orientation' as a reflection of individual basic attitude toward a matter of time in different cultures. Particularly, long-term orientated individual is more concerned with future and regard 'persistence, thrifty and triumph over adversity' as a basic principle for life, while short-term orientated individual is more likely to show more concern for present and regard 'enjoy life, maintain present status, live in the moment' as a basic principle. Bearden et al. (2006) defines cross-cultural time orientation as a cultural value related to time: long-term orientated individual looks at time from the holistic perspective and respects tradition and effects of individual current behaviors in the future. However, short-term orientated individual looks at time from individual perspective and thus only concerns about short-term result.

Under this formal concept's guidance, Bearden et al. (2006) also developed a useful scale for cross-cultural time orientation. It is these efforts that inspire the empirical research of cross-cultural time orientation in the relevant studying areas. For example, Bearden et al. (2006) found that long-term orientated individual possesses a higher moral standard. Yoo et al. (2002, 2011) also found that long-term (vs. short-term) orientated consumer will be more repugnant to corporate immoral business operation. Further study of Zin (2013) has shown that long-term orientated consumer cared more about effects of operating activities on environment. As a further supplement, Hofstede et al. (2010), Minkov and Hofstede (2012) suggested that long-term orientated individual care more not only about his own behavior's impact on future, but also effects of others' and organization's on society.

In essence, CSR reflects that enterprise has realized long-term interest on itself as well as positive influence on society through sacrificing short-term interest (Murph & Schlegelmilch 2013, Sen & Bhattacharya 2001). From the perspective of cross-cultural time orientation, this kind of behavior is more in line with the value of consumers' in the long-term orientated cultural background. According to the value from fit theory (Aaker & Lee 2006, Avnet & Higgins 2006), when consumers receive information in accordance with his basic value, they will make more positive evaluation; otherwise this information will not have significant impact on consumers' evaluation decision. Thus:

H1: Compared to short-term orientated consumer, long-term orientated consumers make more positive evaluation on corporate taking social responsibility actively.

According to the above, long-time orientated consumers give more positive remarks on enterprises' taking CSR because this corporate behavior is consistent with consumers' inner cultural values. In consequence, we consider that consumer-enterprise identification serves as a mediating mechanism of this effect. Consumer-enterprise identification refers to some deep and committed connection between consumer and enterprise and also a reflection of market phenomenon that consumers consider corporate image as indispensable to define their own identity (Bergami & Bagozzi 2000, Bhattacharya & Sen 2003). To explain why consumer-enterprise identification serves as a mediator, the following two questions are posed: Q1. Do long-term time orientated consumers have a higher level of consumer-enterprise identification to enterprise taking CSR? Q2. Dose consumer-enterprise identification improves consumers' evaluation of enterprise?

For the first question, a great number of researches have proved that whether consumers are identified with enterprise largely depended on similarities and consistency between consumer's value and corporate value (Bhattacharya & Sen 2003). The extent of identification increases with the similarity. Long-term orientated consumers' core value is about highlighting the future. And enterprise taking CSR is exactly paying attention to the future of brand and society (Marin & Ruiz 2007). Thus, in the eyes of long-term orientated consumers', corporate image is consistent with their core value, thereby possibly having a higher level of identification. Next, the second question has been solved by existing research. The answer, very clearly and fully justified, is 'yes'. For instance, Bhattacharya and Sen (2003), Press and Arnould (2011) found that the more the consumer-enterprise identification is, the better evaluation of enterprise, purchase intention, verbal communicating intention there are. In summary, thus:

H2: Consumer-enterprise identification mediates the relationship between time orientation and brand evaluation as well as the relationship between Corporate Social Responsibility (CSR) and brand evaluation.

If cross-cultural time orientation affects consumer's evaluation of enterprise based on CSR information indeed, then this effect is mediated by consumer-enterprise identification. In the case of consumer having no identification with target enterprise, taking CSR actively can increase long-term orientated consumer's identity with enterprise, but cannot increase the short-term orientated consumer's. Thus, long-term orientated consumers' evaluation can be improved while short-term orientated consumers' evaluation will not be affected. But when consumers have been identified with target corporate at a high level, long-term orientated consumer's identification with enterprise will not be increased further due to ceiling effect (Lewis-Beck et al. 2004). In consequence, long-term oriented consumers which have the same high level of identification with short-term oriented consumers will result in both of them evaluating enterprise positively. Consequently, the effect of time orientation is attenuated, or even disappeared. Generally speaking, consumers' preexisting identification is a boundary condition for corporate time orientated effect. This point corresponds with the basic principle of method proposed by Spencer et al. (2005), Zhao et al. (2010), using moderating method to examine mediating effect. Thus:

H3: Consumers' preexisting identification with the company moderates the effect of time orientation. Specifically:

H3a: Long-term orientated consumers show higher evaluation on enterprise's taking CSR than short-term orientated consumers when consumers do not have a preexisting identification with the company.

H3b: If the consumers have already been highly identified with the company, long-term orientated consumers make the same high level evaluation on enterprise's taking CSR like short-term orientated consumers.

We present three studies to test the above hypotheses. Study 1 measures consumers' cross-cultural time orientation through manipulating

CSR to verify H1. Study 2 further manipulates time orientation to provide a casual foundation and tests the mediating effect through measuring consumer-corporate identification to support H2. Study 3 manipulates consumer-corporate identification to verify H3.

3 STUDY 1

3.1 *Method*

Two hundred and nineteen participants (148 males, 71 females, average age = 22.64 years) complete the study on one website which is designed to collect data online. We adopt mixed experimental design, a 2 (corporate social responsibility: yes vs. no) × 2 (time orientation: long-term vs. short-term) design. CSR information is manipulated between subjects while time orientation is measured by scales. Participants are randomly assigned to two study groups. Then, subjects are told to read a material about a computer brand. After that, all subjects are asked to provide a brand evaluation. Finally, they provide their time orientation measurement as well as some demographic variables.

Corporate Social Responsibility Based on recommendation in the research of Brown & Dacin (1997), Sen & Bhattacharya (2001), it is through company's information to manipulate CSR. Specifically, subjects in the CSR condition are provided with some information about this computer brand's taking social responsibility, caring social welfare, aiding charity and protecting environment besides the related production and competence information. However, subjects in the no CSR condition are exposed to the same production and competence information and are told that this company aims to seek the shareholders' maximal profits, build market share and increase sales. As a manipulation check, participants then rate company more specifically on following three items: "this company is very concerned about social welfare", "this company possesses a strong moral sense", "this company support charity actively" (1 = "very unfavorable", 7 = "favorable"). We take these items' mean value as test indicator of CSR ($\alpha = 0.76$).

Time Orientation The scale assesses subject's time orientation using Bearden et al.'s (2006) cultural time orientation scale, which consists of eight items: "I work hard for future success", "I don't mind giving up present enjoyment for more bright future", "I plan for my own future" and so on. This scale can not only be applied to assess the differences of time orientation between cultures, but also be used to assess individual time orientation within a culture, which is consistent with this study's situation (Bearden et al. 2006, 2007). These eight items are of good internal consistency coefficient ($\alpha = 0.83$) and its mean value is taken as individual time orientation. Higher mean value indicates a stronger individual long-term orientation. On the contrary, lower mean value indicates a stronger individual short-term orientation.

Brand Evaluation This study measures brand evaluation in a scale on the following three items: "My overall impression of this computer brand is positive", "I think this computer brand is a very excellent company", "I think the product of this computer brand is worth having a try" (1 = "very unfavorable", 7 = "favorable"). The mean value of these three items represents participant's brand evaluation ($\alpha = 0.85$).

3.2 *Results*

Manipulation Check To check whether the manipulation of CSR is successful or not, we apply the analyses of variance (ANOVAs), the manipulation test variable of CSR as dependent variable as well as the manipulation of CSR as independent variable. The results show that compared with those in the control condition, subjects' CSR perceptions are more favorable in the CSR condition (MCSR = 4.74 vs. MNO = 4.33, $F(1, 217) = 4.09$, $p< 0.05$). The success of manipulation is reflected in this significant effect.

Hypothesis Testing According to H1, only long-term orientated consumers can give a higher evaluation to enterprise which takes CSR while short-term orientated consumers would not. We ran a regression to test our prediction, with the dependent variable being brand evaluation. The independent variables were CSR, long-term orientation and the interaction terms of two variables. The main effect was significant and positive ($\beta = 1.79$, $t(215) = 2.25$, $p< 0.05$), stating that consumers tend to give higher rating to enterprises' taking CSR in general. However, the main effect of time orientation is non-significant ($\beta = 1.79$, $t(215) = 2.25$, $p< 0.05$). More importantly, the interaction between the two was positive and significant ($\beta = 0.39$, $t(215) = 2.70$, $p< 0.01$), which illustrates that the stronger consumers' are long-term orientated, the more positive evaluation they will give on enterprises' taking CSR. Following the suggestion of Fitzsimons' (2008), this interaction term should be analyzed further and thus a new categorized variable of time orientation is created through adding or subtracting one standard deviation to run the analyses of variance (ANOVAs). The results are consistent with the results of regression. Taking CSR (M = 4.92) leads to a relatively more positive evaluation on enterprises than enterprises without taking CSR (M = 4.07) for those whose time orientation is long-term ($t(215) = 3.86$, $p< 0.001$).

However, the effect of CSR on brand evaluation disappears for those whose time orientation is short-term ($M_{High\ CSR} = 3.93$ vs. $M_{Low\ CSR} = 4.07$, $t(215) = -1.03$, $p > 0.31$). The results support H1.

3.3 *Discussion*

The results of study 1 are consistent with H1 to verify the theoretical framework of this paper primarily through manipulating CSR and measuring cultural time orientation. However, there are some limitations which need further studies to replenish. First of all, time orientation is measured by scale not to be manipulated, so that causation of time orientation can not be supported to some extent. Second, although this study examines the effect of time orientation and CSR on brand evaluation, the internal mechanism is ignored. Third, only Computer Company is considered. In order to know whether this kind of relationship can be applied to all kinds of companies, it is necessary to examine this effect in other industries. In study 2 we designed an experiment to further solve the above problems.

4 STUDY 2

4.1 *Method*

A total of 117 consumers (average age: 36.68, SD = 13.51, 54 females) participated in this study through an online database. This study uses a 2 (Information of Corporate Social Responsibility: Yes vs. No) × 2 (Time Orientation: long-term vs. Short-term) between-subjects design. Subjects are allocated randomly to four conditions. After time orientation and CSR being manipulated, subjects will complete the measurement of consumer-company identification. Finally, subjects are asked to rate some demographic variables and manipulate tests.

Time Orientation We use the method, which is learned from Kopalle et al. (2010), to manipulate cultural time orientation. Specially, in the long-term orientated condition, subjects are exposed to the following information: "Many people focus too much on present happiness and ignore future life. These people spend too much money on luxurious life while they are failing to consider long-term consequences. In contrast, others choose to sacrifice today's happiness and gain a happy life later. Thus, focusing on future benefits us. For instance, compared with other students, a long-term orientated student decides to study rather than play games. Hence, they are more likely to achieve high grades and get access to prestigious universities. Thus, they can find a favorable job and enjoy a happy life." After finishing reading this, subjects are required to recall and describe three of their own life experiences, in which they have sacrificed temporary happiness to achieve long-term happiness. However, in the short-term orientated condition, subjects read a passage which acclaimed that enjoying present life is the truth of life and recall three related life experiences based on that. Participants complete the scale to check the manipulation of time orientation which is developed from Kopalle et al. (2010), whose mean value is taken as an indicator of manipulation testing ($\alpha = 0.89$).

Consumer-company identification The scale which is used to measure company identification is developed from Elsbach & Bhattacharya (2001): "I am similar to brand image of this company", "I am willing to follow the company's dynamic information", "I will discuss the advantages and disadvantages of this company's products with others". The mean value of these items is used as consumers' identification of this drinks company ($\alpha = 0.88$).

In addition, for increasing external validity, we choose a company from another industry—Drinks Company.

4.2 *Results*

Manipulation Check First of all, we apply ANOVAs to check whether the manipulation of CSR is successful, with the manipulation test variable of CSR as dependent variable ($\alpha = 0.72$) and the manipulation of CSR, the manipulation of time orientation and the interaction term of the two as independent variables. The results showed that only the effect of CSR manipulation is significant ($F(1, 113) = 17.07$, $p < 0.001$). Subjects who have read the information of CSR (M = 3.99) thought that this company has a stronger sense of social responsibility than those who have no access to information (M = 3.99). Hence, the manipulation of CSR is successful. In addition, the main effect of time orientation and interaction effect are non-significant ($Fs < 1$), which states that manipulation of CSR have no effect on consumers' time orientation.

Hypothesis Testing In order to test H1 further, the ANOVAs shows that although the main effect of CSR ($F(1, 113) = 1.62$, $p > 0.20$) and time orientation ($F(1, 113) = 3.44$, $p > 0.07$) is non-significant, the interaction effect of the two is significant ($F(1, 113) = 5.88$, $p < 0.05$). Furthermore, long-term orientated participants tend to have a higher evaluation on companies with CSR than those without CSR (MCRS = 5.06 vs. MNO CRS = 4.21, $t(113) = 2.63$, $p < 0.01$). This suggests that long-term orientated consumers are more likely to support companies which actively take social responsibility. However, there is no difference among short-term orientated consumers (MCSR = 4.08 vs. MNO

CSR = 4.34, t (113) = –0.81, $p > 0.41$). Short-term orientated consumers are not sensitive to CSR and thus show no difference in supporting company whether it takes CSR or not. These results are consistent with study 1 and support H1 further.

The Effect of Mediation In line with H2, consumer-company identification mediates the relationship between time orientation and consumer's response to CSR. We run the regression analysis derived from Baron & Kenny's (1986) method to test mediated relation. First, we run the regression analysis, with brand evaluation as dependent variable, manipulation of CSR, manipulation of time orientation and interaction term of the two as independent variables. The interaction effect is significant ($\beta = 1.11$, t (113) = 2.42, $p < 0.05$). Second, we run the regression analysis, with consumer-company identification as dependent variable, manipulation of CSR, manipulation of time orientation and interaction term of the two as independent variables. There is also a significant effect of interaction ($\beta = 1.74$, t (113) = 4.12, $p < 0.001$). Finally, we run the regression analysis, with brand evaluation as dependent variable, manipulation of CSR, manipulation of time orientation, the interaction term of the two and consumer-company identification as independent variables. As a result, the interaction term isn't significant ($\beta = 0.41$, t (113) = 1.40, $p > 0.16$), while the effect of consumer-company identification on brand evaluation is significant ($\beta = 0.88$, t (113) = 14.13, $p < 0.001$). In conclusion, these results suggest that consumer-company mediate the effect of time orientation and corporate social responsibility on brand evaluation (Sobel's z = 3.98, $p < 0.001$).

Moreover, the problem existing in the Baron and Kenny's method has been found by many studies. According to Preacher & Hayes (2008), Zhao et al. (2010), we employ the bootstrapping approach to test the mediating effect of consumer-company identification again (bootstrapping frequency 1000, 95% confidence interval). The results show that consumer-company identification mediates the relationship between CSR and brand evaluation (indirect effect = 1.28, SE = 0.24, 95% CI: 0.85 to 1.81) among those who are long-term orientated. However, the mediating effect disappears when subjects are short-term orientated (indirect effect = –0.17, SE = 0.26, 95% CI: –0.66 to 0.37). This suggests that there is no difference in identification of company (CSR vs. No CSR) among those who are short-term orientated. Above all, H2 is supported fully.

4.3 *Discussion*

Based on study 1, study 2 further examines the effect of time orientation and CSR on brand evaluation. First, time orientation and CSR are manipulated directly, which demonstrates that the relation between the two and brand evaluation is a casual relationship rather than a correlational relationship simply. Then, this study uses a different method to examine the mediating effect of consumer-company identification by directly measuring consumer-company identification. Finally, this study duplicates the conclusion drawn by study 1 in different industries, which increases the external validity.

It is not hard to see that only when companies actively take social responsibility, will long-term orientated consumers give a higher evaluation than short-term orientated consumers. And for those indifferent to CSR, long-term and short-term orientated consumers' brand evaluation remain low. Upon that, in the study 3, we will only consider that companies actively take social responsibility to simplify experiment design. According to Spencer et al. (2005), Zhao et al. (2010), the method used to examine the mediating effect through measuring mediating variable has some limitations. Moreover, from independent variable to mediating variable, from mediating variable to dependent variable, these causal chains couldn't be demonstrated in essence. Therefore, the most effective method is directly manipulating mediating variable through experiment. The objective of study 3 is further examining the effect of time orientation and CSR on brand evaluation and mechanism.

5 STUDY 3

5.1 *Method*

153 business school undergraduates (average age: 23.54, SD = 4.94, 86 females) have participated in this study in a silent computer lab. This study employs a 2 (consumer-company identification: High vs. Control) × 2 (Time Orientation: long-term vs. Short-term) between-subjects design. The process details are as follows. Subjects first are allocated randomly to four conditions. Next, they receive the manipulation of time orientation, consumer-company identification and CSR. Finally, manipulation testing and some demographic variable are measured. This study's method to manipulate time orientation is the same as that for study 2's. And CSR is manipulated in the same way as study 1 through Computer Company. Only one condition having information about CSR is included.

Manipulation of consumer-company identification We learned from Bergami & Bagozzi (2000), Sen & Bhattacharya (2001) to manipulate consumer-company identification. To be specific, subjects are told that consumers would usually identify with some company. This identity is the result of the similarity and overlap between consumers' self-awareness and consumers' cognition of company. Generally, when consumers think some company's image is very close to his self-image, he will be highly identified with this company. Next, in high identification condition, subjects are required to consider the computer company in the following description as a computer brand which they are highly identified with. However, in the control condition, subjects are required to consider this computer company as some famous computer brand. To test this method of manipulation, we test it by the scale of consumer-company identification in the study 2. The scores obtained by scale are used as testing indicator of consumer-company identification's manipulation ($\alpha = 0.89$).

5.2 *Results*

Manipulation Check First, we run ANOVAs to verify if the manipulation of time orientation is successful. The results show that only main effect of time orientation is significant ($F(1, 149) = 95.76$, $p < 0.001$) and subjects in the long-term condition think that long-term interest is more important ($M_{long\text{-}term} = 5.44$ vs. $M_{short\text{-}term} = 4.29$), which demonstrates that the manipulation of time orientation is successful. Besides, the main effect of consumer-company identification's manipulation ($F < 1$) and the interaction effect ($F(1, 149) = 1.29$, $p > 0.24$) is non-significant, which demonstrates that the manipulation will not affect subjects' time orientation. For checking manipulation of consumer-company identification, ANOVAs yields a significant main effect of manipulation ($F(1, 149) = 93.17$, $p < 0.001$). In the high identification condition, subjects show higher identification ($M_{High} = 5.28$ vs. $M_{Control} = 3.93$). Thus, the manipulation of consumer-company identification is successful. In addition, the main effect of time orientation ($F(1, 149) = 1.75$, $p > 0.20$) and interaction effect are both non-significant ($F < 1$).

Finally, ANOVAs which are used to check the manipulation of CSR reveal that all the factors will not significantly impact consumers' judgments of CSR ($Fs < 1.97$, $p > 0.16$). To further illustrate the fact that the manipulated CSR information is positive, single sample t-test (comparing variable: indicator of CSR's manipulation, comparing factor: 4, median of 7-scale) has yielded a significant result ($t(153) = 14.31$, $p < 0.001$). Subjects think that company actively took CSR ($M = 5.17$). So, manipulation of CSR is successful.

Hypothesis Testing In line with H3, the manipulation of consumer-company identification can moderate the relationship between time orientation and brand evaluation. We run ANOVAs to verify this hypothesis, with brand evaluation ($\alpha = 0.77$) as dependent variable, the manipulation of consumer-company identification, the manipulation of time orientation and interaction term of the two as independent variables. Although the main effects of consumer-company identification ($F(1, 149) = 6.49$, $p < 0.05$) and time orientation ($F(1, 149) = 7.63$, $p < 0.01$) are significant, we pay more attention to the significant interaction effect of the two ($F(1, 149) = 9.87$, $p < 0.01$). The further analysis reveals that in the control condition, long-term orientated subjects make a higher evaluation on company taking CSR ($t(149) = 4.01$, $p < 0.001$), which is same as the previous two studies. That is, long-term orientated consumers will have a more positive brand evaluation on those who actively take CSR. However, in the high consumer-company identification condition, the difference disappeared (M long-term $= 4.92$ vs. M short-term $= 5.03$, $t(149) = -.42$, $p > 0.79$). This illustrates that consumers who have a high identification with company would have a positive evaluation on company whether company taking CSR or not. Above all, H3 is supported fully.

5.3 *Discussion*

Study 3 not only verifies the effect of time orientation and CSR on brand evaluation, but also verifies the mechanism and demonstrates the effect of consumer-company identification from causation.

6 CONCLUSION AND DISCUSSION

This paper suggests that on account of different time orientation, consumers in different culture will respond differently to the same CSR information. To be specific, in the long-term orientated culture, consumers will have a higher evaluation on company which actively takes CSR; while in the short-term orientated culture, consumers will not have a higher evaluation. The reason why time orientation yields this effect is because long-term orientated consumers are likely to have higher consumer-company identification to company which actively takes CSR while short-term orientated consumers will not. This paper, through three studies which 500 representative consumers have participated in, supports the previous hypotheses. More importantly, time orientation, CSR are manipulated using different methods and the hypotheses are tested in different industries. The results are consistent, which demonstrate that the study has strong external and internal validity.

Accordingly, the contributions this paper makes to relevant fields are as follows. First, this research sheds light on the phenomenon why consumers in different cultures respond differently to the same CSR information. As above, the existing researches just illuminate this phenomenon by description and fail to explain how cultural differences can affect consumers' response through causal chain and point out which dimension of culture lead to this effect (Deng 2010, Murphy & Schlegelmilch 2013). This research offsets the deficiency of existing researches.

Second, this research also contributes a lot to the field of consumer-company identification. An enormous amount of researches find that company can achieve consumers' identification through taking corporate social responsibility (Bhattachaya & Sen 2003). However, a growing number of researches find that it is not possible to adopt this kind of consumer-company identification to fit each case. For example, Sen & Bhattacharya (2001) found that only when consumers' attitudes toward corporate activities of CSR are positive, can corporate activities of CSR increase consumer-company identification. In a word, this research enriches the results of this field further, illuminating that cross-cultural time orientation can affect consumer-company identification and only when consumers are long-term orientated, can CSR increase consumer-company identification.

Then, this paper's findings have important implications for the field of cross-cultural consumer behavior. Compared with other four dimensions, cross-cultural time orientation, to a large extent, is ignored by current cross-cultural researches (Gerecke & House 2013). However, as Hofstede & Bond (1988) have pointed out, cross-cultural time orientation might be the most important dimension ever since, because only this dimension can predict economic development and education achievement. Bearden et al. (2006) proposed that current researches should spend more time on paying attention to time orientation so that we can thoroughly understand cross-cultural consumer behavior. Therefore, this research directly examines the effect of cross-cultural time orientation on consumer' response to CSR, which fills the gap of time orientation in the current researches.

Besides, this research reveals something valuable for corporate marketing and management. For example, consistent with these findings, we suggest that multinational enterprise should pay more attention to social responsibility in the long-term orientated country, because consumers in these countries respond more actively to CSR. Besides, for taking CSR, evoking consumer's long-term orientation can lead to a better reward. For companies which perform poorly in CSR, evoking consumer's short-term orientation can be beneficial. A great amount of researches have demonstrated that consumers have both long-term and short-term orientated, which can be influenced by some situational factor. At last, this research suggests that actively taking CSR should be a competitive edge tool when weak brands compete with strong brands. Generally speaking, consumers are less likely to identify with weak brands, so taking CSR can be very helpful to achieve more consumer-company identification for weak brands.

ACKNOWLEDGEMENTS

This research is supported by the National Science Foundation of China "Corporate Social Responsibility and Consumers' Brand Evaluation: Experimental Research Based on Cognitive Differences and Evaluation Mode" (Project number 71072044).

REFERENCES

Bergami, M. & Bagozzi, R.P. 2000. Self-categorization, affective commitment and group self-esteem as distinct aspects of social identity in the organization. *British Journal of Social Psychology,* 39(4): 555–577.

Biehal, G.J. & Sheinin, D.A. 2007. The influence of corporate messages on the product portfolio. *Journal of Marketing,* 71(2): 12–25.

Brown, T.J. & Dacin, P.A. 1997. The company and the product: corporate associations and consumer product responses. *The Journal of Marketing,* 68–84.

Creyer, E.H. 1997. The influence of firm behavior on purchase intention: do consumers really care about business ethics?. *Journal of consumer Marketing,* 14(6): 421–432.

Deng, X. 2012. Understanding Consumer's Responses to Enterprise's Ethical Behaviors: An Investigation in China. *Journal Of Business Ethics,* 107(2): 159–181.

Drumwright, M.E. 1996. Company advertising with a social dimension: The role of noneconomic criteria. *Journal of Marketing,* 60(4).

Du, S. et al. 2007. Reaping relational rewards from corporate social responsibility: The role of competitive positioning. *International Journal of Research in Marketing,* 24(3): 224–241.

Ellen, P.S. et al. 2006. Building corporate associations: consumer attributions for corporate socially responsible programs. *Journal of the Academy of Marketing Science,* 34(2): 147–157.

Gift, M.J. 2013. Cross-cultural perceptions of business ethics: Evidence from the United States and China. *Journal of business ethics,* 114(4): 633–642.

Hofstede, G. 1980. Motivation, leadership, and organization: do American theories apply abroad. *Organizational dynamics,* 9(1): 42.

Hofstede, G.H. 2001. Culture's consequences: Comparing values, behaviors, institutions and organizations across nations. *Sage.*

Ho, F.N. et al. 2012. A global analysis of corporate social performance: The effects of cultural and geographic environments. *Journal of business ethics,* 107(4): 423–433.

Klein, J. & Dawar, N. 2004. Corporate social responsibility and consumers' attributions and brand evaluations in a product–harm crisis. *International Journal of research in Marketing,* 21(3): 203–217.
Menon, S. & Kahn, B.E. 2003. Corporate sponsorships of philanthropic activities: when do they impact perception of sponsor brand?. *Journal of Consumer Psychology,* 13(3): 316–327.
Murphy, P.E. & Schlegelmilch, B.B. 2013. Corporate social responsibility and corporate social irresponsibility: Introduction to a special topic section. *Journal of Business Research,* 66(10): 1807–1813.
Oberseder, M. et al. 2013. CSR practices and consumer perceptions. *Journal of Business Research,* 66(10): 1839–1851.
Pracejus, J.W. & Olsen, G.D. 2004. The role of brand/cause fit in the effectiveness of cause-related marketing campaigns. *Journal of Business Research,* 57(6): 635–640.
Press, M. & Arnould, E.J. 2011. Legitimating community supported agriculture through American pastoralist ideology. *Journal of Consumer Culture,* 11(2): 168–194.
Sen, S. & Bhattacharya, C.B. 2001. Does doing good always lead to doing better? Consumer reactions to corporate social responsibility. *Journal of marketing Research,* 38(2): 225–243.
Sen, S. 2006. The role of corporate social responsibility in strengthening multiple stakeholder relationships: A field experiment. *Journal of the Academy of Marketing science,* 34(2): 158–166.
Spencer, S.J. et al. 2005. Establishing a causal chain: why experiments are often more effective than mediational analyses in examining psychological processes. *Journal of personality and social psychology,* 89(6): 845.
Torelli, C.J. et al. 2012. Doing poorly by doing good: Corporate social responsibility and brand concepts. *Journal of Consumer Research,* 38(5): 948–963.
Valor, C. 2008. Can consumers buy responsibly? Analysis and solutions for market failures. *Journal of Consumer Policy,* 31(3): 315–326.
Yoo, B. & Donthu, N. 2002. Culture's Consequences. *Journal of Marketing Research,* 3: 388–389.
Yoon, Y. et al. 2006. The effect of Corporate Social Responsibility (CSR) activities on companies with bad reputations. *Journal of Consumer Psychology,* 16(4): 377–390.
Yoo, B. et al. 2011. Measuring Hofstede's five dimensions of cultural values at the individual level: Development and validation of CVSCALE. *Journal of International Consumer Marketing,* 23(3–4): 193–210.
Zhao, X. et al. 2010. Reconsidering Baron and Kenny: Myths and truths about mediation analysis. Journal of Consumer Research, 37(2): 197–206.

Engineering Management and Industrial Engineering – Leung (Ed.)
© 2015 Taylor & Francis Group, London, ISBN 978-1-138-02773-2

The influence of retailer's sales effort in dual-channel competition supply chain

G.J. Ji & X.M. Chen
School of Management, Xiamen University, Xiamen, Fujian, China

ABSTRACT: In dual-channel supply chain, the manufacturer opens a direct channel and sells directly to customers over the Internet to compete with its own retailers, which results in two types of conflicts: Vertical channel conflict between manufacturers and retailers, as well as horizontal channel conflict between direct and retail channel. To avoid these channel conflicts, retailer's sales effort is proposed in this paper. We present the Stackelberg and Bertrand competition models in dual-channel supply chain and compare them, finding that the customer acceptance of the direct channel has great effect on the profits of the manufacturer and retailer under these two scenarios. Our study also indicates that when the retailer adopts sales effort, the manufacturer and retailer will benefit (i.e., show an increased profit) under these two scenarios. Given this, the manufacturer should encourage the retailer to adopt sales effort to realize mutual benefits.

1 INTRODUCTION

Forman et al. (2009) pointed out the benefits of electronic commerce to consumers, including lower prices, greater selection, and greater convenience by eliminating travel costs. It is precisely because more and more people realize this point, also more and more companies are opening direct channels for selling products to customers, in addition to their retailer channels. Research shows that more consumers are embracing multiple channels to satisfy their shopping needs. According to one survey reported in *The New York Times*, about 42% of top suppliers (e.g., IBM, Estee Lauder, and Nike) in a variety of industries have begun to sell directly to consumers over the Internet (see Chiang and Chhajed, 2003). As a result, the interest conflicts of manufacturers and retailers become more complex in dual-channel supply chain. On the other hand, businesses today are facing a generation of increasingly sophisticated customers, namely, they become more and more rational and mature. Therefore, dual channels have attracted many scholars and gradually become a mainstream.

Recently, Bernstein et al. (2008) find in an oligopoly setting that a dual-channel structure, although giving rise to an equilibrium channel structure, does not necessarily imply higher profits for the firms. And channel conflict is the biggest deterrent to the manufacturer's decision to adopt the dual-channel business model. There are a number of ways for manufacturers to alleviate these channel conflicts. Some manufacturers have halted direct sales in order to avoid channel conflict, and the typical example is Levi Strauss & Co. (Chiang and Chhajed 2003, and Collett 1999). Cattani et al. (2006) studied the viability of an equal-pricing strategy in alleviating channel conflict. Mukhopadhyay et al. (2008) proposed one way of eliminating the possibility of this channel conflict, where the retailer is allowed to add value to the product to differentiate its offering to the customers, and they found that the manufacturer suffering from the asymmetry of the information-will always benefit from more information. Mukhopadhyay et al. (2008b) use value-added retailer to alleviate the possibility of channel conflict in a mixed channel hi-tech supply chain. Dan et al. (2012) examined the optimal decisions on retail services and prices in a centralized and a decentralized dual-channel supply chain using the two-stage optimization technique and Stackelberg game. They found that retail services strongly influence the manufacturer and the retailer's pricing strategies. Based on the consumer utility theory in this paper, however, we study retailers who introduce value-added services to reduce channel conflicts and encourage consumers to purchase more under the Stackelberg and Bertrand competition models.

2 THE MODEL SCENARIO

We consider a simple supply chain made up of one manufacturer and one retailer. At the same time, the manufacturer sells products on the Internet. Therefore, customers can purchase products either through the retail channel or through the direct channel. Let p_r and p_d denote the unit price at the retail and direct channel, respectively. In addition, let the wholesale price be charged by the manufacturer to the retailer be w.

Customers are heterogeneous in the valuation of the product and we denote the consumption value by v, and for analytic simplicity, assuming that it is uniformly distributed within the consumer population from 0 to 1. A customer with valuation v would derive a net customer surplus of $v-p_r$ by buying the product. Similarly, the product is sold at the direct channel price (p_d) and the value of the product is βv, parameter $\beta\,(0 < \beta < 1)$ is called the customer acceptance of the direct channel. In this paper, we address the retailer's sales effort to make customers purchase more and the levels of effort is denoted by e. Then we use the functional form $ke^2/2$ as the cost of exerting effort.

3 SCENARIO A: STACKELBERG MODEL

In this section, we suppose that the manufacturer and retailer are modeled using the familiar Stackelberg game theory. Thus, the moves of manufacturer and retailer follow: the manufacturer will act as the leader by announcing direct channel price p_d first, and the retailer follows by announcing the retail price p_r, knowing the leader firm's already declared strategy. We further assume that there is no marginal cost of product and the manufacturer and retailer are rational.

3.1 *Without retailer's sales effort*

The customer with product valuation would derive the utility $u_r = v-p_r$ by buying the product from the retail channel, and $u_d = \beta v-p_d$ from the direct channel. Customers decide whether to buy a product, and where to buy it through analyzing product availability, choosing the option that maximizes their utility. Thus, if $u_r = u_d$, buying from the retail channel is indifferent to buying from the direct channel. If $u_r > u_d$ and $u_r > 0$, consumers would prefer the retail channel and if $u_r < u_d$ and $u_d > 0$, they would prefer the direct channel. In this paper, we study each channel in demand, so if and only if $u_r < u_d$ and $u_d > 0$, namely, $0 \le (p_d/\beta) < v < (p_r - p_d)/(1-\beta)$, both of these two channels are in demand. Then the demand of the retailer and the direct channel are, respectively,

$$q_r = 1-\frac{p_r - p_d}{1-\beta},\quad q_d = \frac{p_r - p_d}{1-\beta} - \frac{p_d}{\beta}. \tag{1}$$

Next, we obtain the profit function of the members of the supply chain as follows:

The retailer's profit function:

$$\pi_r = (p_r - w)q_r = (p_r - w)\left(1-\frac{p_r - p_d}{1-\beta}\right) \tag{2}$$

The manufacturer's profit function:

$$\begin{aligned}\pi_m &= wq_r + p_d q_d \\ &= w\left(1-\frac{p_r - p_d}{1-\beta}\right) + p_d\left(\frac{p_r - p_d}{1-\beta} - \frac{p_d}{\beta}\right)\end{aligned} \tag{3}$$

It is straightforward to get the optimal policies by backward induction:

$$p_r^A = \frac{4+4w-5\beta+\beta^2}{8-4\beta},\quad p_d^A = \frac{\beta(-1-2w+\beta)}{2(-2+\beta)},$$
$$q_r^A = \frac{-4+4w+\beta}{4(-2+\beta)},\quad q_d^A = \frac{1}{4}. \tag{4}$$

3.2 *Retailer's sales effort*

Similarly, we get the utility $u_r = v - p_r + \theta e$ and $u_d = \beta v - p_d$, and parameter θ is the effort sensitivity. As a result, we get the demand of the retailer and the direct channels are, respectively,

$$q_r^e = 1-\frac{p_r - p_d - \theta e}{1-\beta},\quad q_d^e = \frac{p_r - p_d - \theta e}{1-\beta} - \frac{p_d}{\beta}. \tag{5}$$

Therefore, the profit function of the retailer is,

$$\begin{aligned}\pi_r &= (p_r - w)q_r - \frac{ke^2}{2} \\ &= (p_r - w)\left(1-\frac{p_r - p_d - \theta e}{1-\beta}\right) - \frac{ke^2}{2}\end{aligned} \tag{6}$$

The profit function of the manufacturer is,

$$\begin{aligned}\pi_m &= wq_r + p_d q_d \\ &= w\left(1-\frac{p_r - p_d - \theta e}{1-\beta}\right) + p_d\left(\frac{p_r - p_d - \theta e}{1-\beta} - \frac{p_d}{\beta}\right)\end{aligned} \tag{7}$$

In such a situation, we get

$$p_r^{eA} = \frac{4+4w-5\beta+\beta^2+4\theta e-3e\beta\theta}{8-4\beta},$$

$$p_d^{eA} = \frac{\beta(-1-2w+\beta+\theta e)}{2(-2+\beta)},$$

$$q_r^{eA} = \frac{4+4w(-1+\beta)+\beta^2+4\theta e-\beta(5+3\theta e)}{4(2-3\beta+\beta^2)},$$

$$q_d^{eA} = \frac{2+8w(-1+\beta)-\beta^2+6\theta e-\beta(1+5\theta e)}{4(2-3\beta+\beta^2)}. \quad (8)$$

4 SCENARIO B: BETRAND MODEL

In this section, the manufacturer and retailer are equal, and they are modeled using the familiar Bertrand game theory. Thus, the moves of manufacturer and retailer follow: the manufacturer decides the direct channel price p_d, while the retailer determines the retail price p_r.

4.1 *Without retailer's sales effort*

Under (2) and (3), we maximizing the manufacturer's and retailer's profits, we get

$$p_r = \frac{1+p_d+w-\beta}{2} \quad (9)$$

$$p_d = \frac{(p_r+w)\beta}{2} \quad (10)$$

Substituting p_d into Equation (9), resulting the optimal prices and demand are given by:

$$p_r^B = \frac{2-2\beta+w(2+\beta)}{4-\beta}, \quad p_d^B = \frac{-\beta-3w\beta+\beta^2}{-4+\beta},$$

$$q_r^B = \frac{2(-1+w)}{-4+\beta}, \quad q_d^B = \frac{-1+w}{-4+\beta}. \quad (11)$$

4.2 *Retailer's sales effort*

Similarly, given in Equation (6) and (7), we get

$$p_r^{eB} = \frac{-2+2\beta-w(2+\beta)-2\theta e+e\beta\theta}{-4+\beta},$$

$$p_d^{eB} = \frac{\beta(-1-3w+\beta+\theta e)}{-4+\beta},$$

$$q_r^{eB} = \frac{2+2w(-1+\beta)+2\theta e-\beta(2+\theta e)}{4-5\beta+\beta^2},$$

$$q_d^{eB} = \frac{1+w(-1+\beta)-\beta-\theta e}{4-5\beta+\beta^2}. \quad (12)$$

5 COMPARISONS AND ANALYSIS

In this section, we will compare the optimal prices and demands in Scenarios A and B.

5.1 *Comparisons*

We will analyse and compare the above-mentioned solutions. These results are given in the next four propositions.

Proposition 1. *Under both cases (A) and (B), the retail price is always lower when the retailer doesn't adopt sales effort compared with sales effort, i.e.,* $p_r^{eA} > p_r^A$, $p_r^{eB} > p_r^B$. *On the contrary, the direct price will be decreased, i.e.,* $p_d^{eA} < p_d^A$, $p_d^{eB} < p_d^B$.

Proposition 2. *Without retailer's effort, in Scenario A, when* $4w < \beta$, *the retail and direct price are always higher than in Scenario B, i.e.,* $p_r^A > p_r^B$, $p_d^A > p_d^B$.

As we saw, Proposition 1 shows that when the retailer adopts sales effort, customers are more willing to buy products from the retail channel, then this allows the retailer to charge a higher price and increase her profitability. On the other hand, the manufacturer set to lower a direct price and online sales may be adversely affected.

Proposition 3. *In scenario A, when the retailer adopts sales effort, its demand will be increased, i.e.,* $q_r^{eA} > q_r^A$. *In scenario B, sales volume in the retail channel will also be increased, while it will be decreased in the direct channel, i.e.,* $q_r^{eB} > q_r^B$, $q_d^{eB} < q_d^B$.

Proposition 4. *Without retailer's effort, in Scenario A, when* $4w < \beta$, *the retail price is higher than in Scenario B, while the direct price is lower in Scenario B, i.e.,* $q_r^A > q_r^B$, $q_d^A < q_d^B$. *Furthermore, in Scenario B, when the retailer adopts sales effort, the total market demand will be higher, i.e.,* $q_r^{eB} + q_d^{eB} > q_r^B + q_d^B$.

It is interesting to see that when the retailer adopts sales effort, the price and the demand in the retail channel will be increased under both cases (A) and (B). As a result, the retailer will make more profits and have more incentive to use sales effort.

5.2 *Numerical analysis*

In this section, we perform a numerical analysis to verify our results and gain more insights into the optimal policies by comparing the profits of the manufacturer and retailer under the two scenarios, respectively. The numerical values and rang of values used in this experiment can be seen in Table 1.

Based on the numerical study, we obtain profits of the retailer and manufacturer as shown in Figures 1–4.

Table 1. The numerical values and rang of values.

w	k	e	θ	β
0.2	0.4	1	0.5	0–0.9

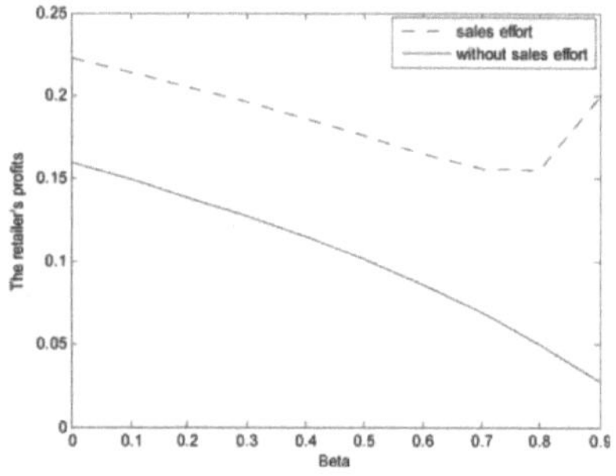

Figure 1. The retailer's profits under Stackelberg model.

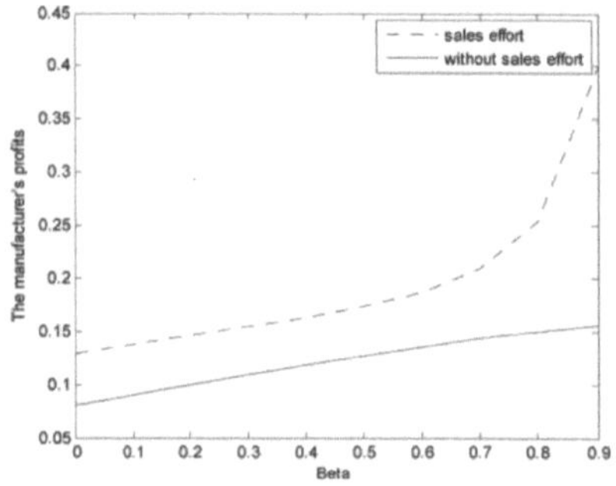

Figure 2. The manufacturer's profits under Stackelberg model.

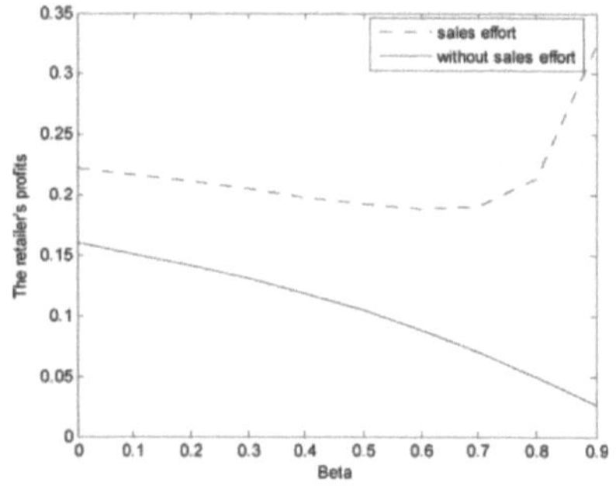

Figure 3. The retailer's profits under Bertrand model.

Figures 1–4 show that the retailer's profits decrease and the manufacturer's profits increase with β without retailer's sales effort under the Stackelberg and Bertrand competition models. That means when the manufacturer focuses on the customer acceptance of the direct channel to increase revenue, the retailer's profit will be decrease. However, it is interesting to see that when the retailer adopts sales effort, the retailer's profit will be increased in some cases.

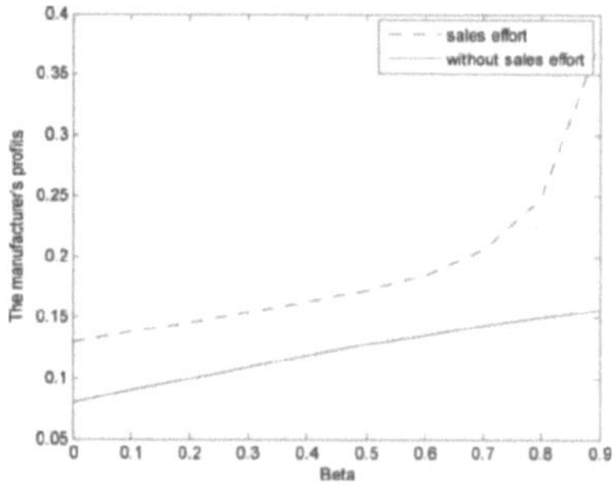

Figure 4. The manufacturer's profits under Bertrand model.

At the same time, we find that when the retailer does not adopt sales effort, the profits are almost equal. Moreover, the profits of the manufacturer and retailer are always larger than that without retailer's sales effort under the two scenarios, and the profits under the Bertrand competition model will add more compared with the Stackelberg model.

6 CONCLUSIONS

In this paper, we study the retailer's sales effort and discuss whether it will eliminate channel conflicts under the Stackelberg and Bertrand competition models. The main results as follow:

1. The profits of the manufacturer and retailer are almost equal under the two scenarios.
2. In some cases, when the retailer adopts sales effort, the manufacturer and retailer will increase profits as the customer acceptance of the direct channel increases.
3. What's more, the retailers have an incentive to adopt sales effort and it is also profitable for the manufacturer.

As a consequence, the channel conflicts will be eliminated by using sales effort. Our model can be extended in many different directions, such as we can also allow the manufacturer to adopt sales effort, instead of the retailer.

ACKNOWLEDGEMENTS

This research was supported by the National Natural Science Foundation of China (NSFC) (Project Nos. 71371159 and 71201138), the National Ministry of Education of China (Humanities and Social Science Project No. 12YJC63026) and the Fujian Provincial Natural Science Foundation (Project No. 2012 J01304).

REFERENCES

Bernstein, F., J.-S. Song, X. Zheng. 2008. "Bricks-and-mortar" vs."clicks-and-mortar": An equilibrium analysis. Eur. J. Oper. Res.187(3):671–690.

Cattani, K., W. Gilland, H.S. Heese, J. Swaminathan. 2006. Boiling frogs: Pricing strategies for a manufacturer adding a direct channel that competes with the traditional channel. Production Operations Management 15(1):40–56.

Chiang W.K, Chhajed D. 2003. Hess J.D. Direct marketing indirect profits: A strategic analysis of dual-channel supply chain design. Management Science. 49(1):1–20.

Collett, S. 1999. Channel conflicts push Levi to halt Web sales. Computerworld. 33(45):8.

Dan B., Xu G., Liu C. 2012. Pricing policies in a dual-channel supply chain with retail services. International Journal of Production Economics. 139(1): 312–320.

Mukhopadhyay S.K, Yao D.Q, Yue X. 2008b. Information sharing of value-adding retailer in a mixed channel hi-tech supply chain. Journal of Business Research. 61(9): 950–958.

MukhoPadhyay S.K, Zhu X, Yue X. 2008a. Optimal contract design for mixed channels under information asymmetry. Production and Operation Management. 17(6): 641–650.

Engineering Management and Industrial Engineering – Leung (Ed.)
© 2015 Taylor & Francis Group, London, ISBN 978-1-138-02773-2

Strategic ambiguity is the biggest risk of "going out" work

Y. Zhang
Wuhan University of Technology, Wuhan, China

ABSTRACT: "Going out" is an important part of China's foreign economic work. The advance strategic planning will play a significant role in guiding the overall situation of the entire work. At present, only some guiding documents are available in China, which cannot meet the requirement of normalization and mandatory standard for such work. Because of the lack of mandatory standards, enterprises tend to make profits by damaging the national interests in actual operation, which is a serious breach of the intended objectives of the "going out" work. In order to promote the "going out" work in a healthy way, strict standards should be explicitly formulated in the advance strategy to standardize sectors such as enterprises and the industry to prevent vicious internal competition and unreasonable administrative intervention. Meanwhile, the interests should be well coordinated and united objective should be set up and finally an effective strategic mode is formed, which is a cornerstone for implementation of the "going out" work.

1 INTRODUCTION

At present, China's economic development has run into its bottleneck of transformation. "Going out" policy can, on one hand, help release part of the domestic development and on the other hand, improve our competitive strength by learning and borrowing advanced management and development experience from foreign peers. As the core of "going out" work, enterprises lead to the actual state that the goods "made in China" represent the goods which are inferior and are of poor quality instead of agglomerating development strength. On one hand, China's economy is in rapid development both in domestic and international markets; on the other hand, overall development strength of Chinese brands in international market has been in continuous overdraft. The most principal cause for the increasingly apparent and intensified conflict is the strategic ambiguity of China's "going out" work.

2 TO CLARIFY REASONS FOR STRATEGIC AMBIGUITY

First of all, the "going out" work belongs to an economic behavior but is also beyond the economic behavior. Economy, military affairs, diplomacy and culture are the extensions of national politics, which should comply with the country's political overall needs. The "going out" work, which is the guiding work for the time being, should be carried out under the premise of political needs. National political needs should be seen as the precondition during preparation of the strategy. Second, red lines should be explicitly stipulated during preparation of the strategy of the "going out" work. That is to say, what should not be done must be clarified, which should be decided and included in standard by the State-owned Assets Supervision and Administration Commission and all industry associations by negotiation. Third, professionals should be assigned to complete professional work in enterprises where "going out" policy is enforced. Especially, some 'window companies' tend to undertake every possible project and find relevant organizations in China to subcontract it from one to another no matter whether they are capable to operate a certain project or business. Consequently, the geometric growth of quality control requirements is caused which easily leads to quality problems under various foreign risk factors and hence damage the image of Chinese brands. Finally, the "going out" work is a systematic and comprehensive economy improvement plan but a new mainland nuggets action. What we need is a peaceful win-win development with mutual benefits rather than "Monoecism" only for profit!

2.1 *Define each relevant interested parties*

The State-owned Assets Supervision and Administration Commission is an institution, which represents national interests. As the preparer of the whole strategy, also an operator of economic refinement of overall national strategies, it plays an irreplaceable role. But due to limitations of its vision, it fails to put forward timely overall strategies and relevant refinement requirements because of its own needs and secondary data.

The industry association is a public institution, which represents the interests of a certain industry. As a participant which forms a connecting link between the preceding and the fully grasping, it has all the overall information of a certain industry and plays a significant role in providing relevant information to the State-owned Assets Supervision and Administration Commission during policy making process. In later period, it could put forward reasonable improvement suggestions regarding whether the policy is suitable for the current economic situation and whether the policy is exercisable in the enterprise, which is conducive to cycle improvement of the overall policy and helps promote overall implementation of the work. [1]

Enterprises represent the interests of a certain executing agency. As the actual performer of strategy, enterprises are the operator of the "going out" work. At the same time, enterprises are directly subject to the specific problems regarding whether the policy is suitable for the actual market requirement and possesses firsthand document and data. Thus, they are, as the provider of information, the basison which the adjustment of the policy is made.

Entrepreneurs represent the interests of the individual. The individual may the entrepreneur himself or any employee working in the enterprise. National interests and enterprise's interests may sometimes damage the interests of the individuals, but the degree is decided by the entrepreneur according to the actual situation of the enterprise. For example, under normal circumstances, daily work time is 8 hours, but some clerks may work for 7 hours and go off work, some other employees taking up special positions may work more than 8 hours and even work until midnight. This should be decided by the entrepreneurs during actual operation and safeguard the interests of individuals' in the enterprise within reasonable scope. If the preparation of the strategy or relevant policy damage individual's interests and the strategy is not exercisable, entrepreneurs must come to safeguard the interests and rights of those who are concerned. [2]

2.2 *Each party gives prominence to their respective interests by stage management of their own*

Because each party represents different interest group and they have different demands, the State-owned Assets Supervision and Administration Commission takes the whole development trend into consideration and is not powerful in control of the factors in specific operational level. Besides, it does not have the need. However, it will certainly ask the enterprise to give up part of the benefits to meet the requirement of the general direction. In addition, it does not exert series of restrictions on entrepreneurs, especially leaders of state-owned enterprises. Consequently, there must be conflicting factors among them.

Industry association may formulate corresponding rules for certain industries to regulate behaviors of relevant enterprises. But a large percentage of the working capital of the association is dependent on the membership dues and investment support of the enterprise. Thus, strict industry standards will result in business development behavior of the enterprises, which eventually leads to dramatic decrease in the membership dues and investment support. At present, some industry associations not only impose corresponding restraints on the enterprise, but also apply to the government for various policies and modify the industry standards, loosen up relevant requirements of the enterprise as the corporate lobbyists in order to "promote" development of the enterprise for the purpose of its own normal running. Such conduct of theirs is in serious violation of the intended meaning of its existence.

In western economics, enterprises are profits-seeking organizations and are profit-driven in nature, which is understandable. However, when Chinese enterprises are implementing "going out" policy, especially various state-owned enterprises, they are representing the image of the whole industry or even the nation, and having certain influence on the politics, diplomacy, culture and military affairs and other aspects of China. Thus, it is strictly prohibited for foreign-related enterprises especially state-owned enterprises to pursue the "savage growth" of the internet economy and corresponding strict control should be conducted at the enterprise level. [3]

Entrepreneurs', actual trader of the enterprises, personal will could influence all details at various operation level of the enterprises. In actual operation, interests would flow correspondingly and finally only orientation of interests would be decided by entrepreneurs. Whether the "going out" strategy can be realized and its abidance or avoidance oftentimes is dependent on the pure thought of the entrepreneurs. In actual operation, entrepreneurs would certainly emphasize interests of the enterprises' and the individuals' and stretch up the policy space for convenience of later operations.

2.3 *Money drives out good money which hinders long-term development*

In accordance with the logic, any routes formulated by the only principle is seeking profit; entrepreneurs care about personal interests most, enterprise cares about individual interests and industry association pursue organizational interests most. Because of this, they deviate from the State-Owned Assets Supervision and Administration, which represents the national interests. As the saying goes where there is a policy, there is a countermeasure. In specific operation level, even as early as in the design period, each party will vigorously develop space conducive to their own development and damage the overall expected requirement. In short, this is internal friction.

The demand of interests of individuals' and organizations' must be squarely looked at during preparation for the strategy of "going out" work. Safeguarding the reasonable interests of each involved party has always been the insurance of the "going out" work. Only by integrating various interests organically and set the unified objective in the same direction, can the preparation of "going out" strategy be seen as successful.

3 TO UNIFY DIRECTION OF INTERESTS FACTORS

3.1 *Establish development path for "people"*

Any rules would be executed by someone. Due to various complicated factors, it is easy to form conflict between individuals and organizations. Even though clear laws and regulations and prompt supervision are available, the behavior of damaging long-term interests of the organizations' by personal interests still cannot be completely eliminated. Therefore, it is the first task to coordinate the personal interests and the organizational interests for maintaining the current strategic objective. [4]

It is the best choice to solve the contradictory between individual and organization by including personal career development planning in the organization's development goals and making individual interests closely integrated with that of the organizations. The stock ownership incentive in the form of futures executed by many foreign enterprises has set a good example in this field. Those enterprises stimulate employees' work enthusiasm by promising the employees that they can buy employee stock if they reach the expected target in their work. Employees who have been working in the enterprise for a longer time and having a higher post can acquire more of the enterprise's stocks; they will have higher consciousness to maintain the interests of the enterprises'. As such, stable and healthy long-term development of the enterprises can be ensured to the greatest extent and impediment to the long-term development caused by seeking of short-term interest can be avoided.

Chinese enterprises and industry associations can learn some lessons from this mode of including the personal career development planning and personal interests in organizational development planning and reformulate a new interests distribution model and system and maintain personal reasonable interests under the premise of maintaining long-term development of the organization.

3.2 *Establish development path for "organizations"*

Both enterprises and industry associations are a form of organizations, which will be developed with the structure and the extension be expanded as the development, function and influence increased. The development of the organization must also be achieved with a clear direction, which must be in conformity with the needs of the country's development and the need. In enterprises, industry associations and other groups which implementing the "going out" policy, the direction is to meet the needs of the development of the current national economy.

The development direction of enterprises must be conducted under the guidance of the industry associations' to prevent vicious competition between enterprises which could decrease the whole nation's competitiveness. Industry associations must be expanded gradually to the extent that they can not only affect the domestic enterprise behavior, set up domestic industry standards, but also affect the overseas enterprises and play a role in gaining international voice. Take FAA as an example, as an institution in USA, the Federal Aviation Administration can grant flying license to all the aircrafts all over the world. Without its permission, all the aircrafts will not be allowed to enter the civil aviation. This is a powerful example of a third-party organization which has the ability to control the whole industry.

Preparation of development plan of enterprises' and industry associations' is an important part of "going out" strategy. Only by integrating the interests of each party's into a strategic direction, can the final goal be achieved. [5]

3.3 *Establish examination and supervision mechanism*

Strict examination and supervision mechanism should be applied to all organizations and individuals. No matter how noble a man is, he may fall into corruption in the environment with no supervision. No measures with differential treatment such as exemption from inspection, submission for inspection and certification should be set up and a series of random checks should be carried out periodically. Punishment should be strengthened and cost of illegal or irregular act should be increased to form the situation [Q] where any individual or organization who breaks the rules will suffer from major irretrievable loss so that they would hold their evil intentions in awe. In this way, we can safeguard interests of the organizations and ensure the strategy does not deviate from the fixed direction.

4 SET UP LONG-TERM HIGH-EFFICIENT STRATEGY

4.1 *Clear behavior boundaries*

Clear behavior boundaries should be applied to each organization. Industry association is an organization that is responsible for safeguarding legitimate rights of the industrys', enterprises' and consumers'. At the same time, they cannot damage the legitimate rights of others because they are protecting the interests of any party. For instance, the reduction of industry standard by some industry associations for the purpose of protecting the interests of the enterprises' is actually a negligent act which can seriously damage the interests of consumers' and the whole industry's long-term development. Temporary easing industry product standards can indeed help enterprises to reduce costs in the short term and create advantages, but it also can make enterprises become dependent and reduce their enthusiasm of upgrading. Consequently, it pushes the customers to overseas competitors while reducing the core competitiveness of enterprises', which will causes severe damage to the long-term development of the industry's.

For enterprises, pursuit of profit becomes the only standard in the absence of clear behavior boundaries. Grabbing short-term interests, walking on the edge of the law and establishing a monopoly environment have become the most convenient and simplest way for profit. These behaviors are completely useless in promoting the industrial transformation and upgrading enterprises. It can only form a development pattern with heavy "protectionism" color. This pattern is comparatively marketable in domestic development in a short period and even relatively competitive. But when it comes to international market, it will become nothing but a joke. In the international market, only by enhancing their core competences, developing comparatively advanced scientific research and technology, reserving talents and industrial capacity, can the enterprises compete with the foreign rivals.

Hence, clear behavior boundaries have become an integral part to the long-term high-efficient strategy because they can promote the suitable development of the organization and advantage training.

4.2 *Set development expected targets*

Phased goal is the most convenient method for testing the strategy realization path. All deficiencies and institutional defects can be fixed in a timely manner during the implementation of the strategy through the establishment of phased goals. Disturbance leading to institutional defects is surely to be met during the implementation of the strategy, because all factors cannot be listed in the early stage of the system formulation. Problems are not terrible itself, what is terrible is that the wrong development plan is still executed after the problem occurs. Timely and effective adjustment is important to ensure long-term goals.

4.3 *Divide development range*

In the actual market development process of different enterprises, market overlapping is surely to be met because the information they acquired is not comprehensive. If the market overlaps in the same field, especially the low-end market with low technology content and low technical level, many enterprises would jostle against each other, which would cause losses on all sides. Obviously, such competition against each other by low price will benefit the owners and cause losses in research and development cost, reasonable profits and development potential of their own. To prevent the loss of profit competition, reasonable division of the development range, especially between the state-owned enterprises and large state-owned enterprises in overseas markets is the priority of this work.

4.4 *Reduce administrative intervention*

Due to historical reasons, our country has set up some "window" enterprises in the early stage of foreign-related business planning, such as foreign economic and trade corporations, import and export enterprises and other related businesses. In planned economy or partially open market economy, the existence of those "window" enterprises is necessary and important. But at present, our country is fostering international multinationals and national brands, such companies with heavy color of protectionism which possesses symmetric advantage by relevant national policies will seriously damage the enterprises which compete with them by their own strength.

Take a foreign engineering market for example. A certain window enterprise who does not have qualifications of enterprises in this industry or even not capable of engineering construction can obtain the advantage of national preferential loans. Then it signs many contracts on EPC engineering projects in this country at a lower price and subcontract those projects to companies in China with relevant engineering implementing competence. It seeks profits of its own by helping the owners in importing relevant equipment and products and lowering the subcontracting reasonable profits of the enterprise. The window enterprise ensures interests of its own and the owner but damages that of companies' with international market development potential and overall development potential of the national economy. In the long run, such kind of enterprises will seriously hinder the healthy development of our country's economy and cause severe damage to the development capacity of related enterprise and China's national brand image.

5 CLARIFY THE STRATEGY TO ADVANCE THE DEVELOPMENT

Always setting up strategy first is a timeless truth. A miss is as good as a mile is the very explanation for the importance of strategy. China's economy is with great volume and differs greatly in region and industry. Besides, there are many participants and organizations. Consequently, conflicts regarding their demand and interests arise. Only by unifying thought and clearing the development path and strengthening supervision and control, can each participant have rules to abide by and operate in the specific scope. Formulating clear strategies will greatly promote the work of "going out", help the development of economic construction and lay a solid foundation for realization of Chinese dream in future.

REFERENCES

[1] Li Guoli, A Brief Discussion on Risk Management of Construction Engineering Supervision, 2009(8): 88–90.

[2] J. Rodney Turner, translated by Qi Anbang etal. Contract Management in Engineering [M], Tianjin: Nankai University Press, 2005.

[3] Hou Lixin, Project Risk Management Factors and Prevention Measures, construction, 2013(3): 145–147.

[4] Tao Feijia, Sun Ming and Zhou Guoqiang, Thinking on Project Risk Management of Construction Enterprise [J], Engineering construction and management, 2008(3):148–149.

[5] Zheng Yanping, A Study on Countermeasures for Construction Engineering Project Risk [J], Shanxi construction, 2001 7, 33:215–216.

Engineering Management and Industrial Engineering – Leung (Ed.)
© 2015 Taylor & Francis Group, London, ISBN 978-1-138-02773-2

Establishment and evaluation of index system for talent flow

H. Lu
Shanghai Finance University, Shanghai, China

L.Y. Xiao
Staffordshire University, Staffordshire, UK

ABSTRACT: Index system described in this thesis is established by taking the north of Jiangsu province as the researching sample and talent migrating as the researching object. Then, the author evaluates the talent migrating in this region with the index system. The research conclusion is shown as below: severe flight of talents with high education and skilled technicians, drain of backup-brain; inburst of labors with low education and qualities; the tendency of talent migrating in the north of Jiangsu province features severe brain drain and deterioration of labor structure.

Keywords: talent migrating; index system; evaluation

1 INTRODUCTION

Today with the rapid development of knowledge economy, human resources are increasingly become the key factor that affecting regional economic development. In September 1995, the World Bank bright out a set of new index system calculating the national wealth based on a report entitled "The Supervision of Environmental Progress—on Work Progress". In this system, a country's total wealth is mainly constituted by three parts—creative wealth, human resources and natural property. This system considers a country's human resources not only as an important part of the country's wealth, but also as the most critical factor of the growth of a country's wealth. The system considers that a country can realize the increase of creative wealth through developing the human resources, improving the natural resources, using efficiently the human resources and increasing the investment of human resources. Therefore, the status of economic development and growth speed of wealth of an area is closely related to the status of the human resources and human resources flow, especially the talents' flow.

At present, some scholars at home and abroad have done a series of researches and discussed about our regional labor force flow, mainly emphasizing on the qualitative analysis test. Evaluation on the regional labor force flow is not only some index comparison, but the evaluations proceeding measurement and integration apply statistic methods. It needs establishing a reasonably objective integrated evaluation system which should include existing talents and reserved talents, static index and dynamic index, and it will found an index system of regional labor force integrative evaluation. This paper based on the contents talents' flow and the full analysis on the content establishes the integrative evaluation index system of regional talents' flow, and evaluates on the talents' flow in the north of Jiangsu province.

2 CONTENTS OF TALENTS' FLOW

2.1 *The definition of related concept*

Among all kinds of concept related to the labor force sources, population resources is the biggest one which means the sum of the population lives in certain social, certain areas. Labor or human resource is the resource which can promote the development of the entire economy and society and consist the sum of population resources. However, human capital is a comparatively abstract concept that emphases on its capital property. In practice it refers to those kinds of laborers possessing invest profits that possess professional knowledge and skills through education, training of labor or human resources, and usually that is our all talents and talent resources. This paper divides our labor resources, also called as human resources, into general labor forces and knowledge labor forces[1]. General labor forces mainly refer to the laborers whose working patterns are manual labors without or less through education, training. Knowledge labor forces refer to the labors taking in complex work with a lot of intelligence, knowledge, skills, abilities and other forms of expression, which is equivalent to the concepts of human capital, talents, human resources and other concepts.

2.2 *The connotation of talents and talents' flow*

The talents of a region include existing talents and reserved talents. Existing talents include the labors with intermediate or above intermediate skills or the labors without the education of junior college or above it, which is above the knowledge labor forces. Reserved talents refer to the college students whose local household registration is not local or the college students who are studying locally with junior college or above education background, including the students in abroad. The talents flow in some region refers to the inflow or outflow of talents within a certain time and certain region, including the changes of existing talents and reserved talents. The existing talents inflow refers to the regional inflow or outflow of the laborers who have the junior or above junior skills, or with the education of junior college or above that. The change of the reserved refers to the returning rate of graduating college students and post-graduates with local household registration and the inflow number of non-local graduating college students and post-graduates.

3 THE ESTABLISHING OF TALENTS' FLOW INDEX SYSTEM

3.1 *The establishing rules of index system*

Establishing the index system of the brain must abide by the following principles[2]:

3.1.1 *The principle of comprehensiveness*

The principle of comprehensiveness refers to the established index system being able to fully reflect the real situation of talents' flow in the region. The established index system not only reflects the static conditions of talents' flow in certain regions, but also reflects the dynamic conditions of talents' flow. It not only reflects the quantity characteristics of talents' flow in certain area, but also reflects the quality characteristics of talents' flow in certain area.

3.1.2 *The principle of dynamic*

The principle of dynamic refers to such a principle that the conditions of talents' flow are constantly changing, and the index system reflecting the regional talents' flow conditions reflects the talents' flow conditions which must change with the objective circumstances. It is dynamic and the index system must be adjustable, not static.

3.1.3 *The principle of scientificalness*

The principle of scientificalness refers to the established index system which must be scientific and reasonable, and the number, weights of the index and the talents' flow conditions reflected by the index system must be comprehensive, reasonable, reliable, primary and secondary.

3.1.4 *The principle of concreteness*

The principle of concreteness refers to every designed index of the index system must be specific, measurable, and the established index system is not abstract and it can reflect the real, concrete situation of talents' flow of some region and the talents' flow of some region reflected and measured by it must have the time limit, within a given time.

3.1.5 *The principle of accuracy*

The principle of accuracy refers to that index system can precisely reflect the talents' flow conditions and its changing tend.

3.2 *Comprehensive evaluation on index system establishment of the brain drain situation*

The establishment of the index system reflecting talents' flow conditions of some certain region must reflect the following aspects of the situation: 1. the static index of existing talent level in a region includes the descriptions of quantity and quality condition; 2. the static index of reserved talents in a certain region; 3. the dynamic index of existing talents in a certain region; 4. the dynamic index of reserved talents in a certain region.

The static indexes reflecting the existing talents conditions mainly are: the national average level of education of the labor force, an average education level of regional labor force, an average education level of regional outgoing labor force, an average number of people having junior college and above education per ten thousand people, an average number of labor force having intermediate and above skills per ten thousand people.

The static index reflecting the reserved talents conditions mainly are: An average number of students accepting junior college and above per year in recent five years.

The dynamic index reflecting the existing talents changing trend mainly are: the number of the inflow talents of a certain region, the number of the inflow labor force with junior college and above education, the number of the inflow labor force with intermediate and above skills; the number of the outflow talents of a certain region, the number of the outflow labor force with junior college and above education, the number of the outflow labor force with intermediate and above skills.

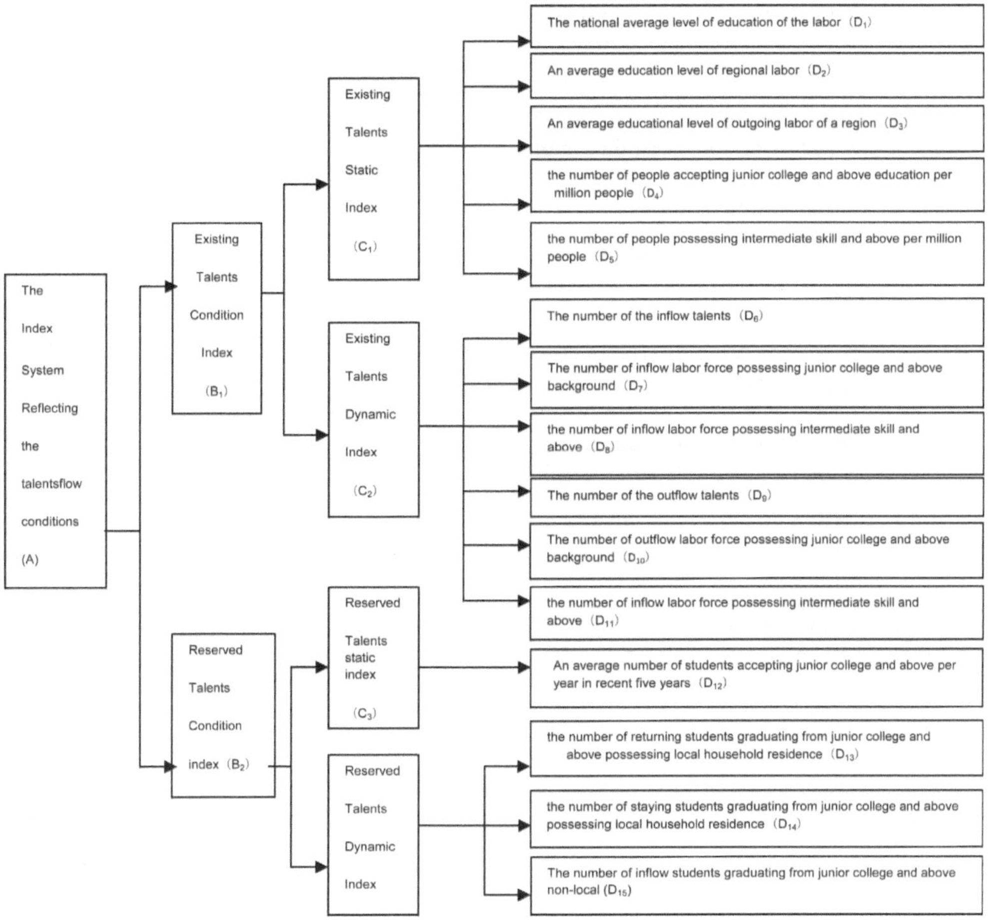

Figure 1. Level structure of the index system reflecting the talents' flow condition of some area.

In general, the dynamic index reflecting the reserved talents changing trend are: In recent five years, with an average number of junior college and above students accepted per year, the number of returning students graduating from junior college and above possesses local household residence are enriching the number of inflow students graduating from junior college and above non-locally. The establishment of talents' flow index system bases on the above analysis, as shown in Figure 1.

4 THE EVALUATION METHODS OF TALENTS'FLOW INDEX SYSTEM

4.1 *The conversion and correction of index system*

To be more precisely and more sensitively reflecting the talents' flow, we do some further conversion and correction on the 15 indexes of Figure 1. Shown as follows, indexes are converted into the following nine indexes. The conversion results as follows:

4.1.1 *The static index of existing talents*

The ratio of regional labor force educational average level and the national average level is: D_2/D_1, ordering $D_2/D_1 = E_1$;

The ratio of regional labor force educational average level and the outgoing labor force educational average level is: D_2/D_3, ordering $D_2/D_3 = E_2$;

The ratio of the educational level of junior college and above per ten thousands people and the national average level: D_4/D_{G4}, ordering $D_4/D_{G4} = E_3$;

The ratio of the level of labor forces with intermediate and above skills per ten thousand people and the national average level: D_5/D_G, ordering $D_5/D_{G5} = E_4$.

4.1.2 *The dynamic index of existing talents*

The ratio of the number of inflow talents and the number of outflow talents of some certain region: D_6/D_9, ordering $D_6/D_9 = E_5$.

The ratio of the number of inflow labor force with junior college and above education and the number of outflow labor force with junior college and above education: D_7/D_{10}, ordering $D_7/D_{10} = E_6$.

The ratio of the number of inflow labor force with intermediate and above skills and the number of outflow labor force with intermediate and above skills: D_8/D_{11}, ordering $D_8/D_{11} = E_7$.

4.1.3 *The static index of reserved talents*

The graduate percentage of the returned working reserved talents and the stay working reserved talents (efficient index of reserved talents) is: $(D_{13} + D_{14})/D_{12}$, ordering $(D_{13} + D_{14})/D_{12} = E_8$.

4.1.4 *The dynamic index of reserved talents*

The ratio of the number of inflow graduate with non-local household register and the number of outflow reserved talents is (dynamic balance status index of reserved talents): $D_{15}/[D_{12} - (D_{13} + D_{14})]$, ordering $D_{15}/[D_{12} - (D_{13} + D_{14})] = E_9$.

4.2 *Evaluation criteria and index score of every index*

Index evaluation standard can divide into 5 ranks[5–8]: 1. a large number of talents outflow; 2. a small number of talents outflow; 3. the balance of the talents inflow and outflow; 4. a small number of talents inflow; 5. a large number of talents inflow. Scores are taken respectively as 1, 2, 3, 4, and 5. The specific evaluation criteria and score of every index are shown as Table 1.

4.3 *The determining of every index weight*

We use the methods of integrated scoring, expert scoring and Analytic Hierarchy Process (AHP) [9–14] to calculate the index weight of layer B, layer C, and layer D. Through the AHP 1–9 scale method and obtaining expert advice, we can compare the relative degree of index and quantify the judgment. In the end, we will tidy the obtained data and construct the judgment matrix. Through the calculation, we get the feature vector, and through the consistency test, that is index weight. If we use $b_{ij} = Z_i/Z_j$ (i,j = 1,2...) to represent the relative importance estimate of the ith index to the jth index. Here Z_i is the ith index weight and Z_j is the jth index weight. According to the calculating value, structural judgment matrix and the judgment matrix of using structure we can do the following homologous calculation:

Calculates the product M_i of each row element b_{ij} in the judgment matrix B:

$$M_i = \prod_{j=1}^{n} b_{ij} \quad i = 1, 2, \ldots, n \tag{1}$$

calculates the nth root A_i of M_i:

$$\alpha_i = \sqrt[n]{M_i} \tag{2}$$

Table 1. The evaluation standard and score of index.

Index attributes	Index	The evaluation standard and score of index: A large number of talents outflow	A small number of talents outflow	The balance of the talents inflow and outflow	A small number of talents inflow	A large number of talents inflow
		1	2	3	4	5
The static index of existing talents	E_1	$0 < E_1 < 1/2$	$1/2 < E_1 < 1$	$E_1 = 1$	$1 < E_1 < 3/2$	$3/2 < E_1$
	E_2	$0 < E_2 < 1/2$	$1/2 < E_2 < 1$	$E_2 = 1$	$1 < E_2 < 3/2$	$3/2 < E_2$
	E_3	$0 < E_3 < 1/2$	$1/2 < E_3 < 1$	$E_3 = 1$	$1 < E_3 < 3/2$	$3/2 < E_3$
	E_4	$0 < E_4 < 1/2$	$1/2 < E_4 < 1$	$E_4 = 1$	$1 < E_4 < 3/2$	$3/2 < E_4$
The static index of existing talents	E_5	$0 < E_5 < 1/2$	$1/2 < E_5 < 1$	$E_5 = 1$	$1 < E_5 < 2$	$2 < E_5$
	E_6	$0 < E_6 < 1/2$	$1/2 < E_6 < 1$	$E_6 = 1$	$1 < E_6 < 2$	$2 < E_6$
	E_7	$0 < E_7 < 1/2$	$1/2 < E_7 < 1$	$E_7 = 1$	$1 < E_7 < 2$	$2 < E_7$
The static index of reserved talents	E_8	$0 < E_8 < 0.2$	$0.2 < E_8 < 0.4$	$0.4 < E_8 < 0.6$	$0.6\ E_8 < 0.8$	$0.8 < E_8 < 1$
The dynamic index of reserved talents	E_9	$0 < E_9 < 1/2$	$1/2 < E_9 < 1$	$E_9 = 1$	$1 < E_9 < 2$	$2 < E_9$

Statistic source: The data in the table according to the survey and scores collected by experts.

Table 2. The index system and every index weight after being converted.

	Layer B		Layer C		Layer D	
Layer A	Index	Weight (W)	Index	Weight (W)	Index	Weight (W)
The index system reflecting a regional talents' flow	Existing talents Index (B_1)	0.7	The static index of existing talents (C_1)	0.4	E_1	0.125
					E_2	0.625
					E_3	0.125
					E_4	0.125
			The dynamic index of existing talents (C_2)	0.6	E_5	0.46
					E_6	0.27
					E_7	0.27
	Reserved talents index (B_2)	0.3	The static index of reserved talents (C_3)	0.4	E_8	1.0
			The dynamic index of reserved talents (C_4)	0.6	E_9	1.0

Data source: Got from the methods of integrated scoring, expert scoring and Analytic Hierarchy Process (AHP).

normalizes the vector $\alpha = (\alpha_1, \alpha_2, ..., \alpha_n)^T$:

$$W_i = \frac{\alpha_i}{\sum_{i=1}^{n} \alpha_i} \quad (i = 1, 2, \ldots, n) \tag{3}$$

And vector $w = (w_1, w_2, ..., w_n)^T$ is the request eigenvector.

Calculates the maximum root characteristic λ_{max} of the judgment matrix B:

$$\lambda_{max} = \frac{1}{n}\sum_{i=1}^{n} \frac{(\mathbf{BW})_i}{W_i} \tag{4}$$

For any pattern of $i = 1, 2, \ldots, n$, $(\mathbf{BW})_i$ is the ith element of vector BW.

Doing the Consistency test:

$$C.I. = \frac{\lambda_{max} - n}{n - 1} \tag{5}$$

$$C.R. = \frac{C.I}{R.I} \tag{6}$$

If it is passed the test, the results obtained of the above calculation will be the weight coefficient.

Through the above calculation, the weights of every index are shown in Table 2. Considering the space limitation, we omitted the calculation process.

4.4 *The evaluation standard of index system*

An ideal status of a regional labor force flow conditions is labor force that can keep a dynamic balance between quantity and quality, or it has the quality improvement and the quantity increasing.

We use N to represent the evaluative result of a regional labor force conditions, and the final score of N is the accumulation of index weight and the product of index standard scoring, and the formula is as follows:

$$N = \Upsilon Y_i W_i \tag{7}$$

If N is between 1 and 2, it will indicate the decrease of talents of some region, manifesting a huge loss of the labor force and a serious brain drain; If N is between 2 and 3, it will indicate the decrease of talents of some region and slight loss of talents; If N is equal to 3, it will indicate the balance of the inflow and outflow of some region, and the labor force manifests a dynamic balance; If N is between 3 and 4, it will indicate the talent increase of some region, manifesting a slight increase of the labor force number; If N is between 4 and 5, it will indicate the talent increase of some region, manifesting a dramatic increase of labor force number.

5 EVALUATION AND CONCLUSION OF TALENTS' FLOW IN THE NORTH JIANGSU PROVINCE

5.1 *Evaluation on talents' flow in the north of Jiangsu province*

1. Establishing the index system of talents' flow in the north of Jiangsu province—done at the third part of this paper.

Table 3. Every index, index scoring and index standard scoring of labor force in the north of Jiangsu province.

Every index of labor force in the north of Jiangsu province	Index scoring	Index standard scoring	Every index of labor force in the north of Jiangsu province	Index scoring	Index standard scoring
E_1	$1 < E_1 < 3/2$	4	E_6	$0 < E_6 < 1/2$	1
E_2	$1/2 < E_2 < 1$	2	E_7	$0 < E_7 < 1/2$	1
E_3	$1 < E_3 < 3/2$	4	E_8	$0 < E_8 < 0.2$	1
E_4	$1 < E_4 < 3/2$	4	E_9	$0 < E_9 < 1/2$	1
E_5	$0 < E_5 < 1/2$	1			

Data source: Household survey data of 2005 from Jiangsu province statistics yearbook (2004).

2. Using the methods of integrated scoring, expert scoring and Analytic Hierarchy Process (AHP) to calculate the index weight, index scoring standard and the index standard scoring, shown as Table 3.

$$N = Y_iW_i = 0.7\times[0.4\times(3.25) + 0.6\times(1)] + 0.3\times[0.4\times(1) + 0.6\times(1)]$$

$$N = \Upsilon Y_iW_i = 0.7\times(1.9) + 0.3\times(1)$$

Index score of existing index Index score of talents dynamic reserved talents dynamic scoring

$$N = \Upsilon Y_iW_i = 1.63$$

6 CONCLUSION

To sum up, we can draw the following conclusions:

1. The index score of existing talents static index, index score of existing talents dynamic index, index score of reserved talents static index, and index score of reserved talents dynamic index of the north of Jiangsu province can be evaluated as: 3.25,1,1,1. The basis of human capital of the north of Jiangsu province is still better, and the educational level of existing labor force in the north of Jiangsu province is higher than the national average level.
2. The average educational level of outgoing labor force in the north of Jiangsu province is manifestly higher than the average level of the north of Jiangsu province ($1/2 < E_2 < 1$, $E_2 = 2$); the number of the inflow labor force per ten thousand people with junior college and above education and intermediate and above skills is further lower than that of local labor force. The quality of labor force in the north of Jiangsu province unbalance seriously, manifesting quality of labor force in the north of Jiangsu province gradually descending.
3. The returning rate and utilization efficient of reserved are seriously low and the drain of human is serious ($0 < E_8 < 0.2$, $E_8 = 1$). Due to the loss of a large number of graduates the development of human source and the huge investment of human investment of the north of Jiangsu province, and the investment profit have been obtained freely by other places.
4. The inflow number of graduates with non-local household register is much lower than the outflow number of local graduates, manifesting as a huge number of reserved talents' loss ($0 < E_9 < 1/2$, $E_9 = 1$).
5. The huge loss of existing talents of the north of Jiangsu province and the low efficiency of reserved.

The inflow number of graduates with non-talents and the huge inflow of labor force from other places with low level, which lead to the further fall of the quality of human source. The rank of labor force loss degree is serious in the north of Jiangsu province.

REFERENCES

[1] Yueyue, Wanghui. Economic Analysis on Labor Force Flow. Market Forum, 2005;3:4–87

[2] S. Cepni. Effects of computer supported instructional material (CSIM) in removing students misconceptions about concepts: "Light, light source and seeing. Energy Educ Sci Technol Part B2009;1: 51–83.

[3] Wang Hecheng, Yaozhong. Study on the Appraisal System of Marketers Business Studies, 2001;10:P82–85

[4] A. Demirbas. Energy concept and energy education. Energy Educ Sci Technol Part B2009;1: 85–101

[5] Yuan Rongqi. Discussion on Relative Questions of Human Resource Management Economy and Management Study, 2004;(10:19–21

[6] K. Turhan, B. Kurt. Computer based medicine education: Improved skin segmentation using YCbCr color space. Energy Educ Sci Technol Part B2010;2:1–20.

[7] Lu Suzhen. Business Accounting of Human Source of Knowledge Economy Condition Business Studies, 2001;1:34–37
[8] M. Dikmenli. Biology students' conceptual structures regarding global warming. Energy Educ Sci Technol Part B2010;2: 21–38.
[9] Zhangqi, Xicun, Ang. Several Methods of Improving the Evaluative Precision on AHP Theory and Practice on System Engineering, 1997;11:83–88
[10] O. Koray, M. S. Koksal, B. Hazer. Simple production experiment of poly (3-hydroxy butyrate) for science laboratories and its importance for science process skills of prospective teachers. Energy Educ Sci Technol Part B2010;2:39–54.
[11] Lu Jiafu, Yi Liangzhong. Vague Level Overall Merit of University Faculty's Teaching Quality. Szechwan Engineering Institute Journal, 1997;2:66–73.
[12] Y. Kaya. The effect of regulator and stiffener exercises for speaking mechanics on the stuttering education and therapy. Energy Educ Sci Technol Part B 2011;3(3):253–268
[13] Wang Yaxiong, Zou Qinghua. Human Capital Based on AHP—Non-Money Value. Statistic Education, 2005:2:47–51.
[14] E. Arslan. The possible contributions of development agencies on Turkish economy. Energy Educ Sci Technol Part B 2011;3(3):285–292.

Engineering Management and Industrial Engineering – Leung (Ed.)

The geographical dispersion of independent directors and the governance of the board

S.Z. Ji
Shanghai University of International Business and Economics, Shanghai, China

ABSTRACT: To improve the corporate governance of SOEs, the government introduces the system of outsider directors. But the outside directors are selected mainly from the retired government officials and corporate executives mainly living in Beijing. Based on the data of independent directors of listed companies, we construct a model with principal component analysis. We find the number of independent directors from the same region has negative correlation with corporate performance. We think the role of outside directors of SOEs is similar to that of the independent directors of listed companies. Therefore we should select the outsider directors from diversified regions. Our study enriches the theory of board governance by highlighting the geographical dispersion of outside directors. This paper demonstrates that the roles of outside directors depend on the interaction between living environment, working experience of the outside directors and inside governance mechanism of the company. Therefore, it is very important to choose the correct types of outside directors to make the board decision-making process better.

Keywords: soft information; independence; monitoring and advising efficiency

1 INTRODUCTION

Since 2001 the independent director system has been brought into China. The independence of the independent directors has been the hot topic discussed by many scholars. Affiliation is defined to represent the independence by most scholars. However, Hermalin and Weisbach (1998) argue that unaffiliated does not necessarily mean independent. Hong wan (2008) points out that geographic proximity has impact on the relation between CEO and independent directors. From the perspective of Chinese culture the relationship among people include relative relationship, geographic relation, business relationship, product relationship, and religious relationship (Lin, 1990). Relative relationship includes family membership and clan membership. Geographic relationship includes neighborhood and hometown membership. Business relationship includes workmates and classmates. Product relationship means some relationship related to the product, such as suppliers and sellers. Religious relationship means people have the same special religious tradition. These five relationships are the cores of interpersonal relationship. The affiliations defined in the independence include only relative and part of business and product relationship. The geographic relationship is totally overlooked, which ancient Chinese paid much attention to and regarded as the second closest relationship only next to relative relationship. One board consisted of directors mostly from the same area is surely different from another board having directors from different areas. The outside directors from the same city as CEO might have closer relationship with CEO than other non-local directors. The affiliation between the local independent directors and CEO will affect the decision and monitoring functions of the board. From the perspective of decision, the local independent directors can get information easily from CEO and can make more contribution to the decision. From the perspective of monitoring, the affiliation will affect the objectiveness of the independent directors and cannot supervise CEO effectively.

2 LITERATURE REVIEW

Research of the board suggests that outside directors generally play two roles: supervising top management and providing resources (Hillman & Dalziel, 2003; Sundaramurthy & Lewis, 2003). Supervising means monitoring the managerial behavior and evaluating the administrative performance. In addition to supervising, outside directors also need to provide advice and counseling to the top management especially in the area of M&A, asset restructuring (Pfeffer & Salancik, 1978). Some scholar found that the number of outside directors had positive correlation with the performance-increasing restructuring programs (Perry and Shivdasani, 2005) and target shareholder gains in a hostile takeover (Cotter et al., 1997). Others found that negative firm performance will lead to a higher number of outside directors on a board (Hermalin and Weisbach, 1988). The results of the empirical research are conflict. In the actual operation of the board, the decision behavior of the directors is heavily influenced by the working experience, cultural background and the relationship between directors. The working experience and cultural background determine the method of information collecting and processing, the formulating of the decision preference, and the social networks they are involved.

The decision preference of the outside directors is influenced by their working experience and cultural background. Tang Qingquan (2008) made the similar conclusion, when he studied on the listed companies in China. He found that the level of education and job position of the independent directors' exerted direct influence on the decision behavior of independent directors, the independent directors presented the risk-aversion preference, and the different working experience background of the independent directors contributed to the better performance of the listed companies. The research by Tang Qingquan brings important enlightenment to our study on the outsider directors of SOEs. Restricted on the limited data of the board in the SOEs, we have no enough data to do empirical analysis. However, the outsider directors system referred to the independent director system when the SASAC launched the pilot project by introducing the board to reform the corporate governance in SOEs. There are many similarities between them. We can study the data from the listed companies and properly adjust conclusion to apply to the outsider directors in SOEs.

3 RESEARCH ASSUMPTIONS

The level of education differs largely between the inside directors and outside directors. The inside directors have comparatively low level of education in China. But the outside directors are always selected from professional institute and considered to be highly educated. It is usually thought people with better culture background might have more rational decision behavior. Wright (1997) found that the students from England tend to express their judgment with probabilities on uncertain occasions while students from East Asia just said they don't know. People with better culture background learn how to analyze from their professional perspectives while people with low level education tend to behave intuitively. The independent directors with master or doctor degree may have more professional training and behave more rationally. When the board behaves more rationally, the firm can get better performance. So we have the first assumption:

The more independent directors with Master or Doctor degree as they are, the better the performance of listed companies is.

The different culture backgrounds influence the decision behavior of directors, especially to the wider difference between eastern and western culture. Some independent directors come from foreign countries and have western culture background. Their decision behavior is different from that of domestic directors. The versatile decision method and different sources of information contribute to the group decision of the board. Most of the foreign outside directors have experience of working in several listed companies abroad as directors or top managers. Obviously, they can bring the advanced idea of corporate governance to the domestic companies. So we have the second assumption:

The more independent directors from abroad, the better performance of listed companies is.

The outside directors assume the role of supervising and resources providing. From the perspective of supervising the outside directors should be independent of CEO and inside directors. The outside directors should have less contact with CEO to ensure their independence. Outside directors from the wider area are less likely to contact with each other and conductive to enhancing the independence. On the contrary, directors from the same region are more likely to contact with each other and may be harmful to the independence. From the perspective of resource providing, outside directors from wider area have different living environment and different methods of processing information. The diversity of decision information is beneficial to the decision efficiency of the board. From another perspective, people depend on social network to collect information when they make decisions. CEO tends to select outside directors from his social network or that overlaps with his. When the directors belong to the same social network, or that overlaps with those of inside directors, the directors will communicate more adequately and the board will be more efficient. So the effect of directors from the same region is mixed. On one hand, the directors from the same region do harm to the independence and diversity of information sources. On the other hand, the directors from the same region can communicate adequately. We don't know which effect is dominant? But we think the number of independent directors from the same region should have evident correlation with the performance of listed companies. So we have the third assumption:

The number of independent directors from the same region has evident correlation with the performance of listed companies.

4 EMPIRICAL ANALYSIS

We choose the data from Guo TaiAn, the largest corporate governance data providing company in China. The SASAC launched the pilot of building the board in SOEs in 2005. We want to use the data of listed companies to apply to the SOEs. So we choose to collect data in 2005. From the data we find each board has more than 2 local outside directors in average, reflecting the listed companies tend to choose the local residents as outside directors. The number of foreign outside directors is very few, average 0.02 each company. The number of high educated outside directors is fewer than expected, average 0.65 each company.

The theorists don't come into agreement about the performance index of the listed companies. ROA, ROE, TOBIN'Q and other financial index were used by different scholars. Each single index has its own limitation. We think it is proper to use the principle component method to integrate the main indexes into a general index through decomposition. We select 19 financial indexes which can be divided into four aspects.

We use the method of principle component to extract 7 factors, which can explain 83.55% of the original 19 indexes.

Firstly, we analyzed the correlation coefficient between the main performance index and different sources of independent directors using Pearson correlation method.

Secondly, we further analyzed the relation between the different sources of independent directors and general performance index with

Table 1. The summary of the different sources of independent directors.

	Average	Variance
SRN	2.28	1.207
HEN	0.65	1.087
FN	0.02	0.195

SRN: the number of independent directors from same region.
HEN: the number of independent directors with master of doctoral degree.
FN: the number of independent directors from foreign countries.

Table 2. The main performance index.

Financial index	Symbol	Definition
(1) Income indexes		
Return on asset	ROA	ROA = net profit/(equity+debt)
Return on equity	ROE	ROE = net profit/equity
Earning per share	EPS	EPS = net profit/outstanding share
GPM	GPM	GPM = gross profit/sales
OPM	OPM	OPM = operating profit/sales
Net profit margin	NPM	NPM = net profit/sales
ADNROE	ADNROE	ADNROE = (earning before extraordinary gains or losses)/equity
CROA	CROA	CROA = the profit from the core business/asset
ROIC	ROIC	ROIC = adjusted after tax operating income/invested capital
(2) The efficiency index		
TAT	TAT	Sales/the average of total asset
FAT	FAT	Sales/the average of fixed asset
IT	IT	Sales cost/average inventory
RT	RT	Sales/average receivable account
(3) Liquidity index		
D/A	D/A	Debt/total asset
CR	CR	Current asset/current debt
QR	QR	Quick asset/quick debt
(4) Cash indexes		
SCR	SCR	Cash collecting/sales
CIR	CIR	Cash income/main income
CFLR	CFLR	Cash balance/current liability

Table 3. Correlation coefficient.

	SRN	HEN	FN
Pearson correlation	−0.112*	−0.059	0.055
Sig. (2-tailed)	0.029	0.160	0.175

*Correlation is significant at the 0.05 level (2-tailed).
**Correlation is significant at the 0.01 level (2-tailed).

Table 4. Regression results.

	Unstandardized coefficients			
Model	B	Std. error	t	Sig.
1. Constant	0.111	0.025	4.425	0.000
SRN	−0.018	0.010	−1.909	0.057
2. Constant	0.075	0.014	5.479	0.000
HEN	−0.011	0.011	−0.995	0.320
3. Constant	0.067	0.012	5.724	0.000
FN	0.056	0.060	0.936	0.350

[a]Dependent Variable: general performance index.
[b]Only choose the companies with EBITDA > 0.

regression method. To correctly analyze the relation between the general performance index and the number of different types of independent directors we delete all the abnormal listed companies with negative EBITDA. We find that only the number of independent directors from the same region has significant correlation with general performance index, significant at 10% level.

5 CONCLUSIONS AND SUGGESTION

From the empirical analysis, we find the number of the independent directors from same region has negative correlation with general performance index. Just as we mentioned above, the number of independent directors has mixed effect on the function of the board. From the perspective of independence and information sources we find the number of independent directors from the same region has negative effect on the function of the board. From the perspective of information communication, we find the number of independent directors has positive effect on the function of the board. The empirical results show that the negative effect dominates the overall effect, reflecting the importance of outside information and independence. There is also an inner link between the independence and the outside information. When the outside directors are independent they can bring more outside information to the board. When the decision information is diversified, the outside directors can make decisions independently and objectively.

As far as the selection of outside directors of SOEs is concerned, the urgent demand of independence and outside information is as important as that of listed companies. So we should avert choosing outside directors from same region. But the SASAC inclines to choose outside directors mainly from the retired government official and corporate executives. Most of them live in Beijing and has similar background and information sources. They will not contribute to the function of the board. The SASAC should select outside directors from wider areas and different background.

ACKNOWLEDGEMENTS

This paper was sponsored by the National Social Science Fund Project (12BGL003), Research Innovation Key Project (12ZS170), Key Discipline Project (J512-01) of Shanghai Municipal Commission of Education, and 085 Project (Z085GSGL14024).

REFERENCES

Huse, M. 1998, Researching the dynamics of board-stakeholder relations, Long Range Planning, 31: 218–26.

Jensen, M.C. and Meckling, W.H. 1976, Theory of the firm: Managerial behavior, agency costs, and ownership structure, Journal of Financial Economics, 2: 305–60.

Morck, R. 2008, Behavioral Finance in Corporate Governance: Independent Directors. Non-Executive Chairs, and the importance of the Devil's Advocate [J]. Journal of Managcnment and Governance 12(2), 179–200.

Scott Plous. The Psychology of Judgment and Dxision Malting. McGraw-Hill, 1993.

Tang qingchuan. 2008, The motivation and effect of the establishment the independent directors in China-the reseach report of National Natural Science Fund Project [J]. Present Economic Management 30(12).

Engineering Management and Industrial Engineering – Leung (Ed.)
 ISBN 978-1-138-02773-2

The mechanism analysis of the interactive development of producer services and clothing industry of world city Beijing

H.T. Han & G.L. Chen
Beijing Institute of Fashion Technology, Chaoyang, Beijing, China

ABSTRACT: To analyze the correlation of clothing industry and the producer services, this article uses the industry relevance theory to analyze the promoting effect of apparel industry to the producer services, and the supporting action of producer services to the apparel industry. Finally it concludes that only two industries' interactive development can better play the role, and enhance its status in the national economy.

Keywords: apparel industry; producer services; industrial interaction

1 INTRODUCTION

The objective of the "twelfth five-year" planning provides the opportunity of upgrading, high-end, internationalization for Beijing apparel industry, and it is also a good opportunity for the development of producer services. The structure upgrading and international competitiveness in Beijing's apparel industry should not only rely on the progress of the apparel industry itself, but also depend on the development of producer services and other related supporting industries. And producer services only developed with the associated industries can better play the role, and enhance its status in the national economy.

2 THE PROMOTING FUNCTION OF BEIJING PRODUCER SERVICES TO THE APPAREL INDUSTRY

Producer services play a great role in promoting transformation, upgrading and developing garment industry, and it can play a role in each link of garment industry chain, such as research and development design, logistics, trade and business, marketing consulting, convention and exhibition dissemination and so on.

2.1 *Research and design services with apparel industry*

R&D and innovation of Fabrics, fiber, dyeing and finishing, etc., have great significance for the transformation and development of Beijing's apparel industry. To make Beijing a fashion capital and follow the international trend; we must enhance the independent innovation ability of Beijing clothing textile enterprises. Therefore, R&D and design services play a great role in promoting the transformation and development of Beijing clothing textile enterprises.

2.2 *Logistics, finance and insurance services with apparel industry*

Logistics, finance and insurance services can help the Beijing clothing enterprises save costs and reduce risk. As a modern logistics enterprise the cost of warehouse and transportation can be reduced effectively; financial services can provide enterprises with funds, helping enterprises analyze the capital cost, turnover, profitability, etc. and make optimal decision; Insurance services help enterprise analyze risk, and provide constructive strategy to avoid or transfer risk and reduce loss.

2.3 *Marketing planning, commercial exhibition and international trade services with garment industry*

Marketing, trade and exhibition services have great significance for brand molding and market development of Beijing clothing enterprises. First of all, marketing planning service through market demand research provide enterprises with effective brand and promotion plan; Secondly, international trade services help enterprises widen sight and provide enterprises with rich resources of international trade through exports and imports. Thirdly, garment industry can also communicate with others and promote its own brand with the exhibition service. At the same time, it will get more resources of supplier and customer to expand the industry chain in both upstream and downstream.

3 THE PULL FUNCTION ON THE PRODUCER SERVICES OF BEIJING APPAREL INDUSTRY

3.1 *The industry associations between apparel industry and the producer services of Beijing*

With the refinement of social division, the producer services inside the garment industry are gradually out of the enterprise's value chain and vertical separation, which is dedicated to provide the production enterprises and enterprises with services. Producer services have a close correlation with apparel industry. First of all, the producer services are needed in garment industry chain to provide services of research and development, technology, logistics, commerce, training and so on; Second, the enterprise's good operation also need management consulting services and financial services, etc. Therefore, the rapid development of Beijing apparel industry inevitably leads to the development of producer services, related to promote the producer services in Beijing technology innovation and industrial structure optimization.

3.2 *Beijing apparel industry pull the spatial agglomeration of producer services*

Producer services interacting with apparel industry can be divided into two forms: one is to match the producer service industry cluster around the clothing industry agglomeration area, which can realize clustered development of producer services and get the advantage of cluster growth. At the same time, providing relevant supporting services for clothing industry cluster makes clothing industry improve efficiently, finally realizing the producer services interactive development with the apparel industry; Second is to develop producer services cluster areas as core, make apparel industry gathered in the peripheral zone, through innovation and continuous development of producer services, and then spread to the apparel industry of peripheral area, promoting the clothing industry structure optimization and upgrading.

4 THE PROBLEM OF BEIJING PRODUCER SERVICES BASED ON THE APPAREL INDUSTRY

As a world-class city, the construction of "fashion" is an important item for Beijing's economic and social development, but the development of producer services in Beijing is still hard to achieve the goal of a world-class city development. In this paper, through investigation and study, it sums up the current development of producer services in Beijing's based apparel industry, which mainly exist the following problems.

4.1 *The overall benefits of Beijing producer services remains to be improved*

The overall benefits of Beijing's producer services based on apparel industry need to be improved. For example, logistics efficiency is generally not high; logistics market demand and market competition is intense; prices and profits of transportation and warehousing fee is falling; some small and medium-sized logistics enterprises are in trouble; the benefit of the logistics enterprises is in a relatively low level; small enterprises face the fierce market competition; and the main indicators growth is generally lower than average productive service enterprise.

4.2 *Producer services in Beijing do not have a high level of specialization*

Compared with the international advanced level, at present the specialization degree of most productive service enterprise is not high, and the service concept is relatively backward, with a low degree of internationalization. Productive service enterprise of Beijing is almost small and medium-sized enterprises. The lack of large enterprise can provide high quality service. Although small and medium-sized enterprises have strong innovation consciousness, the level of service and technology content is low, so it is difficult to meet the high standard demand of customers, especially the orders of multinational service outsourcing. Service industry in China is gradually opening to the outside world, and international service companies begin to come into Beijing market and enable the market competition in domestic enterprises more severe. Many enterprises have their competitive disadvantages.

4.3 *The unbalanced industry structure of producer services in Beijing*

According to the current situation of producer services in Beijing, each industry's development is not balanced. From the total value added of Beijing producer services in 2011, circulation services, information services, financial services, business services, and technology services, is respectively 26.07%, 18.38%, 27.27%, 14.30%, 13.98%. Although business service and information sector has developed rapidly in recent years, it can be seen that the traditional financial industry is still in the largest proportion of service industry, and there is a great contrast with the developed countries and regions in the structure of producer services, which are mainly trade services, software industry, the securities industry, consulting and other emerging industry. The industrial structure of producer services in Beijing has few competitive edges, and its internal development among various industries is uncoordinated.

5 DEVELOPMENT STRATEGIC PLANNING OF BEIJING PRODUCER SERVICES BASED ON THE APPAREL INDUSTRY

5.1 *Government level*

In order to accelerate the development of producer services in Beijing, first, it is urgent to transform government functions, strengthen the service consciousness, improve the service level and efficiency, and improve the market environment; Second, it requires to further improve and perfect the institutional environment of producer services, such as market access system, improve the industry's standards and norms, vigorously promote the establishment of the legal system of producer services, strengthen market supervision, and maintain the market order; Third, it must promote its cooperation with the surrounding areas, expand the market space of the producer services in Beijing, accelerate integration of Beijing producer services with the surrounding clothing industry, and develop the pattern of labor division and cooperation.

5.2 *The industry level*

The rapid development of producer services depends on the rationalization of the internal structure Based on the current reality, it needs to formulate specific plan and related policy to be the leading department for improving the service industry structure, and select key industries and form a good industrial structure.

First, from the aspects of financial resource allocation, cross-regional business development, background services and financial products innovation, it needs to develop the financial services and improve the financial environment in Beijing; Second, it needs to make business services bigger and high-qualityand actively take the advantage of knowledge-intensive enterprises in capital, thus advancing internationalization and specialization in business service of Beijing. Meanwhile, it must improve the intermediary and advisory services and build up the business service industry cluster in Beijing. Third, it needs to comprehensively develop the science and technology services. Beijing needs to further accelerate the process of marketization, specialization and socialization, effectively playing the demonstration effect, diffusion effect and agglomeration effect of science and technology service industry so as to enhance the level of science and technology. Fourth, it must actively develop the information service, promote the development of information service and the cultural creative industry, and expand the products and market of information service. Fifth, it must improve the development of modern logistics industry. At present Beijing should strengthen the system planning and promote effective operation mechanism, make full use of logistics resources, widen the channel of the logistics, and develop the international logistics. By applying the supply chain management and information technology, it can enhance the level of informatization and specialization of logistics enterprises.

5.3 *Corporate level*

Enterprise is part of the producer services. The development of producer services in Beijing is dependent on the growth of each enterprise.

First, from the aspects of system, technology and organizational structure, the productive service enterprise should seize the opportunity and the challenge to innovate and raise their competitiveness. Second, it should support the small businesses among producer services to solve the management problems. At the same time, it needs to encourage the productive service enterprise merger and reorganization to promote the strategic alliance of producer services. Finally, the productive service enterprise should actively change to adapt to the needs of the development of Beijing for world city construction. By changing ideas and increasing investment in technology innovation, it must focus on human resources development, accelerate the information and scientific management, and improve the competition ability.

6 DEADLINE

In general, this paper makes an analysis on the correlation between Beijing producer services and apparel industry, and based on this, it puts forward the strategic planning countermeasures about the development of the producer services in Beijing. Producer services in China is still in its infancy, and the current statistical yearbook of producer services statistics is less. Therefore, it cannot accurately grasp the development of producer services in Beijing as well as the effect on clothing industry's growth. This paper is still far from mature, and will be further improved in the later study.

ACKNOWLEDGEMENT

It is supported by the construction fund project on off-campus training base for graduates in Beijing Institute of Fashion Technology.

The project of construction of innovative teams and selection and development of excellent talents for Beijing institute of fashion technology (PTTBIFT).

REFERENCES

[1] Greenfield H., Manpower and the Growth of Producer Services [M], New York: Columbia University Press, 1966.

[2] Browning H.C., Singelmann J., The Emergence of a Service Society national Technical Information Service [M], Springfield Virginia, 1975.

[3] Herbert G. Grubel, Michael A., Walker: Service Industry Growth: Cause and Effects [M]. Vancouver: Fraser Institute, 1989, 279.

[4] Coffer J., The Geographies of Producer Services [J]. Urban Geography, 2000(2).

[5] Ghosh A., Sarkar H., An Input-Output Matrix as a Spatial Configuration [J]. Economics of Planning, 1970(1–2):133–142.

Engineering Management and Industrial Engineering – Leung (Ed.)
© 2015 Taylor & Francis Group, London, ISBN 978-1-138-02773-2

Profit allocation for joint development projects based on co-operative game theory

C.W. Liao
Department of Civil Engineering and Hazard Mitigation Design, China University of Technology, Taipei, Taiwan

Y.H. Perng & H.S. Hsu
Department of Architecture, National Taiwan University of Science and Technology, Taipei, Taiwan

ABSTRACT: Extravagant profits exist in land development projects after the approval of an urban renewal plan. Nevertheless, a lot of vacant property at a location is not economically usable. Joint development of private parcels is a method to strengthen the economic base of a location. In addition, the government offers joint development projects strong incentives to achieve the objectives of an urban renewal plan. A landowner may earn more profit if he collaborates with others in a joint development project. However, lesser effort has been made on the allocation of the profits of land development. The hedonic price model is employed in identifying land price functions for individual parcels of land and multiple parcels in a joint development project based on historical sales data. Profit can also be reasonably allocated to each landowner in a joint development project by using the Shapley value and nucleolus.

1 INTRODUCTION

In an urban renewal plan, two Independent appraisals are required for each parcel to be acquired [1]. There are quite a few increases in the second appraisal after the approval of an urban renewal plan. In other words, extravagant profits exist in land development projects after the approval of an urban renewal plan. Therefore, criminals, with the connivance of politicians and bureaucrats, can intervene to make a lot of money [2].

Although vacant property is abundant at a location, it is not economically usable. Assembling small parcels to a size suitable for development may be costly. Once assembled, a site often requires expensive demolition, environmental cleanup, and buildup. A sustainable economic base for a location can be created only through private, for-profit initiatives and genuine competitive advantage. Therefore, joint development of private parcels is a method to strengthen the economic base of a location. In addition, the government offers joint development projects strong incentives to achieve the objectives of an urban renewal plan.

However, lesser effort has been made on the allocation of the profits of land development. Traditionally, a broker may perform negotiations for the purchase of the parcel after the urban renewal plan has been approved [1]. The total price for a parcel, pursuant to the first appraisal before the approval of an urban renewal plan, is paid to the landowner. The increases between two appraisals do not go to the landowner, but to the broker and the developer. Sometimes, extensive litigation follows.

Moreover, despite the desperate need for new projects, joint development is much more difficult because of strict urban regulations, expensive building costs, and the uncertainty that the regulatory process creates. As a result, only if landowners make sure that joint development is profitable and they are in agreement on allocating profits can joint development be carried out.

The hedonic price model is presented herein to identify the land acquisition price of each parcel to be acquired in the urban renewal plan. Game theory is then applied to investigate the possibility of improving profitability by joint development of several private parcels. Nucleolus and Shapley Value are also suggested for allocating profits among the landowners.

2 LAND APPRAISAL

Two Independent appraisals are required for each parcel to be acquired. The first appraisal shall provide the basis for the initial estimate of the value of the parcel to be purchased. The second appraisal shall be prepared after the approval of the urban renewal plan. The determination of the proposed acquisition price shall be based on review of the appraisals [1].

The hedonic price model is employed in determining land acquisition prices. The functional form for the hedonic equation is not determined theoretically. Rather, it is determined empirically based on historical sales data.

2.1 *Hedonic price model*

The theory of hedonic prices is based on the hedonic hypothesis that goods are valued for their utility-bearing attributes or characteristics. Hedonic prices are defined as the implicit prices of attributes and are revealed from observed prices of differentiated products and the specific amounts of characteristics associated with them. A hedonic price model provides a framework for the analysis of differentiated goods such as housing and land, whose individual attributes do not have observable market prices. It explicitly recognizes the complexity and heterogeneity of a property. It posits a relationship between the price of the property and its attributes and it can be used to explore the effects of different attributes on the price structure of the property [3]. Consequently, property markets are analyzed in terms of implicit trait markets in which consumers and suppliers trade off among these various attributes. A typical hedonic price function is expressed as equation (1).

$$V = f(z_1, z_2, \ldots, z_n) \qquad (1)$$

where V is the market price of the property, $z_1, z_2, \ldots, z_n$ are the attributes of the property.

A hedonic function should include only attributes that are costly to produce and yield utility to residents. When empirically tested by a regression, the regression coefficients measure the implicit prices of the individual characteristics of the land.

2.2 *Attributes affecting land acquisition price*

Much has been discussed about the attributes affecting land acquisition price. The attributes generally can be classified as two categories: neighborhood attributes, and locational attributes. The two attributes include ease and speed of transportation in the area of the parcel, distance to the central business district, population density, land use district, time-of-sale, ethnic clustering (homogeneity), governmental zoning, neighborhood road conditions, pupil-teacher ratio in elementary schools, ratio of hospital beds to population, parcel size, width of road abutting the parcel, quality of neighborhood dwelling units, etc.

Joint development project parcels discussed in this article are adjacent to each other. Aside from parcel size and width of road abutting the parcel, the characteristics of adjacent parcels mentioned above, are all the same. In addition, the government offers private sectors strong incentives to assemble adjacent parcels into joint development projects in order to achieve the objectives of urban renewal plans—the effective utilization and development of urban lands, the revitalization of urban functions, the improvement of urban living environments, and the increase in public benefit. Therefore, only three variables including parcel size, width of road abutting the parcel, and government incentives for joint development projects, are taken into account in this article. These considerations yield the following hedonic model:

$$\ln Y = a + b_1 \ln A + b_2 W + b_3 G \qquad (2)$$

where $\ln Y$ represents the natural logarithm of the ratio of land acquisition price to parcel size, $\ln A$ is the natural logarithm of parcel size, W is width of road abutting the parcel, G is a dummy variable which equals 1 if the project conforms to a condition of government incentives and 0 otherwise, a is an error term, and b_1, b_2 and b_3 are the respective coefficients to be estimated.

2.2.1 *Parcel size*

Determining the correct relationship between land acquisition prices and parcel size is critical to our understanding of the land market and has significant practical implications. Therefore, it is necessary to probe into the price-size relationship to confirm the niche for joint development projects.

Many empirical studies find that parcel size affects land acquisition price. Equation (3) is typical:

$$\ln V_i = \alpha \ln A_i + \beta Z_i + u_i \qquad (3)$$

where V_i is total sales price or value for parcel i, A_i is the size of the parcel, Z_i is a vector of explanatory variables other than parcel size, u_i is an error term, and α and β are coefficients to be estimated.

The price-size relationship is linear if $\alpha = 1$, convex if $\alpha > 1$, and concave if $0 < \alpha < 1$. Most studies that use equation (3) or a version with $\ln(V/A)$ as an estimating equation find concavity. Only one study finds that the price-size relationship is convex for small parcels and is concave at large values.

2.2.2 *Width of road abutting the parcel*

The increase in width of road abutting the parcel will lead to the enhancement of sale price of land. Also, the building height limitations stipulated by relevant statutes of different countries are related to width of road abutting the parcel. According to relevant statutes of Taiwan, the building height must not exceed 1.5 times as the width of the road abutting the parcel, plus 6 meters. In addition, the floor area may increase when width of road abutting the parcel is larger.

2.2.3 *Government incentives*

The government offers strong incentives to assemble adjacent parcels into a building lot. In accordance with relevant statutes of Taiwan, additional building height and floor area is permitted when parcel size conforms to statutory standards.

3 CO-OPERATIVE GAME THEORY IN BRIEF

3.1 *Game theory in general*

The participation of landowners in a joint development project is similar to players' participation in a game. Non-cooperative and co-operative games are two types of games discussed in Game theory [4]. Non-cooperative games deal with a situation where a decision-making unit in a market chooses its strategies independently and treats the others as competitors. In co-operative games, a group of decision-makers decide to undertake a project together in order to achieve business objectives: for example, to increase total revenue (profit maximization) or market share or to decrease total costs (cost minimization). And all partners joining the project regard the others as collaborators.

Game theory assumes that a player evaluates various outcomes in terms of the utility derived from them. Two classes of co-operative games are TU-games in which utility can be transferred, and NTU-games in which utility cannot be transferred. In this article, the cooperation of landowners is considered as a TU game.

3.2 *Profit allocation in co-operative games*

There are two major issues in a co-operative game: coalition formation and distribution of profit gained through co-operation.

In a co-operative game, different members of various coalitions will obtain different profits. If the players can obtain more profits by cooperating together and acting independently, they will certainly seek to establish a coalition.

Every player wants to obtain maximum profits in the coalition. The adopted method for allocating profits among the members will affect whether various members remain active in the coalition. The allocation problem may be solved in many ways, but an allocation scheme that prescribes, somehow, a solution for the allocation problem should satisfy desirable criteria such as rationality, efficiency and others. The Shapley value and the nucleolus are the two best known concepts which fit the aforementioned criteria.

3.3 *The shapely value*

The Shapley value is the expected marginal amount that a player contributes to a coalition [5]. This method is based on the potential fairness in distributing the total profit achievable by the coalition.

Considering a TU game (N,v), N = {1, …., n} is the set of players, and v:2 $^N \rightarrow$R is a characteristic function satisfying $v(\phi) = 0$, where ϕ is the empty set. For a coalition $T \subseteq N$, $v(T)$ is the total payoff that players in coalition T can jointly obtain if this coalition is formed. The Shapley value ($\phi_i[v]$) of every member $i \in T$ in the TU game (N,v) is:

$$\phi_i[v] = \sum_{\substack{T \subset N \\ i \in T}} \frac{(m-1)!(n-m)!}{n!}[v(T) - v(T - \{i\})] \quad (4)$$

where m is the number of members of coalition T, n is the number of all the members in grand coalition N, $T - \{i\}$ is the coalition, not including member i. If a player i co-operates with the coalition, which consists of members $T - \{i\}$, he receives the amount $v(T) - v(T - \{i\})$, the marginal amount which he contributes to the coalition, as payoff. Then the Shapley value $\phi_i[v]$ is the expected payoff to player i. Under the randomization scheme, $(m-1)!(n-m)!/n!$ is the probability of player i joining coalition $T - \{i\}$.

3.4 *The nucleolus*

The nucleolus is a solution concept of profit allocation problems in co-operative game theory. The set of acceptable distribution more than any distribution is called the nucleolus. The nucleolus is based on the idea of the 'excess'. Before obtaining the nucleolus, the concept of excess is introduced below.

For the n-player game (N,v), let T be a coalition, and $x = (x_1, x_2, \ldots, x_n)$ a payoff vector. Then the excess of the coalition T associated with x is

$$e(T,x) = v(T) - \sum_{i \in T} x_i \quad (5)$$

The nucleolus is the solution that minimizes the maximum ε; that is,

$$\min_{x \in X} \max_{T \subset N} e(T,x \quad (6)$$

which can be formulated as a Linear Programming (LP) problem

minimize ε

subject to,

$$e(T,x) \leq \varepsilon, T \subset N, x \in X \quad (7)$$

where X is the set of all allocations.

4 CASE STUDY

A joint development project is selected to explore the potential benefits of co-operation. Three parcels A, B and C are assembled to a size suitable for an urban renewal plan. The sizes of parcels A, B, and C are 400, 800, and 400 square meters respectively. Roads abutting the lot are as shown in Figure 1, including a main commercial street about 18 meters wide, a second commercial street about 10 meters wide, and a service street about 8 meters wide. The land acquisition price function is derived from the hedonic price analysis and presented as equation (8).

$$\ln Y = 4.92 + 0.16 \ln A + 0.01W + 0.12G \qquad (8)$$

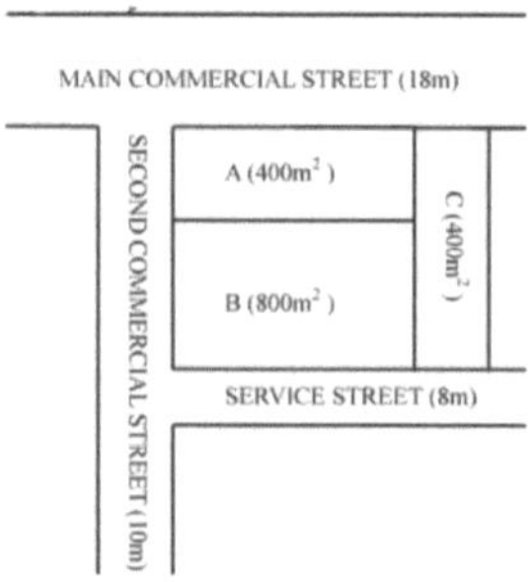

Figure 1. Map of A, B, C.

Three parcels may be sold independently or jointly as: {A}, {B}, {C}, {AB}, {AC}, {BC} or {ABC}. {A}, {B} and {C} represent the land prices for A, B and C respectively, if each parcel is sold independently; while {AB}, {AC}, {BC} and {ABC} depict the land prices for parcels A and B, A and C, B and C, and so on, respectively. Land prices are regarded as the profit and can be calculated as shown in Table 1. It is evident that the overall values increase by 37.99% if the three parcels are sold jointly instead of independently.

Both the Shapley value and nucleolus are applied to examine possible profit allocation schemes. The profit allocation scheme suggested by the Shapley value can be computed as in Table 2. Profit increases for parcels A, B and C are 45.00%, 34.13% and 39.35%, respectively, if they are sold jointly. On the other hand, using a linear programming program, nucleolus can be calculated as in Table 3. Profit increases for parcels A, B and C are 40.59%, 38.23% and 34.71%, respectively, if they are assembled into a lot. Moreover, the sum of the Shapley value of all parcels is the same as the sum of nucleolus, and it is larger than the value before the parcels are assembled.

The proposed method is promising, with the condition that the profit allocation scheme should be communicated and agreed before all parcels are assembled into a lot.

Table 1. Land acquisition prices and variations.

Land	Unit price (Thou.NT/M^2)	Total price (Thou.NT)	Total price increase (Thou.NT)	Total price increase ratio (%)
{A}	400	160,000	–	–
{B}	430	344,000	–	–
{C}	380	152,000	–	–
{AB}	540	647,714	143,714	28.51
{AC}	426	340,489	28,489	9.13
{BC}	513	615,329	119,329	24.06
{ABC}	566	905,221	249,221	37.99

Table 2. Profit allocation scheme suggested by the Shapley value.

Land	Profit before coalition (Thou.NT)	Profit after coalition (Thou.NT)	Profit increase ratio (%)
{A}	160,000	231,998	45.00
{B}	344,000	461,418	34.13
{C}	152,000	211,805	39.35

Table 3. Profit allocation scheme suggested by nucleolus .

Land	Profit before coalition (Thou.NT)	Profit after coalition (Thou.NT)	Profit increase ratio (%)
{A}	160,000	224,946.00	40.59
{B}	344,000	475,521.50	38.23
{C}	152,000	204,753.50	34.71

5 CONCLUSIONS AND SUGGESTIONS

A two-step procedure is presented in this study to demonstrate the potential benefits for joint development of several private land parcels. First, the hedonic price model is employed in identifying the land acquisition price of each parcel based on historical sales data. Then, both the Shapley value and nucleolus approaches are suggested for examining profit allocation schemes among the landowners.

The case study reveals an attractive profit increase for the landowners, if the parcels are jointly developed. Assembling small parcels into a lot may require an extra negotiation costs. A land acquisition price, for each parcel, can be acquired by other methods, aside from the hedonic price model. Similar methods may also be useful in setting up co-operation among landowners, developers, and a broker for development.

REFERENCES

[1] Massachusetts Department of Housing and Community Development. Massachusetts Department of Housing and Community Development's Regulations: Urban Renewal Regulations, Massachusetts Department of Housing and Community Development, Boston MA, 1996.

[2] Minar, P., and Lysa, J. "Security of Tenure in Mumbai." Tenure Security Policies in South African, Brazilian, Indian and Sub Saharan Cities: A Comparative Analysis, Centre for Applied Legal Studies, University of Witwatersrand, Johannesburg, 1999.

[3] Harding, J.P., Knight, J.R., and Sirmans, C.F. "Estimating Bargaining Effects in Hedonic Models: Evidence from the Housing Market." Real Estate Economics, 31(4), 601–622, 2003.

[4] Curiel, I. Co-operative Game Theory and Applications: Co-operative Games arising form Combinatorial optimization Problems, Kluwer Academic Publications, Boston MA, 1997.

[5] Shapley, L.S. "A value for n-person games." Annals of Mathematics Studies, 28, 307–317, 1953.

Engineering Management and Industrial Engineering – Leung (Ed.)
© 2015 Taylor & Francis Group, London, ISBN 978-1-138-02773-2

Family-friendly, role stress and innovation behavior: Assessing their effects on social enterprise employee base on market orientation

Z.H. Lian
Tan Kah Kee College, Xiamen University, Zhangzhou, China

B. Wang
Faculty of Business, City University of Macau, Macau, China

ABSTRACT: The purpose of this paper is to explore the mediating effect of role stress and the moderation effect of market orientation on the relationship between family-friendly and innovation behavior of social enterprise employees. In order to exam the model we defined, SEM and SPSS are used to analyze the data collected from the pairing sample of 500 employees and 100 supervisors in social enterprise on the Chinese mainland. Results demonstrate that family-friendly can affect innovation behavior through mediator of role stress. In addition, market orientation moderating impacts family-friendly and role stress, and also plays moderation effect between role stress and innovation behavior. This paper demonstrates the internal relationship between family-friendly, market orientation, role stress and innovation behavior on the basis of characteristics of social enterprise employees and offers ways for employers to improve organizational performance of social enterprise.

Keywords: family-friendly; role stress; innovation behavior; market orientation

1 INTRODUCTION

1.1 *Background and motivation*

In recent years, the new organizational philosophy called Social Enterprise is emerging in China rapidly, showing its influence through a variety of ways, and becoming the top popular discussion topic of academics and practitioners. However, most of the organizations' establishment purpose is to converging the force of vulnerable groups. Despite the enthusiasm, they are lacking of the experience of management and all kinds of resources. They are difficult to survive and heavily rely on external donations and government financial promotion service. Towards this end, the emphasis of commercializing market in order to solve social service and innovation performance becomes the focus that the social enterprise employees are mostly concerned about. Meanwhile, highly market-oriented organization atmosphere is causing more and more employees' work-family conflict as well as role stress. Building Family-Friendly policy is an important factor to alleviate the social enterprise employees' role stress and increase their innovation behavior.

1.2 *Literature*

Based on the view of the Social Exchange Theory and Social Justice Theory, organization's investment for employees can promote their contribution and performance at work (Tsui, Porter & Tripoli, 1997). The Working-Family welfare and other goodwill policies are the best way that an organization can invest for their employees. In addition, the study found, the family accident vacation can increase the employees' work satisfaction, and then increase the working efficiency (Galinsky & Stein,1990). Sands, J., & Harper, T. (2007) indicated that family-friendly policies can relieve the employees' working-life and family-life stress, improve their working behavior, and then promote the organization's overall performance.

Role stress is a mediating effect between the staff perception environment (Market orientation, Occupational characteristics, Family-Friendly culture etc.) and job performance (Zhihua Lian 2013). Yochum. D. (1997) pointed that the middle and small enterprise should construct a more family-friendly culture to effectively relieve the staff's occupational stress, and then reduce the quitting behavior of employees. The scholar Janis and Lazarus emphasized that different personal knowledge will cause different levels of stress. The anxiety generated by role stress is the main reason of working performance decline. If we can reduce the staff's role stress, we can reduce their anxiety, and then promote the staff's job performance. Similarly, if we strengthen the support of the staff's emotion,

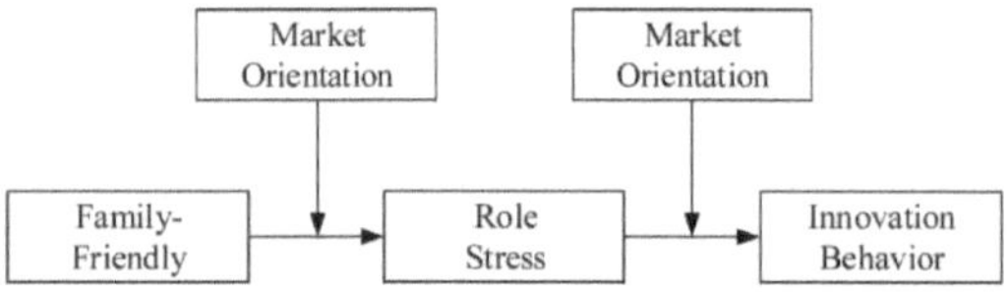

Figure 1. Conceptual model.

tools, appraisal and work-family relationship, promote the staff's self-efficacy and other psychological perception in time, then the negative effect of the stress will be greatly reduced, or even turned into an encouraging and promoting effect, and then positively affect job performance. And this is just the view this paper describes. In summary, we propose the following hypothesis: The role stress is the mediating variable between family-friendly relationship and innovation behavior.

According to the study by Sin et al. (2003) and Kirca et al. (2005), there are three mainstream theories in market orientation research: the measure and pre-variable impacting of market orientation; market orientation and mediating effect of organizational performance and moderating effect of environment/situational variables. Siguaw. J, Brown, G. & Widing. R. (1994) constructed the empirical model of market orientation, customs orientation and working attitude. Through the sales staff survey, they verified that the market orientation can significantly affect the employee's work attitude. Zablah A.R. et al. (2012) discovered that the market orientation can affect the staff's job behavior through frontline staff's stress. If we can improve the staff's psychological welfare, we can promote their work performance. Therefore, we further hypothesize that market orientation not only moderates the relationship between family-friendly and role stress, but also adjusts the relationship between role stress and innovation behavior (see Fig. 1).

2 METHODS

2.1 *Survey*

A cross-sectional research design was used in this study. In order to know if this model can be supported in the sample of the social enterprise employees, we used convenience sampling and occasional sampling method and tested the hypotheses with the sample survey. Researchers obtained data on employees in social enterprise in mainland China, and collected it from the pairing sample of 500 employees and 100 supervisors. We received 556 questionnaires (92.6 percent), among which 500 are valid questionnaires (83.3 percent). All subjects who joined in this study did on a voluntary basis with ensured confidentiality.

2.2 *Measure*

Family-Friendly. FF are employer-sponsored programs and policies that are designed to help employees manage work and personal life demands (Glass & Finley, 2002; Lobel, 1999). They generally include flexible work schedules, dependent care assistance, leave arrangements, counseling and referral services. In this study, we assess family-friendly on the basis of the FF Search Activities Index, which has a list of 9 FF search activities developed by Hammer et al.(2005). The Cronbach's α for the FF scale in the current study was 0.885.

Role Stress. Role conflict and role ambiguity are among the most widely studied role stress variables (Bettencourt and Brown, 2003; Brown and Peterson, 1993; Rizzo et al., 1970). Generally, they are negatively related to job outcomes. Role stress was measured by the Perceived Role Stress Scale (House et al. 1979). The scale consisted of three items answered on five-point Likert-type scale from 1 (not at all) to 5 (completely). The Cronbach's α for the role stress scale in the current study was 0.912.

Market Orientation. Market orientation, as a corporate culture, characterizes an organization's disposition of delivering superior value to its customers continuously (Slater and Narver, 1994). This variable is measured by a nine-item scale developed by Han, Jin k. et al. (1998). In this study, the α coefficient for the scale was 0.92.

Innovation Behavior. The degree of participants' performance creation as a team member was assessed using supervisor ratings of 6 items from Janssen, O. (2008) innovation scale. The response format was the same as the one used for measuring citizenship behavior. Cronbach's alpha was 0.94.

3 RESULTS

3.1 *Common method variance analyses*

When self-report questionnaires are used to collect data at the same time from the same participants, Common Method Variance (CMV) may be a concern. We test CMV loads all items from each of the constructs into an exploratory factor analysis to see whether one single factor does emerge or whether one general factor does account for a majority of the covariance between the measures; if not, the claim is that CMV is not a pervasive issue. The expected ex post model showed good fit indexes (RMSEA = 0.050; NFI = 0.919; CFI = 0.953; GFI = 0.909; IFI = 0.954).

Nevertheless, this model did not yield a significantly better fit to the data than ex ante CMV. Table 1 presents Δdf = 581–571 = 10, ΔX^2 = 893.214–585.239 = 307.975, P < 0.000, indicating no significant variance.

3.2 *Hierarchical regression-mediating effect*

To identify potential mediator of the relationship between family-friendly and innovation behavior, three multiple regression models were followed next to test for mediation (Baron and Kenny, 1986). First, a significant relationship was established between family-friendly and innovation behavior. Then a significant association was shown between family-friendly and role stress. At the third step, we use innovation behavior as the criterion variable in a regression equation; both family-friendly and role stress were entered. The Family-friendly can still find the significant relations with innovation behavior, but the influence became much weaker, thus partial mediation of role stress is indicated, which is shown in Table 2.

Table 1. Deal with CMV in the ex post statistical analyses.

CMV	Ex-ante	Ex-post
x^2	585.239	893.214
df	571	581
x^2/df	1.046	1.527
CFI	0.996	0.953
NFI	0.943	0.919
RMSEA	0.015	0.050
IFI	0.996	0.954
GFI	0.921	0.909

Table 2. Mediating multiple regression results.

	Model 1		
Variables	IB	RS	IB
Gender	–0.035	–0.031	–0.076
Age	0.035	–0.006	–0.027
Education	0.207***	0.051	0.231***
Tenure	0.164***	0.071	–0.182***
Marital status	–0.080	0.017	0.008
FF	0.076***	0.225***	0.069*
RS			0.025***
R^2	0.085***	0.123***	0.097***
Adj-R^2	0.074***	0.113***	0.085***
F	2.836***	24.938***	1.576***

Notes: Significant at: *p < 0.05; ***p < 0.001; n = 500; IB-innovation behavior; FF-family-friendly; RS-role stress.

Table 3. Moderating hierarchical multiple regression results.

	Model 2		Model 3	
Variables	RS		IB	
Gender	–0.031	–0.027	–0.067	–0.060
Age	–0.006	–0.002	–0.032	–0.035
Education	0.051	0.059	0.209***	0.211***
Tenure	0.071	0.074	–0.152***	–0.151***
Marital status	0.017	0.024	0.022	0.022
FF	0.225***	0.228***		
RS			–0.047*	–0.064
MAR	–0.012*	–0.006	0.278***	0.276***
FF*MAR		0.087*		
RS*MAR				0.062*
R^2	0.069***	0.076***	0.168***	0.172***
Adj-R^2	0.056***	0.061***	0.157***	0.158***
F	12.445***	9.656***	22.973***	16.039***

Notes: Significance at: *p < 0.05; ***p < 0.001; n = 500; IB-innovation behavior; FF-family-friendly; RS-role stress; MAR-market orientation; FF*MAR-family friendly and market orientation; RS*MAR-role stress and market orientation.

3.3 *Hierarchical regression-moderating effect*

As shown above in Table 3, we first test the moderating effect of market orientation between family-friendly and role stress from models 2. In the control of the respondent's gender, age, education, tenure, marital status, the interaction of the preference of family-friendly and market orientation is significant (β = 0.087, p < 0.05); the moderating effect of family-friendly and market orientation is identified, which means once market clime has stronger preference to chase for role stress, their family-friendly effort will lead to stronger fit perception. At last we test the interaction of the preference of role stress and market orientation and innovation behavior, which is significant (β = 0.062, p < 0.05); the moderating effect of role stress and market orientation is identified, which means once market culture has stronger preference to chase for innovation behavior and independence, their role stress effort will weaken their fit perception in organization.

4 CONCLUSIONS

This study extends the literature by examining the consequences of family-friendly on role stress, market orientation and innovation behavior of social enterprise employees. As hypothesized, family-friendly variables can affect role stress and also has direct and indirect effects on innovation behavior mediated by role stress. Moreover, we have deepened and expanded our understanding of the relationship between family-friendly and role stress, role stress and innovation behavior by the moderating effects of market orientation from one aspect to another. In general, this paper contributes to the ongoing efforts to understand the strategic management aspect of family-friendly and innovation performance.

5 LIMITATIONS AND FUTURE WORK

In the current study, due to resource limitations, only social enterprise workers in southern China are selected for the study population. Therefore, current study results cannot be extrapolated to describe the experiences or circumstances of social enterprise employees in other areas. In addition, all variables are concerned with perception or self-behaviors, which is difficult to be assessed by others; that may cause single source bias. We need to consider this bias in research design and try to survey team member and their customer in further study to get more accurate data and avoid common variance.

ACKNOWLEDGMENT

At the point of finishing this paper, I'd like to express my sincere thanks to all those who have lent me hands in the course of my writing this paper. First of all, I'd like to take this opportunity to show my sincere gratitude to my leader, Ms. Xue Yuan. Secondly, I'd like to express my gratitude to my friends who offered me references and information on time. Without their help, it would be much harder for me to finish my study and this paper.

REFERENCES

Baron, R.M. and Kenny, D.A. 1986, The moderator-mediator variable distinction in social psychological research: conceptual, strategic and statistical considerations, *Journal of Personality and Social Psychology*, Vol. 51, pp. 1173–82.

Bettencourt, L.A. and Brown, S.W. 2003, Role stressors and customer-oriented boundary-spanning behaviors in service organizations, *Journal of the Academy of Marketing Science*, Vol. 31 No. 4, pp. 394–408.

Brown, S.P. and Peterson, R.A. 1993, Antecedents and consequences of salesperson job satisfaction: meta-analysis and assessment of causal effects, *Journal of Marketing Research*, Vol. 30, pp. 63–77.

Galinsky, E., & Stein, P.J. 1990. The impact of human resource policies on employees:Balancing work/family life. *Journal of Family Isues,* 11(4), 368–383.

Glass, J., & Finley, A. 2002. Coverage and effectiveness of family-responsive workplace policies. *Human Resource Management Review*, 12, 313–337.

Hammer, L. B., Neal, M. B., Newsom, J. T., Brockwood, K. J., & Colton, C. L.2005. A Longitudinal Study of the Effects of Dual-Earner Couples' Utilization of Family-Friendly Workplace Supports on Work and Family Outcomes. *Journal of Applied Psychology*, 90(4), 799–810.

Han, J. K., Kim, N., & Srivastava, R. K.1998. Market orientation and organizational performance:Is innovation a missing link?. *Journal of Marketing*, 62(4), 30–45.

House, J. S., Wells, J. A., Landerman, L. R., McMichael, A. J., & Kaplan, B. H.1979. Occupational Stress and Health among Factory Workers. *Journal of Health and Social Behavior*, 20(2), 139–160.

Janis, I. L. 1958. *Psychological Stress.*. New York: Wiley.

Janssen, O., & Huang, X. 2008. Us and Me: Team Identification and Individual Differentiation as Complementary Drivers of Team Members' Citizenship and Creative Behaviors. *Journal of Management*, 34(1), 69–88.

Kirca, A. H., Jayachandran, S., & Bearden, W.O. 2005. Market orientation: A meta analytic review and assessment of its antecedents and impact on performance. *Journal of Marketing,* 69(2), 24–41.

Lian Zhihua, 2013, Relationship model research between market orientation, job characteristics and job performance base on the moderation of family-friendly: job stress as a mediation. *Macau University of Science and Technology*, Macau.

Lobel, S. A. 1999. Impacts of diversity and work-life initiatives in organizations. In G. N. Powell (Ed.), *Handbook of gender and work* (pp. 453–476). Newbury Park, CA: Sage.

Rizzo, J.R., House, R.J. and Lirtzman, S. 1970, Role conflict and ambiguity in complex organizations, *Administrative Science Quarterly*, Vol. 15, pp. 150–63.

Sands, J., & Harper, T.2007. Family-Friendly Benefits and Organizational Performance. *Business Renaissance Quarterly,* 2(1), 107–126.

Yochum, D.1997. Small firms get more family friendly for workers Efforts to reduce employees' stress, attract higher quality people cuts turnover, absenteeism. *Tampa Bay Business Journal,* 17(28), 18.

Sin, Y.M., Tse, C.B., Yau, H.M., Chow, R., & Lee, S.Y. 2003. Market orientation and business performance: A comparative study of firms in mainland China and Hong Kong. *European Journal of Marketing,* 37(6), 910–936.

Siguaw, J., Brown, G., & Widing, R. 1994. The influence of the market orientation of the firm on sales force behavior and attitudes. *Journal of Marketing* (31), 106–116.

Slater, S. F., & Narver, J. C. 1994. Does competitive environment moderate the market orientation-performance relationship? Journal of Marketing, 58(1), 46–55.

Tsui, A. S., Porter, L. W., & Tripoli, A. M. 1997. Alternative approaches to the employee-organization relationship: Does investment in employees pay off? *Academy of Management Journal,* 40(5), 1089–1121.

Zablah, A. R., Franke, G., Brown, T., & de Bartholomew. 2012. How and When Does Customer Orientation Influence Frontline Employee Job Outcomes? A Meta-Analytic Evaluation. *Jouranal of Marketing,* 76(3), 21–40.

Engineering Management and Industrial Engineering – Leung (Ed.)
© 2015 Taylor & Francis Group, London, ISBN 978-1-138-02773-2

The study of the pull effect of international air transport on entry-exit tourism

G.C. Liu
Management and Economics of Tianjin University, Tianjin, China

D.L. Liang
School of Economics and Management of Civil Aviation University of China, Tianjin, China

ABSTRACT: International air transport is the main mode of transport utilized by entry-exit tourists, and can greatly stimulate the development of international tourism. The research is based on the relevant date from 2012. Co-integration analysis and the least squares method are used. Granger causality test examines the pulling effect which is exerted to entry-exit tourism by the international transport of our country. The results show that there exists a significant positive correlation. The results of the study play an important role on the collaborative development of China's civil aviation and tourism.

1 INTRODUCTION

With the rapid development of economic globalization and aviation internationalization in the 21st century, the proportion of disposable income continues to increase. The majority of passengers are prone to go overseas. As the feature of greater space distance and longer time, time and space efficiency have became the modern sense of region connection and the convenience of network. Air transport has become the main transportation for the entry-exit tourists. Regarding to the fact that aviation has a great influence on the tourism, the future study of the relation between each other's development will be of great value. According to industry statistics bulletin from CAA, the proportion of entry-exit passengers is increasing year by year in the passenger structure of our country. Under the international environment and popularity of civil aviation, the number of visitors traveling will continue to increase.

Yang Guishan (2011) analyzed the relationship between air transport and air travel by using qualitative method and recommended how to boost collaborative development from the perspective of policy. Chen Huiying (2002) used the method of partial least squares regression to analyze the relationship between integrated transport and tourism and considered that air passenger has great impact on international tourism, but has little influence on railway traffic. It will stimulate the development of exit tourism and largely increases aviation passengers. Zhou Bei, Li Yanna who ordered the existing airport layout and levels of development of tourism got some conclusions: regional differences in tourist flows generated by tourism resource development can actively change the level of aviation network structure. The layout of the airport and tourism resources currently interactive development is not yet clear. It still has great room for improvement. Wang Zhaofeng (2012) deeply analyzed the collaborative evolution and difference between southwest inbound tourist flow and air transport network from the correlation of passenger flow, spatial structure, connection strength of network joint about inbound tourist flow and air transport. The results showed that the significant correlation existed in inbound tourism flow of southwest and air transport. And interactive effects were obvious, but the promotion that inbounds tourism flow affected air transport was more.

2 THE DEVELOPMENT OF CIVIL AVIATION AND TOURISM

International air transport is the basic industry of the national economy with large external economic benefits. The number of passenger air routes characterizes the universality of regional association. In other words, with the number of air routes rising up, the range of radiation and the association between different regions will be wider and the number of tourists will grow accordingly. The civil aviation of our country has developed dramatically from 1978 to 2013 due to the reform and opening up policy. During this period, the number of international air routes has grown sharply from 12 to 427 and the navigable city international of scheduled flights have covered 118 of 50 countries. A statistic report released by the International Civil Aviation Organization reveals that the turnover of international passenger air transport has increased by 7.4% mainly due to the strong demand from emerging economies for both business and leisure travel. Then this report predicted that China will be one of the most quickly developing countries in terms of air passenger transportation. Generally speaking, passenger traffic is a symbol of the connection between two regions. That means the connection between two regions will be tighter and the number of tourists will increase if the passenger traffic rise up. As is clearly shown in the following graph, the international air transport of China is growing steadily every year because of the quickly developing economy and tighter connection with other countries.

Tourism is not only one of the largest industries in the world but also one of the fastest-growing industries in the service sectors and one of the power sources to promote the global economy. According to a study conducted by the World Tourism Organization, the number of international tourists is increasing rapidly by the rate of 4.3% every year and the number will reach 1.6 billion in 2020. As can be seen from Figure 1, the number of entry-exit passengers rise up steadily every year in general and this tendency has become more obvious after China's accession to WTO. The whole change process can be divided into three stages. First, from 1992 to 2001, the increase of entry-exit tourism was slow because of the weak demand for travel due to the limitation of economy and policy. Second, from 2001 to 2007, Chinese economy experienced a fast development and the demand for travel grew up rapidly after China's accession to WTO. As a result, the entry-exit tourism went into a high-speed growth period. Third, from 2008 to this day, the number of entry-exit passengers fell in 2008 due to the global financial crisis but the figure began to recover in 2009. Although the absolute number shows a tendency to increase steadily by the large, the growth rate fluctuates obviously over the same period.

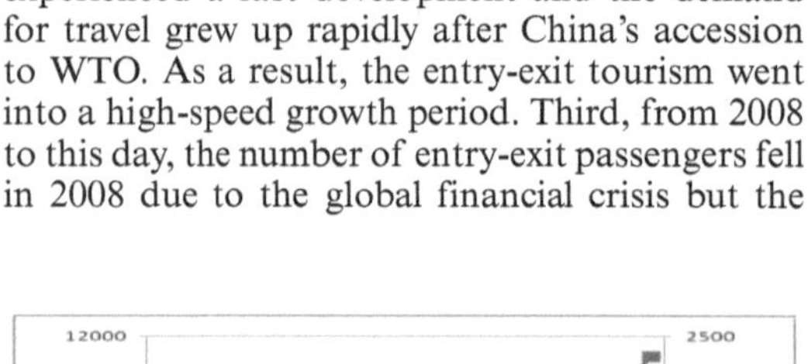

Figure 1. Caption relationship of entry-exit tourism and international passenger traffic.

3 EMPIRICAL ANALYSIS

3.1 *Data sources and methods*

This paper is based on the date of "Statistical Yearbook of China Tourism" "China Statistical Yearbook" and "a statistical viewing the Civil Aviation". The international passenger traffic of China aviation, C, and entry-exit tourism are regarded as object using the econometric software Eviews carries on the corresponding analysis through time series of unilateral root tests, OLS regression analysis and Granger causality analysis.

3.2 *Times series of unilateral root tests*

Firstly, it is based on the cointegration of Econometric analysis. Using unit root test method examines the long-term stable development relationship of the air transport industry and inbound-outbound passenger flow. Secondly, after the unit root test of international passenger traffic ((ln(LK))), international routes ((ln(HX))), the number of entry-exit tourism ((ln(LY))), under the premise of stable data, the paper uses cointegration analysis to analyze the date. Under the principle of minimum AIC, after the first round of the ADF test, ln(LY), ln(HX) and ln(LK) are not stable, after the differentially processed, new indicators are obtained. Test results are shown in Table 1.

$$\Delta \ln(LY) = \ln(LY_t) - \ln(LY_{t-1})$$
$$\Delta \ln(HX) = \ln(HX_t) - \ln(HX_{t-1})$$
$$\Delta \ln(LK) = \ln(LK_t) - \ln(LK_{t-1})$$

As can be seen from Table 1, at the 1% level of significance, unit root are presenting in the variable ln(LY), ln(HX) and ln(LK), and time series data are showing the instability, but the unit root about

Table 1. Test results of stationary time series variables.

Variables	Test mode 1	ADF	ADF thresholds	Conclusion
ln(LY)	(c,t,0)	–0.608251	–3.80854	Unstable
ln(HX)	(c,t,0)	–1.263632	–3.80854	Unstable
ln(LK)	(c,t,0)	–1.268294	–3.80854	Unstabe
Δ ln(LY)	(c,t,m)	–4.827633	–3.80854	Stable
Δ ln(HX)	(c,t,0)	–5.177635	–3.80854	Stable
Δ ln(LK)	(c,0,0)	–5.817189	–3.80854	Stable

the associated difference in 1% level of significance does not exist, all are stability data, appearing the first-order single whole time series. Therefore, ln(LY), ln(HX) and ln(LK) meet the premise of cointegration.

3.3 *OLS regression analysis*

After the unit root test showed a single whole sequence order, we can further test the long-term equilibrium relationship between variables. In model, LY represents for entry-exit tourism, HX international routes, LK International passenger traffic of China aviation, and LK, HX are explanatory variables, LY is dependent variable, u is other distractor, the paper uses logarithmic regression equation to reduce and eliminate the effects of heteroskedasticity and other factors, the results are showed as follows:

$$\ln(LY_t) = 0.2162 + 0.8922\ln(HX_t) + 0.4757\ln(LK_t)$$

$$T\ (-0.802731) \quad (3.312816) \quad (2.218835)$$

$$P\ (0.4326) \quad (0.0039) \quad (0.0396)$$

$$R^2 = 0.979 \quad \overline{R}^2 = 0.977542 \quad F = 436.2840$$

R^2 and F are respectively 0.9775,436.2840. We can know that the regression equation has a high overall level of significance, and thus, the estimated econometric model has better explanatory power. Coefficient of ln(HXt) is 0.8922, standing that international routes gains e, the number of entry-exit tourism will add 8922.Under the confidence level of 10%, international routes are not significant. The higher correlation of HX and LK may the main reason. After we control LK, HX plays little influence on LY, this may make LK act on HX indirectly. In order to analyze the theory among them, we examine the relationship among the three via Granger causality test.

3.4 *Granger causality test*

Based on the above analysis, we can use Granger causality test to further test. After lag order is determined, the results are showed in Table 2.

From the Table 2, under the significance level of 5%, ln(LY) and ln (LK) haven't bidirectional Granger causality, p are respectively 0.5725, 0.1239, it shows that both don't promote the development of each other clearly, that is to say, but the more is LK, the passengers of traveling may increase. ln(HX) is not Granger cause ln(LY), p is 0.8568, accepting the null hypothesis, but that is not saying that HX is not promoting LY. The gain of HX broaden the increasing needs of tourism market. ln(HY) is Granger cause ln(HX), p is 0.0435, refusing the null hypothesis, indicating that LY can promote the development of international routes.

Table 2. Results of Granger causality test.

The null hypothesis	DOF	P	Conclusion
ln(HX) is not Granger cause ln(LY)	2	0.8568	Accept
ln(LY) is not Granger cause ln(HX)	2	0.0435	Refuse
ln(LK) is not Granger cause ln(LY)	2	0.5725	Accept
ln(LY) is not Granger cause ln(LK)	2	0.1239	Accept

4 CONCLUSION AND RECOMMENDATION

The following are conclusions we can get from the above analysis:

1. Based on the analysis of time series data, a long-term stable equilibrium relationship is showed in the exit-entry tourism passenger flow with the number of international airline routes and civil aviation of China's international passenger traffic.
2. The relationship existed in the exit-entry tourism and China's civil aviation passenger traffic is a significant correlation, and a positive correlation relationship. In other words, if the number of international airline routes and international civil aviation passenger traffic are on the rise, there will be a dramatic increase in numbers of entry-exit passengers. The result of Granger Causality Test suggests that the improvement of exit-entry tourism is beneficial to the increase in international civil aviation passenger traffic.

With the proceeding of economy globalization and integration, the mutual promotion of civil aviation and tourism will be more and more significant. In order to make full use of the positive role of civil aviation in promoting the development of tourism, some essential measures should be taken:

1. Establish a long-term strategic cooperation mechanism between the civil aviation and tourism. Open special exit-entry tourism routes and provides specific civil aviation services to make civil aviation passenger transport and exit-entry tourism go forward hand in hand.
2. Make out reasonable allocation of civil aviation transport capacity and draw up rational planning of route network with tourism demand as guidance in order that the development of civil

aviation and tourism complement each other and win together.

3. Provide encouragement, policy support and guidance for airlines to adjust and optimize the industrial structure so that the allocation of tourism and aviation resources can be optimized and the integration of tourism and aviation resources can be maximally realized.

REFERENCES

[1] Cejas R R M. Tourism growth versus airport environmental capacity: An application of Ramsey pricing to Spanish tourist airports [J]. Transportation Research Part D:Transport and Environment, 2010, 15(3): 175–178.

[2] X.G. Wu. Correlation Analysis of the Development of Tourism and Economic Growth in China[J]. Journal of Applied Technology College of Shanghai, 2003, 3(4):238–241.

[3] X.J, Yang, D. NIU, B.H. WU. Travel behavior and evaluation of spatial patterns, [J]. Journal of Economic geography, 2000.

[4] H.Y. Chen. Analysis of the relationship between transportation and tourism using the partial least squares regression method [D]. Shanghai Maritime University, 2002.

[5] B. Zhou, Y.N. Li. The analysis of the interaction of civil aviation transportation airport layout and development of tourism resources in China [J]. Journal of Industrial and Commercial University Of Chongqing, 2004(6): 78–82.

[6] X.K. Wang. Transport in the mechanism research of tourist flow. [J] Research and discussion, 2010.

[7] J.J. Sun, G.N. Sun, L.F. Wang. Chinese Tourism Association in third industries driving analysis since 1982[J]. Journal of Progress in Geography, 2011, 30 (8):1047–1055.

[8] D.X. He, C.L. Huang, S. B. Liang. Correlation analysis of China development of tourism and air transport industry. [J]. Journal of Tropical Geography, 2007.

[9] G.S. Yang. The coordinated development of tourism and air transport aviation. Transport aviation, 2011.

[10] H.D. Hu, L.M. Zhao. Study on the correlation of passenger transport industry of China's tourism industry and civil aviation and Enlightenment. Business management, 2011.

[11] J.F. Wu, R.P. Ren, L.N. Han, Y.Q. Wang. The influence of characteristics of international aviation and China network structure of inbound tourism. [J]. Journal of Economic geography, 2012.

[12] Z.F. Wang. Analysis of evolution and the coordination of differences of inbound tourism flow and air transport network—taking the southwest area as an example. Geographical research, 2012.

Engineering Management and Industrial Engineering – Leung (Ed.)
© 2015 Taylor & Francis Group, London, ISBN 978-1-138-02773-2

Analysis and anticipation of inbound tourism market for Jiangmen

S.Y. Luo & J.H. Xiao
School of Economics and Management, Wuyi University, Jiangmen, Guangdong, China

ABSTRACT: The number of inbound stay-overnight tourists is a significant indicator of measuring the tourism strength of a place, while accurate analysis and anticipation of inbound stay-overnight tourists plays a vital role in the formulation and implementation of related tourism policy measures. In this paper, a deep analysis is conducted according to the horizontal data of inbound tourism market in Jiangmen from 2007 to 2013, and Grey GM (1,1) Residual Error Modified Model is established. According to the prediction results, the inbound tourism market develops vigorously in Jiangmen and the number of stay-overnight tourists would be keeping a fast growth. Furthermore, some new thoughts and suggestions were given to enlarge the inbound tourism market of Jiangmen.

1 INTRODUCTION

Inbound tourism is a significant constituent part of the tourist industry, playing a vital role in improving the quality of tourist industry, adjusting the structure of tourist industry, increasing the foreign currency earnings and enhancing the international exchange[1]. With the development of economic globalization, people's life quality and cultural level keeps improving constantly, and the diversification, individuation and internationalization of tourism demand tendency grows more and more evident. Moreover, the inbound tourism market also displays strong vitality. According to the *Tourism Spotlight in 2013* released by World Tourism Organization, in the past six decades, the international tourism developed vigorously, and tourists reached an unprecedented scale. The number of international tourists throughout the world increased to 278 million in 1980, 528 million in 1995, till the 1,035 million in 2012, which was 40.4 times of the number in 1950 with an annual growth of 6.4%. According to the *Statistical Report of World Tourism in 2013* released by UNWTO, even when the global economy was stagnant, the international tourists in 2013 still increased by 5%, when compared to the 2012, reaching 1.087 billion. Moreover, it is estimated that the number of international tourists in 2014 will keep increasing by 4% to 4.5%. Furthermore, China has kept its status as the third largest inbound tourist destination since 2010. Considering the increasingly flourishing international tourist market, coastal tourist areas in China propose measures of constructing international coastal tourist attraction, and the correct analysis and anticipation of inbound tourists is a significant premise of the healthy development of international tourist destination. Jiangmen City is the 'first hometown of overseas Chinese', a great tourism resource city in Guangdong, the garden in the backyard of the Pearl River Delta, as well as the west gateway of Guangdong. In the recent decades, the inbound stay-overnight tourists received by Jiangmen present a rapid growth trend, achieving outstanding achievements. However, as for the entire province, it still lags behind. In order to draw the attention from related tourism sector, as well as the confidence in developing international coastal tourist resort, the current status of inbound tourism market in Jiangmen is analyzed in detail, and grey system theory is applied for constructing the grey GM (1,1) residual error modified model. Finally, new thoughts and suggestions for the inbound tourism development are proposed according to the current status analysis and anticipation results.

2 CURRENT STATUS OF INBOUND TOURISM MARKET IN JIANGMEN

It can be seen from the statistics (Fig. 1) of inbound stay-overnight tourists (tourists from Hong Kong, Macao, Taiwan and foreign countries) received from 2007 to 2013 in Jiangmen that the number of inbound stay-overnight tourists in Jiangmen maintained a steady growth, with an annual average growth rate of 23.45%. Tourists from Hong Kong, Macao and Taiwan have always been the main body of the inbound tourist market, and its proportion always remains above 84%. As for the total number, tourists from Hong Kong, Macao and Taiwan received in Jiangmen increased from 514,700 in 2007 to

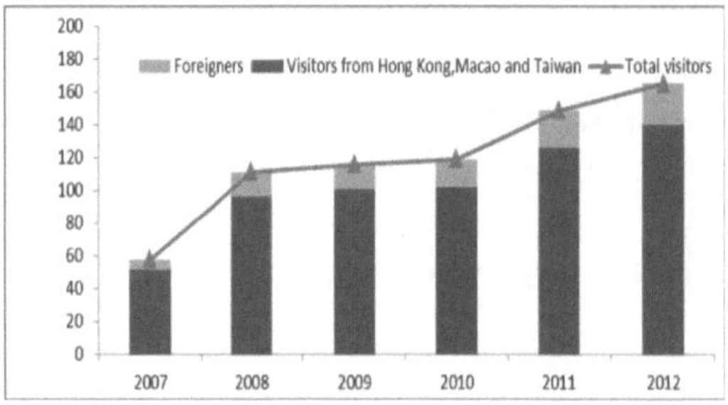

Figure 1. Variation trend of inbound stay-overnight tourists in Jiangmen (Unit: 10,000 people).

1,398,400 in 2012, increased by 1.87 times. Seen from the growth rate, the number of tourists from Hong Kong, Macao and Taiwan has always been in stable growth trend, in which the growth rate in 2008 was the highest, about 87.03%, but it grew slowly in 2009 and 2010 due to external factors. In 2011, the inbound tourism market revived, receiving 1,256,200 tourists from Hong Kong, Macao and Taiwan, increased by 30.04% when compared with the number in 2008. In 2012, it received 13,984 tourists from Hong Kong, Macao and Taiwan, suggesting the recovery of the normal development track.

The proportion of foreign tourists among the Jiangmen's inbound stay-overnight tourists is relatively small, ranging from 10% to 15%, but the growth rate was 32.79% in 2007 and 2012, faster than the annual average growth rate of tourists from Hong Kong, Macao and Taiwan, and it reached 147.33% in 2008. The number of foreign tourists in 2012 was 252,500, increased by 3.13 times when compared with the 61,300 in 2007, fully suggesting that the potential of foreign tourist market is huge, and the proportion of foreign tourists in the inbound tourist market in Jiangmen presents a rapid growing trend. Consequently, Jiangmen shall take this opportunity to develop the foreign tourist market.

It can be seen from the variation trend of total inbound stay-overnight tourists that it maintained a growing trend in the past years. In 2007, Jiangmen received 575,900 inbound stay-overnight tourists. However, it suffered from the international financial crisis in 2008, and the global economy was weakening. As a result, other tourist attractions were impacted to some different degrees with decreasing inbound tourists, but Jiangmen achieved a good development momentum, and the inbound stay-overnight tourists reached 1,113,900, enjoying the maximum growing rate, about 93.44%. In 2009 and 2010, the global economy was in the recovery period, and the inbound stay-overnight tourists increased by 3.34% slowly. In 2011, the inbound tourism market recovered, receiving 1,486,400 inbound tourists, increased by 33.43% when compared to 2008. Seen from the inbound stay-overnight tourists in Jiangmen over the past years, it has made certain achievements, but according to the *'Tenth Five-Year' Plan of Tourist Industry in Jiangmen* issued by its government, the inbound tourism still lags behind, and it ranks poorly among cities and provinces in the Pearl River Delta. Consequently, it is considered that it is quite necessary to predict and analyze the future inbound stay-overnight tourists in Jiangmen, so as to be aware of the potential in developing the inbound tourist and enhance the confidence in developing the inbound tourist market.

3 PREDICTION OF INBOUND SATY-OVERNIGHT TOURISTS IN JIANGMEN

3.1 *Construction of grey GM(1,1) residual error modified model*

Grey prediction method is an approach of analyzing, modeling, solving and predicting the grey system with uncertain factors, and it has already been widely applied in various aspects, such as the economic management, social system, agriculture, military, etc. The basic principle of grey prediction method is to generate the number sequence with the accumulation and reduction of original data and establish corresponding differential equation model, for carrying out quantitative analysis of the development trend of economic system. With low demands on data, the grey prediction model can weaken the randomness of the system and improve the precision of prediction. It is especially applicable for the modeling of little data, and it is quite advantageous in the short-term prediction. It has been discovered by referring to substantial document literatures that firstly, there are various factors impacting the development of tourist market, and it cannot conduct accurate quantitative evaluation for each factor, but it can establish the grey model for tourist market with grey system theory, dig the internal principles of the original data and make accurate evaluation. Secondly, grey GM(1,1) prediction model has been widely applied in the inbound tourism market, but it has been proved by researches that[2] some of the grey prediction results can reach the required precision, while some are characterized by huge mistakes, over 10%, and it is difficult to grasp the prediction error. Therefore, residual error modification is conducted for the grey prediction model for compensating the error caused by uncertainties and improving the precision of prediction. Thirdly, research achievements targeting at the inbound tourism market in Jiangmen are nearly blank. On that basis, according to the *Statistical Yearbook of Jiangmen* and the official website of tourist administration, the inbound stay-overnight tourists from 2007 to 2012 are counted (Table 1), the GM(1,1) prediction model of inbound stay-overnight tourists from 2007 to 2012 was constructed. And then, the prediction precision is improved through residual error modification. Eventually, the total inbound tourists in Jiangmen are predicted.

Step 1: the original series of inbound tourists in Jiangmen $X^{(0)}=[x^{(0)}(1), x^{(0)}(2), \ldots, x^{(0)}(n)]$; new series are generated from primary accumulation $X^{(1)}=[x^{(1)}(1), x^{(1)}(2), \ldots, x^{(1)}(n-1)]$, in which,

$$x^{(1)}(k)=\sum_{k=t}^{t} x^{(0)}(k) t=1,2,\ldots,n-1 \tag{1}$$

The GM(1,1) model of accumulated generation series $X^{(1)}$ shall be generated, and the corresponding differential equation is:

$$\frac{dx^{(1)}}{dt}+ax^{(1)}=\mu \tag{2}$$

where, μ is the constant input of the system, named as endogenous control grey level, and a is the development coefficient. This differential equation shall be worked out with the least square method, and the least square estimation is obtained:

$$\hat{\alpha}=\begin{pmatrix}\hat{a}\\ \hat{\mu}\end{pmatrix}=(B^T B)^{-1} B^T X_n \tag{3}$$

where, B is the (n−1)-order two-dimensional accumulation matrix, $X_n=\left(x^{(0)}(2), x^{(0)}(3), \ldots, x^{(0)}(n)\right)^T$ is the constant term. The estimated value $\hat{a}$ and $\hat{\mu}$ shall be brought into the differential equation for obtaining the prediction model of $X^{(1)}$:

$$\hat{x}^{(1)}(t+1)=\left[x^{(1)}(0)-\frac{\hat{\mu}}{\hat{a}}\right]e^{-\hat{a}t}+\frac{\hat{\mu}}{\hat{a}} \tag{4}$$

The $X^{(0)}$ prediction model is obtained with primary reduction for $X^{(1)}$:

$$\hat{X}^{(0)}(t+1)=\hat{X}^{(1)}(t+1)-\hat{X}^{(1)}(t) \tag{5}$$

Table 1. Original data of inbound tourists in Jiangmen (Unit: 10,000 people).

Year	2007	2008	2009	2010	2011	2012
Tourists	57.59	111.39	115.89	119.04	148.64	165.09

And when $t=1,2,\ldots,n-$, $\hat{x}^{(1)}(t+1)$ is the fitted value; when $t>n$, $\hat{x}^{(1)}(t+1)$ is the predicted value.

According to the above equation and data in Table 1, the GM(1,1) prediction model of inbound stay-overnight tourists in Jiangmen is:

$$\hat{x}^{(1)}(t+1)=900.114524e^{0.109815t}-842.524524$$
$$\hat{x}^{(0)}(t+1)=568.9809e^{0.109815t} \tag{6}$$

Step 2: accuracy check. Residual check and posterior ratio inspection is conducted for the original series, and the precision of the model shall be judged by referring to the model precision check list[3]. If both are acceptable, the established model can be applied for prediction.

The predicted inbound tourists from 2008 to 2012 in Jiangmen, corresponding residual error and posterior ratio shall be calculated with the inbound stay-overnight tourist GM(1,1) prediction model, and the results are shown in Table 2. It can be learnt that the average relative error is 4.04782%, and it is acceptable in the residual error check. The posterior ratio $C=0.1833<0.35$, the small error probability $P=1>0.95$. The precision of the model is excellent, and the posterior ratio is acceptable. Although the residual error and posterior ratio check of the model passes, the precision of the model is not high enough. There was a huge discrete in 2010, and the maximum reached 11.0995. In order to obtain higher precision, residual error modification is conducted for the inbound stay-overnight tourists grey GM(1,1) model.

Step 3: residual error modification. According to the GM(1,1) predicted model (equation 6) of inbound stay-overnight tourists in Jiangmen, the predicted series $X^{(1)}$ is obtained. Suppose $t_0=3$, the modeling residual end-piece of

$$\varepsilon^{(0)}(t)=\left|x^{(0)}(t)-\hat{x}^{(0)}(t)\right| \tag{7}$$

can be obtained, and its primary accumulation production series is:

$$\varepsilon^{(1)}=\left[\varepsilon^{(1)}(t_0), \varepsilon^{(1)}(t_0+1), \ldots, \varepsilon^{(1)}(n)\right] \tag{8}$$

Corresponding GM(1,1) model is established for the production series $\varepsilon^{(1)}$, and corresponding grey GM(1,1) model is established, the prediction model of primary series $\varepsilon^{(1)}$ is:

$$\hat{\varepsilon}^{(1)}(t+1)=-18.496746e^{-0.826268t}+19.211705 \quad t\geq 3 \tag{9}$$

The predicted value of $\varepsilon^{(1)}$ prediction model is added into the prediction model of $\hat{X}^{(1)}$

Table 2. Inspection table of GM (1,1) model and model after residual error modification.

Year	No.	AV	PV	PVM	Res	ResM	Rel	RelM
2008	2	111.39	104.478	111.2777	6.912	0.1123	6.2052	0.1008
2009	3	115.89	116.605	112.8494	−0.715	3.0406	0.6169	2.6237
2010	4	119.04	130.1395	124.59	−11.0995	−5.55	9.3242	4.6623
2011	5	148.64	145.2451	149.0672	3.3949	−0.4272	2.284	0.2874
2012	6	165.09	162.1039	161.3412	2.9861	3.7488	1.8088	2.2708
GM(1,1)			Post-test odd: C = 0.1833			Small-error probability P = 1		
After residual error modification			Post-test odd: C = 0.0975			Small-error probability P = 1		

Note: AV = Actual Value; PV = Predicted Value; PVM = Predicted Value after Modification; Res = Residual error; ResM = Residual error Modification; Rel = Relative error; RelM = Relative error after Modified.

(equation 6), and the residual error modified model obtained is:

$$\hat{x}^{(1)}(t+1) = -37.587852e^{-0.37299t} + 42.490329 \quad t \geq 3 \tag{10}$$

The model is modified with the three stimulation value $t = 4, 5, 6$, and it can be seen from the modified precision check list (Table 2) that the precision of the model is excellent, and the average relative error is 1.989%, the maximum absolute residual error is 10.4011, the posterior ratio is 0.0975, which are all smaller than that before the modification. It can be concluded that the modified GM(1,1) model receives higher precision, and it can be applied for the short-term prediction of inbound stay-overnight tourists in Jiangmen.

3.2 *Prediction of grey residual error modified model*

According to the prediction model of inbound stay-overnight tourists in Jiangmen:

$$\hat{x}^{(1)}(t+1) = -37.587852e^{-0.37299t} + 42.490329$$

The prediction results of inbound stay-overnight tourists from 2013 to 2016 are shown in Figure 2. According to the real data released by the government of Jiangmen, the number of inbound stay-overnight tourists in 2013 is 1,695,900, while the number predicted with grey GM(1,1) residual error modified was 1,778,300, and the accuracy rate is as high as 95.37%. The following conclusion can be drawn from Figure 2:

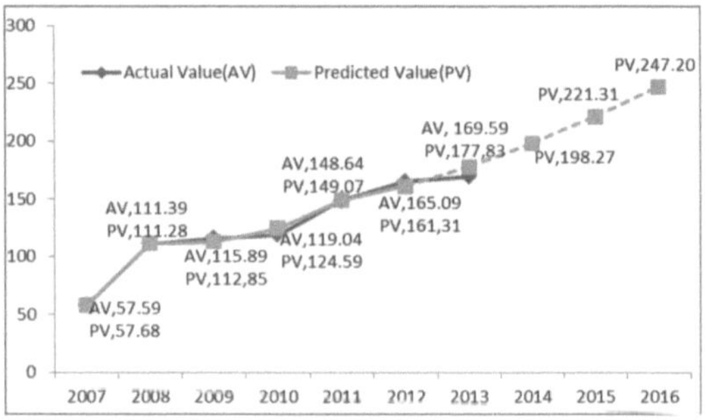

Figure 2. Fitted curve of the actual and predicted inbound stay-overnight tourists in Jiangmen.

1. The actual and predicted value receives quite good fitting results, and large discrete only occurred in 2010 and 2013.
2. Jiangmen has already obtained certain stable inbound tourist base, and it will have a bright future if there are no other influencing factors. The inbound stay-overnight tourists in Jiangmen will maintain a stable growth trend, and in 2015 the number of inbound stay-overnight tourists will break through two million.
3. In front of the rapid growth of inbound stay-overnight tourists, the inbound tourist market will have a brilliant future, and it may become the new engine driving the economy in Jiangmen.
4. In this paper, the dynamic changing process of inbound stay-overnight tourists is regarded as a grey process, and corresponding grey GM(1,1) residual error modified model is established. It has been agreed that the practice has a high reliability, and it can provide scientific basis for the tourism decision of related departments.
5. It shall be pointed out that the inbound tourism market is a complicated system. It is considered as a grey system, and according to historic data, the prediction results obtained with the grey residual error modification model is scientific but limited.
6. The development coefficient of the predicted model in this paper $a = 0.37299 \in (0.3, 0.5]$, suggests that the model is only applicable for short-term prediction, and it can only reflect the trend of long-term prediction. Therefore,

the predicted value of 2014–2016 is a relatively conservative estimation.

4 SUGGESTIONS FOR THE DEVELOPMENT OF INBOUND TOURISM IN JIANGMEN

4.1 *Consolidate the current passenger source, attract new passenger source, and adjust the market structure constantly*

Generally speaking, at present, Jiangmen has already obtained certain stable inbound tourist foundation. In order to consolidate the current regular customer market, Jiangmen shall further dig the distinctive culture, such as 'overseas Chinese' culture, watchtower culture, coast defense culture, etc. To package the tourist attractions with culture, set up the tourism features in Jiangmen, create the tourist brand, strengthen the promotion among tourists from Hong Kong, Macao and Taiwan, as well as overseas Chinese, and then enlarge its reputation and influence in domestic and foreign tourist market. Moreover, it can also cooperate with related tourist agencies in Hong Kong, Macao and Taiwan, and make long-term mechanism of regional tourism cooperation. What's more, the proportion of foreign tourists is relatively low, and it shall attach importance to the expansion of new tourists. For instance, the customer market of Southeast Asia makes rewarding mechanism to encourage the travel agencies in countries of tourists to bring tour group to Jiangmen, or carry out regional cooperation with countries of tourists, so as to transport tourists to each other. Eventually, it shall regulate the market structure, develop high-demand, high-consumption and high-accomplishment inbound tourism, strengthen the macro-control and configuration force in recreation, reception, traffic, commerce, marketing, tourism, etc. and try to form a diversified market structure.

4.2 *Construct the international tourist island, and intensify the development of international tourist products*

At present, the international tourism prevails all over the world, and island tourism is quite popular among international tourists for its unique geographical condition, thus becomes a hot choice among the international tourism. Successful international tourism islands in the world, such as American Hawaii, Japanese Okinawa-jima, South Korean Jeju, and so on show that international tourism island is a new tourist body which can help driving the local economy. It can not only attract more international tourists and generate more tourist income, but also stimulate the domestic demand and enlarge the social employment. Jiangmen is provided with the condition of developing international tourism island, such as the island tourism resource, regional conditions, traffic environment, development condition, inbound tourist base, etc. Consequently, the construction of international tourist island will become the new engine of developing the economy in Jiangmen, and it will also provide a new thought for the development of economic society in Jiangmen. Therefore, it shall fully discuss the feasibility of constructing international tourism island, make related policies and measures, and promote the process of constructing international tourism island with development planning. According to the statistics, there are 107 tourist attractions and tourist sites, which are relatively low in level, and it is difficult to attract foreign tourists. In order to enlarge the development scale of international tourist products, it shall realize the internationalization and informatization of operation, management and service, so as to form a competitive international tourist product system with good reputation throughout the world.

4.3 *Strengthen the travel service promotion, and improve the international reputation*

Reputation plays an immeasurable role in the development of sightseeing place, and distinctive tourist image may improve the reputation of the sightseeing places, which is one of the key factors in attracting the tourists. However, as long as it is a successful tourist image lies in the promotion and advertising. Jiangmen can improve from three aspects: firstly, setting up the tourist image 'first hometown of overseas Chinese, and enchanting scenery in Jiangmen, paying attention to the overall package, and promoting with three combined methods, including the media advertising, festival advertising and theme advertising by integrating location of Jiangmen in tourists, cities and provinces, as well as the features of tourism resource, so as to reach the brand effect. For instance, the successful holding of overseas Chinese carnival in 2013 received a good advertising effect, and improved the reputation and influence in both China and foreign countries. Secondly, taking the combined measures of bringing in (reward mechanism for the media in places of tourists, and agencies) and walking out (advertising in countries with tourists, and participating in international tourist conference actively) to enlarge the influence. Thirdly, inviting overseas Chinese or stars with international influence, such as Andy Lau, Eric Tsang, etc. to be the tourism image ambassador in Jiangmen.

4.4 *Deepen the talent cultivation system reform, improve the talent quality*

The development of inbound tourist market sets more demands on the high-quality tourism talents, but the individual and overall quality of tourism employees is relatively low, only with high school or junior college education level, and insufficient talents reserve, and it lacks high-quality tourism talents. On that basis, Jiangmen can conduct talent cultivation system reform from two aspects, firstly, establish and perfect the tourist resource development and management mechanism, encourage enterprises to focus on the management and training of talents and improve the professional quality and overall quality. On the other hand, by relying on the local colleges and universities, it shall establish related tourism majors for adapting to the international demands, construct practice base with local tourist enterprises, arrange the schooling resources of schools ad enterprises, build the linked multi-discipline and multi-level open education platform and cooperate with other colleges and universities out of Jiangmen city, cultivate tourism talents and transport them to Jiangmen or bring in professional talent with preferential policies.

REFERENCES

[1] Hu Xue-jun, Shan Yuan-yuan. 2008. Analysis and Anticipation on Inbound Tourism Market of Hangzhou. Journal of Guilin Institute of Tourism (6): 370–374.

[2] Huang Yinzhu, Yang jianming, Chen Yali. 2009. Forecasting of Inbound Tourism Demand Using Grey Dynamic Models in Fujian Province. Journal of Beijing International Studies University 07: 42–47.

[3] Liu Yu-feng, Sun Hu, Yuan Zhi-hua. 2008. Analysis and Anticipation of Overseas Tourism Market to Shanxi. Journal of Arid Land Resources and Environment 01: 113–117.

[4] Deng Ju-long. 1990. Tutorial of Grey System Theory. Wuhan: Press of Huazhong University of Science and Technology.

[5] Zhu Xiao-hua, Yang Xiu-chun, Cai Yun-long. 2005. Forecasting Models of Tourism Passenger Based on The Grey Theory—A Case of The International Tourism Passenger Source of China. Economic Geography 02: 232–235.

Engineering Management and Industrial Engineering – Leung (Ed.)
© 2015 Taylor & Francis Group, London, ISBN 978-1-138-02773-2

Point-to-point transportation analysis for domestic travel in Malaysia

F.I. Romli & H. Rashid
Department of Aerospace Engineering, Faculty of Engineering, Universiti Putra Malaysia, Serdang, Selangor, Malaysia

ABSTRACT: Point-to-point travel refers to the trip taken from the initial departure point to the final destination point, which can be further categorized as either direct or non-direct. Direct point-to-point travel indicates that the trip is made with a single transportation mode from beginning until the person reaches the final destination. On the other hand, the non-direct type of point-to-point travel involves several modes of transportation throughout the journey. Comparison analysis is made between the available modes of domestic transportation such as buses, trains, commercial flights and cars to determine the preference and the market potential for direct point-to-point travel means. Several evaluation criteria including time on transport, time of waiting, cost of transport, cost of other expenses, travel comfort, flexibility of departure point and privacy are used in this study. Based on the result from case studies of domestic travel in Malaysia, it shows that direct point-to-point means of transportation represented by cars generally have higher benefits in comparison to other means of transportation, especially in terms of time and travel costs. This signifies a big potential in introducing a better means of direct point-to-point transportation than current cars like the dual-mode Personal Air Vehicle (PAVE) for domestic travel in Malaysia.

1 INTRODUCTION

Transportation has brought interest in the discussion of feasible development due to the constant growth in personal mobility and its correlated environmental costs (Frändberg & Vilhelmson, 2011). Today, there are several different modes of transportation that are used by the majority of people in their daily life to travel from one point to another. Their choice of transportation mode depends on few personal considerations. For example, some people are mainly concerned on how much they have to pay for the transportation regardless of the time taken to travel while others may think the opposite. It has been suggested that travel demand is typically constrained by the time budget while the choice of travel mode is frequently dictated by the traveler's financial budget (Hahn, 2006).

In general, modes of transportation can be classified as direct or non-direct point-to-point. The direct point-to-point travel, or sometimes also known as direct door-to-door travel, is defined as a direct travelling from one point to the destination point using only a single mode of transportation. This means there is no need to change transportation mode throughout the journey. For instance, domestic travel between the cities within Peninsular Malaysia using a personal car is considered as direct point-to-point since the traveler can depart with his/her car directly from any departure point and arrive at the intended destination point with the same car in ideal circumstances. There is no need to use another different mode of transportation throughout the journey. In most parts of the world, the cars and motorcycles are the only currently available modes of transport that have the capability for direct point-to-point travel, depending also on available facilities between the travel points. It is anticipated that future personal travels will emphasize more on having a doorstep-to-destination journey at any time, in any weather and in as little travel time as possible (Marchman et al, 2002). This notion puts heavy emphasis on future capability of direct point-to-point travel.

In Malaysia, numbers of motor vehicles have increased over the years. From 2009 to 2010, cars have recorded the highest increment at 608,840 units whereas motorcycles are second with 501,677 units, followed by taxis (5,512 units), buses (2,568 units) and other transport vehicles like aircraft and trains with 21,510 units (Ministry of Transport Malaysia, 2010). It can be implied from the statistics that cars and motorcycles are the preferred mode of transportation for domestic travel in Malaysia. This can also be interpreted that many Malaysians prefer to have the capability for direct point-to-point travel. In support of this notion, this paper aims to further examine direct point-to-point domestic travel in Malaysia.

2 MODES OF DOMESTIC TRAVEL IN MALAYSIA

In Malaysia, there are few types of domestic transportation means that include land, maritime and air transportation. The examples of land transportation include both those on road and rail such as cars, motorcycles, buses and rail vehicles like trains, Light Rail Transit (LRT) and monorail. The maritime transport is usually associated with cargo that involves import and export of goods, apart from passenger transportation, similarly for air transportation. These transportations have utility fees based on the number of trips they generate, providing a direct connection between demands for transportation facilities and costs of constructing and maintaining them (Junge & Levinson, 2012). Other than that, walking and bicycling are also considered as transport means. They are among active transportation means that involve physical activities specifically for travel. Many travelers typically use the active transportation for short-distance travel due to common factors such as costs and indirect improvement of their health.

All the above transportations have their own link or network within Malaysia and they can be used interchangeably at points of operational intersections. For domestic on-ground travel in Malaysia, people can use rail vehicles or buses as their mode of transportation. In short, the monorail is used for public transportation within Kuala Lumpur while the only funicular railway line is in Pulau Pinang. LRT that is used in Kuala Lumpur has two lines of railway. The first one is the Kelana Line that covers the areas between Gombak and Kelana Jaya, while the other one is the Ampang Line that covers areas between Sentul Timur and Sri Petaling. Meanwhile, KTM Intercity provides alternative choice for passengers with a variety of couch seating categories and destinations including Peninsular Malaysia. KTM Commuter connects Kuala Lumpur city areas with the surrounding suburban areas such as Tanjung Malim, Rawang, Seremban, Sungai Gadut and Pelabuhan Klang. In addition, there are more than 100 bus companies that provide transportation services in Malaysia. A famous bus company is the state-run Transnasional that provides the most extensive coverage of bus routes service in Malaysia. The other companies like Alisan Golden Coach Express, LionStar Express, Unititi Express, KKKL, Delima Express, Konsortium Express, Sri Maju Yakin BG Express and Grassland, to name a few, also provide services for domestic travel. For air transportation services, domestic flights within Peninsular Malaysia are provided by the full-service national airline carrier that is Malaysia Airlines and several low-cost airlines such as Air Asia, FireFly and Malindo Air.

For ground transportation facilities, there are built highways that connect the roads in Malaysia. In this study, the North-South Expressway (NSE) has been used as the main facility to travel within the Peninsular Malaysia. The NSE runs from Bukit Kayu Hitam in Kedah, near the Malaysia-Thailand border, all the way to Johor Bahru at the southern part of Peninsular Malaysia. This expressway links many major cities and towns in the western Peninsular Malaysia, acting as the 'backbone' connection line of the west coast of the peninsula. It provides a faster alternative to the old state routes, thus reducing the travelling time between various towns and cities. On the other hand, available airports in Malaysia consist of international airport, domestic airport and also airstrip. The airports are built and located throughout Malaysia including Peninsular Malaysia, Sabah and Sarawak. The international airport, which is for flights arriving and departing from/to other countries, has larger facility and also longer runway compared to the domestic airport. Meanwhile, small airstrips offers runway and simple facilities like fueling equipment.

3 COMPARISON STUDY

Comparison between the available modes of transportation for domestic travel in Malaysia is done using the Technique for Order Preference by Similarity to Ideal Solution (TOPSIS). The key idea here is that the best alternative is the one that is as close as possible to the positive ideal solution and as far as possible to the negative-ideal solution. Positive ideal is a set of best-possible values for each of the evaluation criteria whereas negative ideal represents the worst-case of each evaluation criterion. In this study, several variables are identified as important evaluation criteria: time on transport, time of waiting, cost of transport, cost of indirect travel expenses like food, drinking water and snacks, flexibility of departure point, travel comfort and privacy. These criteria are further defined in Table 1 and a typical rating scale used in the evaluation process is as presented in the following Table 2.

Two case studies are done under the assumption of the normal road traffic conditions for this research paper. This is a rather significant assumption as traffic conditions can severely affect the evaluation criteria listed in Table 1. The first case is designed as a domestic travel from the Kuala Lumpur Convention Centre (KLCC), Kuala Lumpur to the Pekan Rabu in Alor Setar, Kedah. In the meantime, the second case considers the travel from the KLCC, Kuala Lumpur to the Johor Bahru City Square in Johor. Seven different weighting scenarios are developed to study

Table 1. Evaluation criteria for TOPSIS.

Criteria	Definition
Time on transport	Time taken for the overall journey, from the departure point up until the final destination point.
Time of waiting	Time spent waiting for the mode of transport to start the travelling journey.
Cost of transport	Total cost of travel due only to the fees or the incurred travel expenses such as toll fees, fuel, etc. of the transport.
Cost of indirect travel expenses	Total cost incurred for the travel excluding the cost of transport. This includes foods, drinks, etc. while waiting at the station.
Flexibility of departure point	Measure of flexibility level to start the travel from any departure point.
Travel comfort	Measure of travelling comfort throughout the journey.
Privacy	Measure of provided privacy level throughout the journey.

Table 2. Rating scale for TOPSIS.

Rating	Explanation
1	Performance level of the evaluation criterion is estimated to be ***very low*** compared to the perceived travel preference.
3	Performance level of the evaluation criterion is estimated to be ***low*** compared to the perceived travel preference.
4	Performance level of the evaluation criterion is estimated to be ***moderately low*** compared to the perceived travel preference.
5	Performance level of the evaluation criterion is estimated to be ***on average*** compared to the perceived travel preference.
6	Performance level of the evaluation criterion is estimated to be ***moderately high*** compared to the perceived travel preference.
7	Performance level of the evaluation criterion is estimated to be ***high*** compared to the perceived travel preference.
9	Performance level of the evaluation criterion is estimated to be ***very high*** compared to the perceived travel preference.

Table 3. Evaluation scenarios for TOPSIS.

Scenario	Description
Travel time	This scenario emphasizes on the shortest travel time taken for the journey. In this particular scenario, both shortest time on transport in the travel and time of waiting for the transport to arrive are the upmost importance. Hence, the weighting for time on transport and time of waiting for the transport are assigned as 0.25 each while the other criteria are assigned 0.1 each.
Cost	This scenario emphasizes on the lowest cost for the journey. In this particular scenario, both lowest cost of transport (fees) and cost of other expenses such as drinking water, meals and others during waiting time are the upmost importance. Hence, the weighting for cost of transport and cost of other expenses time are assigned as 0.25 each while the other criteria are assigned 0.1 each.
Departure point flexibility	This scenario emphasizes on the flexibility of travel for the journey. In this particular scenario, the best case is to be able to start the journey straightaway from the initial departure point without the need to go to a fixed hub. Hence, weighting for departure point flexibility is given as 0.25 while the other criteria are assigned 0.125 each.
Comfort	This scenario emphasizes on the comfort level throughout the travel for the journey. In this particular scenario, it is based on the comfortability of the passenger while on transport, time of arrival either day or night and how they feel when arrived at night. Hence the weighting for travel comfort is assigned as 0.25 while the other criteria are assigned 0.125 each.
Privacy	This scenario emphasizes on the level of privacy provided throughout the travel for the journey. In this particular scenario, factors such as seating arrangement (or group seating) and type of seating play a big role. Hence, the weighting for privacy is assigned as 0.25 while the other criteria are given 0.125 each.
Balance	In this scenario, all evaluation criteria are taken to be of equal importance. Hence, all criteria are assigned with equal weighting of 0.1428 each.
All rounded	In this scenario, it is taken that time on transport, cost of transport, flexibility of departure point and comfort are the important criteria with assigned weighting of 0.175 each while the other are assigned with 0.1 each.

the effects of the different travelers' preferences. This is essential to cover the diversity aspects of decision-making process in selecting available travel options and highlight the robustness of the travelling means.

For instance, if travel time is the most important selection criterion for the traveler in choosing his/her transportation option, this condition can be reflected with a higher weighting for time on the transport and also waiting time for the transport in comparison to the other evaluation criteria. These different scenarios are further described in Table 3.

4 RESULTS AND DISCUSSIONS

In the first case study, it involves travel trip from the Kuala Lumpur Convention Centre (KLCC) to Pekan Rabu in Alor Setar, Kedah. For the purpose of dictating the ticket fare and timetable of the available public transportation services, it is taken on Wednesday, 23rd April 2014 and the departure time is assumed to be at 10.20 a.m. Based on the evaluation criteria listed in Table 1 and using the rating scale as outlined in Table 2, the assessment is done for this travel trip for different main transportation modes for all seven different evaluation scenarios (and importance weightings) in Table 3. For each scenario, the results of closeness rating in TOPSIS for the different transportation modes are presented in Table 4.

A closer look at the results can be made by looking at the web chart presented in Figure 1. From the figure, it is observed that car consistently obtains a high closeness rating in most of the different evaluation scenarios considered in the TOPSIS evaluation. It places second on only two scenarios, which are the travel time and cost, while appears on top in all other scenarios.

Table 4. Closeness rating for case study 1.

Scenario	Bus	Train	Car	Airplane
Travel time	0.6620	0.1756	0.6621	**0.6872**
Cost	**0.7728**	0.5131	0.7343	0.4530
Departure point flexibility	0.5751	0.3630	**0.8130**	0.4942
Comfort	0.5788	0.3532	**0.8121**	0.5272
Privacy	0.5706	0.3535	**0.8209**	0.4728
Balance	0.6367	0.3701	**0.8012**	0.5292
All rounded	0.6150	0.4390	**0.7706**	0.4872

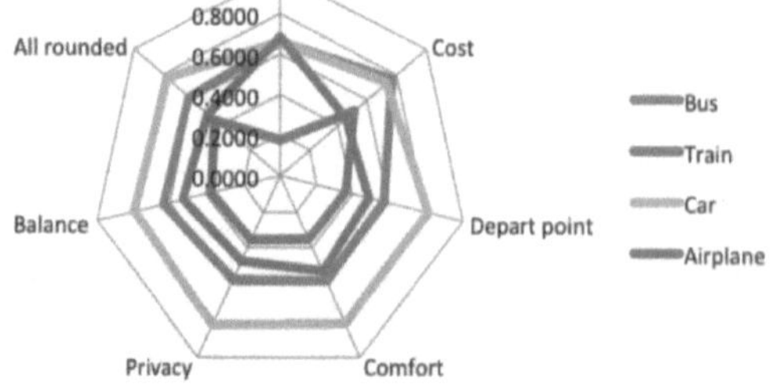

Figure 1. TOPSIS results for case study 1.

In the time travel scenario, an aircraft surely has the shortest travel time over the majority of the distance involved in the trip. Nonetheless, as reflected by a small score difference between taking flight and driving a car for the trip, the latter has indeed countered the aircraft's advantage with the benefits of not having to wait for the transport. The travel trip can be started straightaway by car but when an aircraft is chosen, the traveler may have to wait at the airport for the scheduled departure time, thus delaying the arrival time at the intended final destination.

On the other hand, in terms of cost, it appears that taking the bus for the trip will incur the least amount of travel costs. In addition to fuel, traveling by car also involves paying the highway's toll fares. Furthermore, the total costs of traveling by bus are being shared with other passengers, as already distributed in the bus ticket price. Nonetheless, the score difference is comparatively small between car and bus for the particular evaluation scenario, indicating that the total costs of using these two transportation means may not differ that much.

Overall, based on the TOPSIS results of first case study, it can be said that direct point-to-point travel as represented by car is the most preferable and robust for the different evaluation scenarios compared to the other considered means of transportation.

Similar evaluation process is done for the second case study, which is a travel trip from Kuala Lumpur Convention Centre (KLCC), Kuala Lumpur to Johor Bahru City Square in Johor. To maintain an appropriate comparison basis with the first case study, the ticket fare and timetable of available public transportation services are determined using the same travel date and time, which is on Wednesday, 23rd April 2014 and with the departure time assumed to be at 10.20 a.m. For each of the evaluation scenarios in this second case study, results of the closeness rating in TOPSIS for different transportation modes considered are presented in Table 5 and its associated web chart is depicted in Figure 2.

Table 5. Closeness rating for case study 2.

Scenario	Bus	Train	Car	Airplane
Travel time	0.4260	0.3412	**0.8423**	0.4456
Cost	0.6602	0.6588	**0.7682**	0.2379
Departure point flexibility	0.4509	0.4662	**0.8340**	0.3186
Comfort	0.4540	0.4496	**0.8330**	0.3559
Privacy	0.4497	0.4461	**0.5503**	0.3042
Balance	0.5006	0.4870	**0.8190**	0.3414
All rounded	0.5430	0.5252	**0.7921**	0.3711

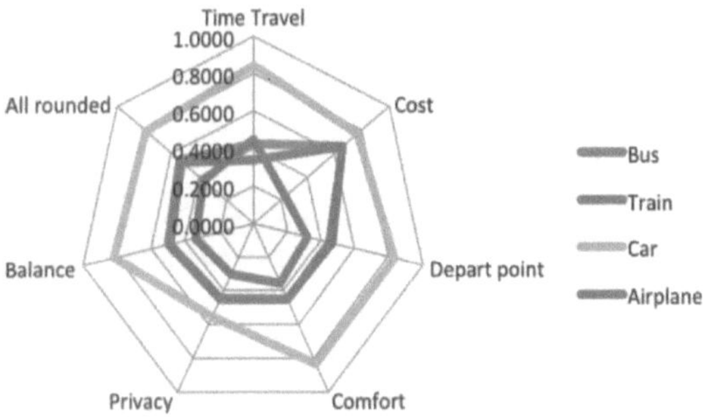

Figure 2. TOPSIS results for case study 2.

From Figure 2, car clearly is the preferred transportation means as it comes on top for all of the assessment scenarios. It should be noted that, while the cost of travel for car is not the lowest, it still beat the other transportation means due to a competitive travelling time. This highlights its high robustness in terms of satisfying the many different aspects of travelling preferences. Another thing to note from results of this second case study in comparison to the first case study is the improvement of score difference between train and bus. The main reason for this is the close availability of the scheduled public train service from KLCC to Johor Bahru with intended departure time. This reduces the waiting time and the overall travel time compared to bus, even though the latter takes much less on-the-road time to travel the distance to the destination. One of the things that is highlighted by this occurrence is the high dependency and influence of departure time for use of public transport. It is not as flexible as using private transportation means since the traveler has to follow the scheduled travel time. Overall, it can be concluded that car still emerge as the best solution for point-to-point travel from KLCC, Kuala Lumpur to Johor Bahru City Square, Johor.

4.1 *Personal Air Vehicle (PAVE)*

Based on the previous results, the preference of cars as the mode of domestic transportation means can be extended to the preference for having the capability of direct point-to-point travel. However, the high use of cars in Malaysia has also led to problems of traffic congestion. For example, in Klang Valley area alone where 21% of the nation's population or about 4.8 million people live, about 84% of them are car owners (Ariffin & Zahari, 2013). As the result, problems of traffic congestion become more severe and this situation leads to wasted travelling time and costs, increased fuel consumption, air and noise pollution, and also elevated accident and fatality rates (Kasipillai & Chan, 2008). Note that previous comparison case studies are conducted under the assumption of a normal traffic conditions. If problems of the ground traffic congestion are factored in, they may outweigh some of the prescribed benefits of travelling in cars. Because the need and preference of a direct pint-to-point travel cannot be simply ignored, there is a need for new alternative mean of transportation that can improve the capability of the current cars.

It has been suggested that application of PAVE in Malaysia could provide a sound solution and also offer economic and social benefits by reducing wasted time and cost from getting stuck in traffic congestion (Romli & Yaakob, 2014a). A dual-mode PAVE, also sometimes simply known as a flying car, technically combines performance elements of both ground and air transportation. The concept of such transportation means is not exactly new as it has already been envisioned a few decades ago in many futuristic, science-fiction movies, cartoons and books. Several design concepts for PAVE have already been researched in the United States and also European countries. For instance, a four-year myCopter project (from 2011 to 2014) has been funded by European Union to study the feasibility of having personal flying vehicles with the concept of city-wide skyway (Czyzewski, 2011). The PAVE concept considered in myCopter project is as shown in Figure 3.

To highlight the potential of PAVE for domestic travel use, the performance comparison between car and PAVE is necessary. As published in (Romli & Yaakob, 2014b), results of cost and travel time comparison are reproduced in Table 6 for travelling case that is almost similar to the one used in the first case study, which is from the Universiti Putra Malaysia (UPM) Campus, Selangor to Pekan Rabu, Kedah.

It could be seen that the expected performance of PAVE in both travel time and cost aspects exceeds

Figure 3. PAVE concept in EU myCopter Project (Czyzewski, 2011).

Table 6. Comparison of car and PAVE for travel trip from UPM Campus, Selangor to Pekan Rabu, Kedah.

Parameter	Car	PAVE
Travel time	4 hours 43 minutes	3 hours 8 minutes
Total cost	$50.00	$30.46

the capability of the current car. Hence, this raises a big potential for PAVE as an alternative transportation to cars for future direct point-to-point travel. It should be noted that the comparison was made in that study using published technical specifications and performance characteristics of Transition PAVE design as published by its manufacturer.

5 CONCLUSION

The purpose of this research is mainly to examine the direct point-to-point domestic travel capability in Malaysia. From the results of TOPSIS assessment for two sample case studies, it can be concluded that under normal traffic conditions the ideal result is to travel by car. This is mainly due to its flexibility and less time taken to travel, even though the overall cost of using a car is usually higher compared to the bus and train. With the limitation of the current cars, especially due to the rising problems of traffic congestion on the road, there is a need for a new alternative mean of direct point-to-point transport. Based on an example comparison between the performance of a car and a PAVE for the same case of a travel trip, it shows that PAVE seems to have a big potential edge to become better direct point-to-point mean of transport in the future.

ACKNOWLEDGEMENT

The authors acknowledge financial support for this project from Ministry of Education, Malaysia under the Fundamental Research Grant Scheme (FRGS)—FRGS/1/2012/TK09/UPM/02/1.

REFERENCES

Ariffin, R.N.R. & Zahari, R.K. 2013. The Challenges of Implementing Urban Transport Policy in the Klang Valley, Malaysia. *Procedia Environmental Sciences*. 17: 469–477.

Czyzewski, A. 2011. Personal Flying Vehicles Project Aims to End Road Congestion. in *The Engineer*. http://www.theengi neer.co.uk [Accessed: June 2014].

Frändberg, L. & Vilhelmson, B. 2011. More or less travel: personal mobility trends in the Swedish population focusing gender and cohort. *J. Transp. Geogr*. 19(6): 1235–1244.

Hahn, A. S. 2006. Next Generation NASA GA Advanced Concept. in *SAE General Aviation Technology Conference and Exhibition*, Wichita, USA.

Junge, J.R. & Levinson, D. 2012. Financing transportation with land value taxes: Effects on development intensity. *Journal of Transport and Land Use*. 5(1): 49–63.

Kasipillai, J. & Chan, P. 2008. Travel Demand Management: Lessons for Malaysia. Journal of Public Transportation. 11: 41–56.

Marchman, J.F., Interatep, N., Skelton, E., Mason, W.H. 2002. An Investigation of CTOL Dual-Mode PAVE Concepts. in *Technical Report to NASA Langley Research Center*, January 2002.

Ministry of Transport Malaysia. 2010. *Statistik Pengangkutan Malaysia.*

Romli, F.I. & Yaakob, M.S. 2014a. Personal Air Vehicle (PAVE) Application in Malaysia. in *International Conference on Advanced Logistics and Transport (ICALT) 2014*. 59–64, 1–3 May 2014, Yassmine Hammamet: Tunisia.

Romli, F.I. & Yaakob, M.S. 2014b. Travel Time and Cost Analysis of PAVE Application in Malaysia. *Applied Mechanics and Materials*. 629: 246–251.

Engineering education and training

Engineering Management and Industrial Engineering – Leung (Ed.)
© 2015 Taylor & Francis Group, London, ISBN 978-1-138-02773-2

Model and design of decision support system for party cadres modern distance education

L.N. Dai, J.X. Guo & J.C. Zhao
The Research Center of Beijing Engineering Technology for Rural Remote Information Services, Institute of Agricultural Scientech Information, Beijing Academy of Agricultural and Forestry Sciences, Beijing, China

ABSTRACT: The work of party cadres modern distance education has been carried out more than 10 years. It is very necessary to know whether the terminal sites have been used or not and whether the terminal sites have developed distance learning on a regular basis or not. They can be known easily if the large data in data warehouse can be analyzed through some analytical models by building decision support system. In this paper, design flow, logical model, data table structure and inference algorithm used in decision support system will be studied in order to provide decision support for the leadership and develop the work of party cadres modern distance education well.

Keywords: decision support system; design flow; logical model; inference algorithm; distance education

1 INTRODUCTION

At first, the work of party cadres modern distance education is developed in rural areas. That is why it is called as rural party cadres modern education. With the development of this work, party cadres modern distance education has been developed in communities, state organs, state-owned enterprises, universities and the "two new" organizations, so it is named as party cadres modern distance education. The building of distance education platform, video education programs and terminal sites are necessary for this work. Video education programs will be made by some designated unit states and uploaded to the distance education platform firstly. Then students go to the terminal site in their rural or community to watch. Terminal sites can be understood by classroom and there are some video receiving equipment, network devices, such as computers, projectors and TV [1]. Because there are so large input of human, material and financial resources, it is very necessary to know whether the terminal sites have been used or not, and whether the terminal sites have developed distance learning on a regular basis or not, even though video programs are more welcome, in order to keep the balance of input and output. Now, building decision support system is very popular, because it can provide decision support for the leadership through analyzing the large data in data warehouse, relying on some analytical models, and developing the work of party cadres modern distance education well.

2 DEFINITIONG AND SYSTEM ARCHITECTURE OF DECISION SUPPORT SYSTEM

2.1 *Definition of decision support system*

Based on the computer environment support, decision support system is one kind of dynamic interactive system, which can support the semi-structured and unstructured decision, allow decision makers to intervene directly, and also can accept decision makers' intuitive judgment and experiences [2,3]. Decision support system's essential characteristic is model driven. Its main purpose is to improve the effectiveness of the decision. Based on all kinds of data and effective models, it can provide effective computer support environment for senior management and decision makers to find the answer of poorly structured problems in a fast and flexible way [2, 4, 5].

2.2 *System architecture types of decision support system*

Decision support system has gained great development from 1970. Now its system architecture concludes three basic types. They are based on databases, knowledge, or data warehouse respectively [6]. System architecture based on databases comes from the structure of two warehouses and three parts, which is studied by R. H. Sprague in 1980. Two warehouses are Database (DB) and Model Base (MB). Three parts are dialogue component, data component and model component [7].

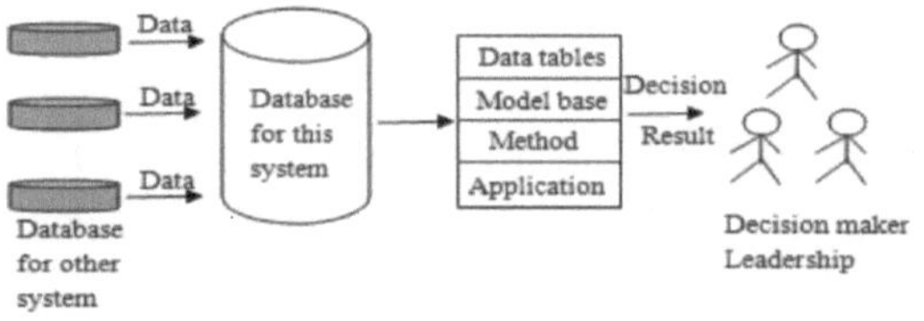

Figure 1. System architecture of decision support system for party cadres modern distance education.

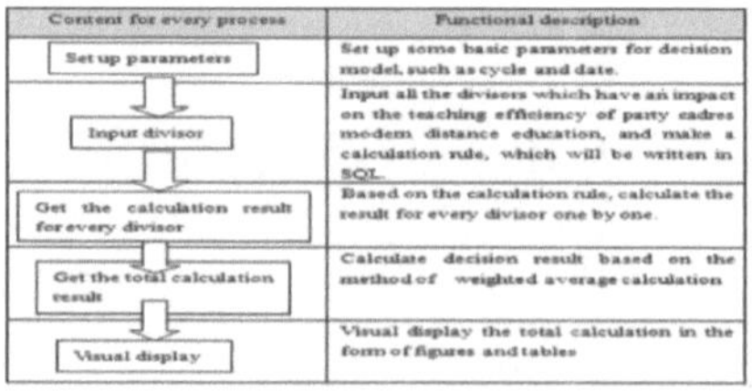

Figure 2. Design process of decision support system for party cadres modern distance education.

System architecture based on knowledge is studied by Bonczek in 1981. It concludes language subsystem, knowledge subsystem and problem processing subsystem. Based on his study, knowledge component is added, which concludes knowledge base, knowledge management system and inference system [8, 9]. System architecture based on data warehouse is used to analyze large data stored in information system by the data mining technology [10].

2.3 *System architecture of this decision support system*

System architecture of party cadres modern distance education is shown in Figure 1. From Figure 1, it can be known easily that the system architecture of decision support system in this paper is built on the database and model base.

The system architecture of decision support system for party cadres modern distance education is: the database of this system receive large data from other system's databases through setting up data interface, than build data tables, decision models, calculation rules, and so on, lastly get the decision result which will be given to decision makers or leaderships.

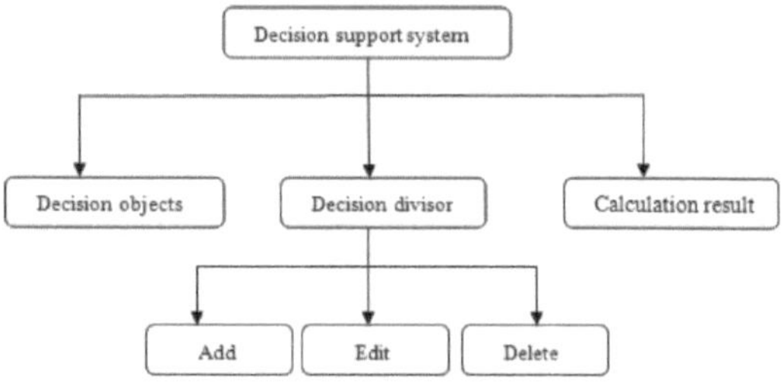

Figure 3. Logic model of decision support system for party cadres modern distance education.

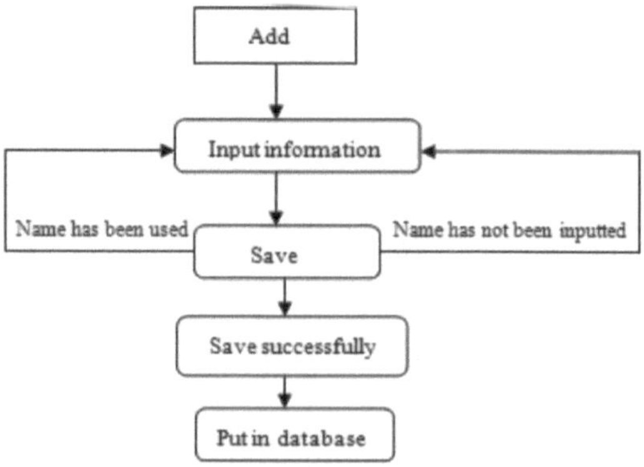

Figure 4. Processing logic for adding decision object.

3 DESIGN AND MODEL OF DECISION SUPPORT SYSTEM

3.1 *Design process and system framework*

Based on the system architecture in Figure 1, the design process of model base can be collected as shown in Figure 2. There are five steps from setting up parameters, inputting divisor, getting the calculation result for every divisor, getting the total calculation result to visual display.

Based on the system architecture and design process of model base, logic model on the front page can be designed easily and clearly, just as shown in Figure 3. There are only three parts. They are decision object maintenance page, decision divisor maintenance page, and calculation result display page. There are also divisors' add, edit and delete function on the decision divisor maintenance page.

3.2 *Function realization*

3.2.1 *Decision objects*

Decision object maintenance page realize the functions of add, delete, enable and disable about decision objects. When a new decision object is added, it is easy to transfer the add event in the script, upload the URL, and open the page in IFRAME. After input some relevant information and save successfully, start SQL batch processing and save the new decision object in the database (see Fig. 4).

3.2.2 *Decision divisor*

Decision divisor maintenance page realize the functions of add, edit and delete about decision divisor.

Add: If there are many decision objects and divisors, it is convenient to replay objects, child nodes and divisors in a data table, transfer JS script and make the form to a tree structure, in order to show a more clear structure and use easily. If the scientific calculation result is expected, when a new divisor is added, it is better to assign weight for it. It is known that the methods for determining weight are weight factor judgment table method, expert intuitive judgment method, hierarchy analysis method, ranking method, Delphi method, etc [11]. The divisor's name and weight can be used or not will be checked by the SQL batch processing and check function (see Fig. 5).

Edit: Devisors' name, rank, weight and calculation rule are all can be edited. To edit calculation rule is the most important and complicated.

Calculation rule will be inputted firstly, than the system will generate SQL automatically. If decision makers want to know how many times some video has been watched by terminal sites, minimum value can be set to 800 times and maximum value can be set to 2000 times according to the teaching plan or

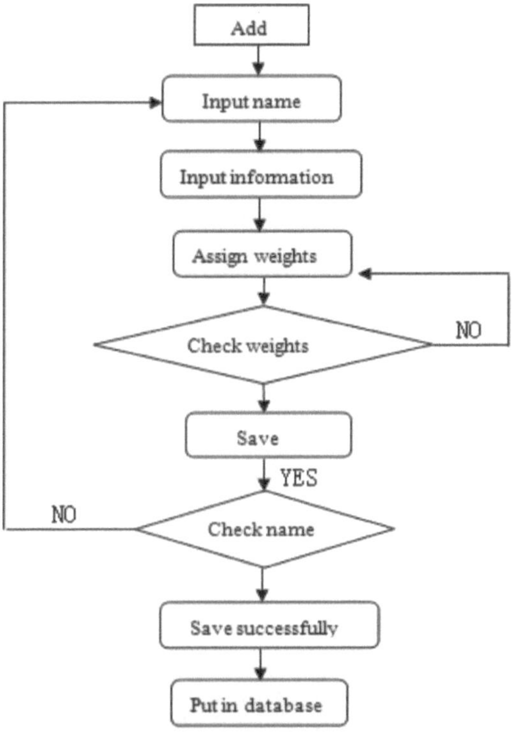

Figure 5. Processing logic for adding decision divisor.

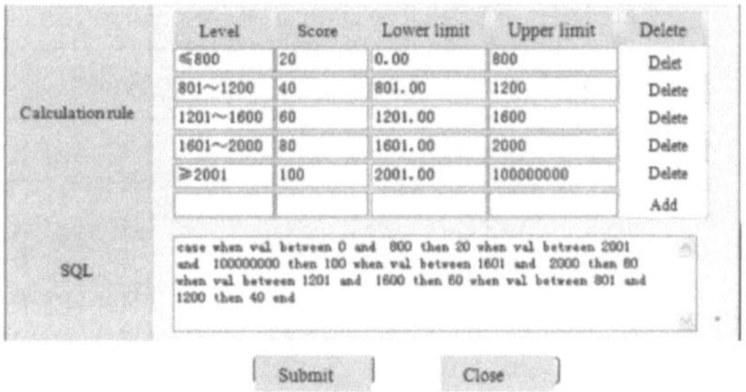

Figure 6. Front edit page for calculation rules.

historical study, just as shown in Figure 6. How to make levels is free, and it will be deleted and added at any time.

Delete: The decision divisors, even the decision objects are all can be closed or deleted, according to the decision change of decision makers or leaderships. When some divisor or object will be deleted, system can transfer the JS script and carry out operation through ID.

3.2.3 *Calculation result*

The calculation result of comparison chart, run chart, year-on-year growth and analysis suggest can be shown through selecting necessary search terms. It is required to use tool strip, transfer JS script, pass parameter and check database in these visual displays.

3.3 *Calculation model*

In this paper, the developmental evaluation method will be used in the decision-making and reasoning process. This method that is made in this paper is: firstly, use modern information technology and network technology means to collect evaluation data for each evaluation indicator; Secondly, through comparing the actual state with the past situation lengthways, manage a large number of evaluation data; finally, calculate the results of evaluation, and make teaching effect clearly using Excel software, mathematical formulas and Likert scale. The procedures are as follows:

Step 1: Based on modern information technology and network technology means, database collection and the collection of electronic questionnaires as two kinds of channels to obtain evaluation data will be used. Many rural remote education platforms of provinces and municipalities take full advantage of the feathers of flexibility and convenience that the network has. Based on .net or .java technologies, a variety of databases can be built to do real-time monitoring and data capture for the whole process of teaching activities. This provides convenient conditions for some assessment indicators to

Table 1. Data table structure for calculation rules.

Column name	Type	Yes/No	Description
DSS_ITEM_ID	NUMBER	Y	ID of decision divisor
DSS_STANDARD	VARCHAR2(100)	Y	Level set
DSS_SCORE_STANDARD	NUMBER	Y	Score reference
DSS_MIN	NUMBER(10,2)	Y	Lower limit
DSS_MAX	NUMBER	Y	Upper limit
ID	NUMBER	Y	Add

collect evaluation data from the database. In addition, for those assessment indicators which could not collect evaluation data directly from the database, we should give play to the convenience of the web-based survey system, by using electric questionnaire method to collect evaluation data from the teaching organizers and participants of rural internet education. We use scoring form to answer questions, in order to make the respondents answer questions commodiously, and to batch processing the huge amounts data.

Step 2: Manage historical evaluation data for each evaluation indicator, and calculate the average value by using simple arithmetic mean method. The computation model is:

$$\overline{X_j} = \sum_{i=1}^{n} X_{ij} / n \tag{1}$$

where $\overline{X_j}$ represents the average value of historical evaluation data of index content of item j, j is equal to 1, 2, ..., m (the same below); X_{ij} refers to the historical evaluation data of index content of item j, i is equal to 1, 2, ..., n (the same below); n stands for the total number of the historical evaluation data of index content of item j.

Step 3: After identical score for the result of computation model is given (1), the current and historical evaluation data need to be compared, and the score of each evaluation indicator should be calculated by using linear equation solving method. The computation model is:

$$Pj = A_j \cdot Y / \overline{X_j} \cdot Q_j \tag{2}$$

where P_j represents the score of index content of item j; A_j refers to the current evaluation data of index content of item j; stands for the given score; Q_j represents the weight of index content of item j.

Step 4: Summarize the result of computation model (2), and calculate the score of evaluation project. The computation model is:

$$W = \sum_{j=1}^{m} P_j / m \tag{3}$$

Table 2. Likert scale for teaching effect.

Over 100 points	80–99 points	60–79 points	40–59 points	Below 39 points
Best	Better	Good	Bad	Worse

where W represents the score of evaluation project; P_j refers to the result of computation model (2), m stands for the total number of evaluation indicators.

Step 5: Classify the result of computation model (3) using Likert scale, and then describe the extent that teaching effect has achieved. If the historical evaluation data would be given 60 points (the given score does not have practical significance, it just compares current teaching effect with the historical, so it can be set arbitrarily), Likert scale for 5 grades to describe the teaching effect. It can be made easily as shown in Table 2.

4 CONCLUSION

The work of party cadres modern distance education has been done more than ten years, but the effect of this work, especially terminal sites' study effect is very difficult to know. Even though a few provinces doing this work have built study effect management system to monitor terminal sites' study, the work of study effect evaluation through digging and calculate large data is almost not done. Based on relevant papers and books study, the decision support system in this paper has been built by BEIJING party cadres modern distance education platform. It will be used in the near future, so applicability and reasonability are needed to be checked in practical application.

REFERENCES

[1] Leadership Coordination Group Office for the Work of Rural Party Cadres Modern Distance Education in the Whole Country. 2008. *Work Book of Rural Party Cadres Modern Distance Education*: 1–88. Beijing: Party Book Press.

[2] Haibin Yuan, Haiwen Yuan. 2009. Decision Support System Model for Fault Comprehensive Evaluation. *J. Huazhong Univ. of Sci. & Tech. (Natural Science Edition)*: Vol. 37, Sup. I, August 2009.
[3] Bonczek R.H, et al. 1981. *Foundations of Decision Support Systems.* NewYork: Academic Press.
[4] Becvar P, et al. 2007. IEEE Transactions on Systems, Man, and Cybernetics, Part C: Applications and Reviews. *Anintelligent Telephony Interface of Multiagent Decision Support Systems*: 553–560 37(4), 2007.
[5] Beliakov G, Warren J. 2001. IEEE Transactions on Fuzzy Systems. *Appropriate Choice of Aggregation Operators in Fuzzy Decision Support Systems*: 773–784 9(6), 2001.
[6] Yun Wu, Guoxiang Zhou, Lei Shi. 2009. Design of the Model Base of the DSS for Vehicle Performance Analysis. *Journal of Hefei University of Technology (Natural Science Edition)*: Vol. 32, No. 1, January 2009.
[7] Erm C. 1988. Decision support systems: a summary, problems and future trends. *Decision Support Systems*: 355–363 4(4), 1988.
[8] Cha J.R, Kim J.H. 2005. Novel Anti Collision Algorithms for Fast Object Identification in RFID System *//Proceeding of the 2005 11th International Conference on Parallel and Distributed Systems*: 63–67 Vol 2, 2005.
[9] Le Gao, Yuanming Wu, Xiaolei Wang. 2007. Frame Length Method based on RFID System. *Microcomputer Information*: 213–215 23(2), 2007.
[10] Rui Lu, Gang Wang. 2006. ALOHA Arithmetic Improvement for Multiple Target Recognition. *Microcomputer Information*: 231–233 22(11), 2006.
[11] LI Y.H. 2011. University Performance Appraisal Index Weight Determination Method under Balanced Scorecard. *Economic Research Guide*: 107, 2011(3).

Engineering Management and Industrial Engineering – Leung (Ed.)
© 2015 Taylor & Francis Group, London, ISBN 978-1-138-02773-2

Study on Beijing residents' learning and reading time in their free time based on the sample survey in 2011

Z.Y. Liu
Beijing Vocational College of Labor and Social Security, Beijing, China

Q.Y. Wang
Renmin University of China, Beijing, China

ABSTRACT: This paper analyzes the learning and reading time of Beijing residents based on the survey data of their life time allocation that was conducted in Beijing in May 2011. The conclusion is that there is a close relationship between the learning and reading time of Beijing residents and their degree, gender, family income, age, and so on.

1 INTRODUCTION

People begin to receive education since childhood. Undoubtedly, school education will benefit us during lifetime, but it cannot fully meet the needs of modern education. The knowledge gained in school is only a very small part, and much more is to be learned in post-school education. In modern times, science and technology advances rapidly and information doubles every day. Scientific knowledge in all areas of society is continuously changing from single aspect to multiple one. It is impossible for us to learn all knowledge at one time, so everyone needs a lifelong learning. Learning is an important means for human to survive and develop, and the only way to get self-development and adapt to their careers. Just as an old saying goes, "It's never too old to learn" this saying should be the concept that people in the new century follow. In order to understand the situation of Beijing residents' learning and reading, we conducted a sample survey of Beijing residents' life time allocation in May 2011. Respondents are asked to select a weekday and a weekend, 24 hours a day, to fill in their specific activities of the two days honestly. Finally, there are effective samples of 1106 people. This paper analyzes Beijing residents' learning and reading time in their free time in 2011.

2 ANALYSIS OF LEARNING AND READING TIME OF BEIJING RESIDENTS

2.1 *The analysis of the learning and reading time of Beijing residents at different educational levels*

Beijing residents' learning and reading time of different educational levels in their free time in 2011 and the percentage of learning and reading time in their free time are shown in Tables 1 and 2.

The statistics show that there is a positive correlation between Beijing residents' learning and reading time and their education levels. With the improvement of education levels, the learning and reading time increases. Moreover, for people at the same education level, the school students spend more time on learning and reading than the graduates. The reasons why higher educated people have longer learning and reading time are as follows: first, they have formed the habit of learning and reading after years of education; second, the majority of higher educated people are engaged in mental work, which requires them to turn to books for help to get problems solved.

Table 1. Beijing residents' learning and reading time of different educational levels in their free time in 2011 (Unit: hours: minutes).

Different educational levels	Learning and reading time		
	Average daily	Week-day	Week-end
Graduate			
Primary school, junior high school and others	00:12	00:08	00:22
Senior high school	00:18	00:14	00:28
Undergraduate and Graduate	00:18	00:13	00:31
In school			
Junior high school	00:28	00:18	00:54
Senior high school	00:36	00:00	02:08
Undergraduate and graduate	00:32	00:25	00:50

Table 2. Percentage of learning and reading time in their free time of Beijing residents at different educational levels in 2011 (%).

Different educational levels	Learning and reading time		
	Average daily	Week-day	Week-end
Graduate			
Primary school, junior high school and others	1.73	2.78	5.41
Senior high school	2.70	5.88	6.55
Undergraduate and graduate	2.97	7.52	7.17
In school			
Junior high school	4.90	20.45	11.16
Senior high school	5.29	0	22.46
Undergraduate and graduate	5.21	12.68	12.00

Therefore, they need more learning and reading. In all groups, the learning and reading time of senior high school students is the longest and the average reading time is 36 minutes a day (accounts for 5.29 percent of their free time). The main reason is the increasing academic pressure and the severe employment situation. That many employers only focus on diploma have exerted an invisible impact on senior high school students who have not entered the society. In order to get admitted to a famous university and find a good job in the future, most senior high school students spend more time learning and reading.

At the same time, according to the data in Table 2, we also find that in terms of the indicator of "Beijing residents' learning and reading time", the learning and reading time at all education levels on weekend is more than that on weekdays. The main reason is that people have more free time on weekend than on weekdays. In terms of the indicator of "percentage of Beijing residents' learning and reading time in their free time", the percentage on weekend is lower than that on weekdays for people at school, and the percentage on weekend is higher than that on weekdays for the graduates. The main reason is that people at school do more other activities after a week's learning and reading, so the proportion of learning and reading time declines on weekends.

2.2 *The analysis of Beijing residents' learning and reading time of different gender*

According to the survey data in Tables 3 and 4, we know that men spend 20 minutes on learning and reading on average everyday (accounts for 2.92 percent of their free time) and women spend 17 minutes (accounts for 2.86 percent of their free time). Men's learning and reading time is 3 minutes more than women's. It shows the pattern of "men dominate outdoor world, while women indoors" has a certain impact on Beijing households. It is generally believed that women do more housework than men, so they have less time on learning and reading.

According to the survey, we know that in terms of the indicator of "Beijing residents' learning and reading time", men and women spend more time on learning and reading on weekend than on weekday, and the difference of learning and reading time between men and women is bigger on weekend than on weekday. In terms of the indicator of "percentage of Beijing residents' learning and reading time in their free time", the difference of learning and reading time between men and women on weekend is smaller than on weekday,

Table 3. Beijing residents' learning and reading time of different gender in their free time in 2011 (Unit: hours: minutes).

Different gender	Learning and reading time		
	Average daily	Weekday	Weekend
Male	00:20	00:15	00:35
Female	00:17	00:12	00:26

Table 4. Percentage of Beijing residents' learning and reading time of different gender in their free time in 2011 (%).

Different gender	Learning and reading time		
	Average daily	Weekday	Weekend
Male	2.92	7.41	7.25
Female	2.86	5.73	6.75

Table 5. Beijing residents' learning and reading time of different income in their free time in 2011 (Unit: hours: minutes).

Different annual income	Learning and reading time		
	Average daily	Weekday	Weekend
<0.99	00:20	00:15	00:32
1.00–1.99	00:15	00:09	00:29
2.00–2.99	00:19	00:12	00:39
3.00–4.99	00:19	00:15	00:29
5.00–9.99	00:19	00:14	00:28
10.00–19.99	00:16	00:12	00:26
>20	00:22	00:16	00:36

which further shows the pattern of "men dominate outdoor world, while women indoors" influences Beijing households.

2.3 *The analysis of learning and reading time of Beijing residents of different income*

Statistics tell us that according to the absolute indicator of "the learning and reading time", the group who has the longest learning and reading time are those with annual income more than 200,000 Yuan, 22 minutes. The group who has the second longest learning and reading time is those with an annual income less than 9,900 Yuan, 20 minutes. The reason is as follows: the majority with an annual income over 200,000 Yuan are company managers. Due to the needs of their work, they spend more time learning and reading every day. While lower income groups generally have shorter working hours, so they have more time on learning and reading.

At the same time, the survey data show that although, the learning and reading time on weekend is more than that on weekday regardless of different income groups, the difference between different income groups is bigger on weekend. 30,000 Yuan is the cut-off point of this indicator of "percentage of the learning and reading time in their free time". Before it, the indicator on weekend is bigger than that on weekday; after it, opposite. The main reason is that high-income groups have more social activities on weekend, which makes their indicator on weekend lower than that on weekday.

2.4 *The analysis of learning and reading time of Beijing residents of different ages*

According to the survey data in Table 7, the learning and reading time of Beijing residents shows the trend of first decreasing and then increasing as age grows. The group who has the longest learning and reading time is those over 60 years old, 26 minutes daily; the group who has the second longest learning and reading time is those less than 19 years old, 23 minutes; the group who has the shortest time is those between 30 to 39 years old, only 15 minutes. The majority of 19 year-olds and younger ones are high school students. Under the pressure of entering universities, they have a longer learning and reading time. The group between 30 to 39 years old, who are the backbone of their work units, are under pressure from family and society, having parents and children to take care and housework to do. Therefore, they have to reduce learning and reading time to cope with the housework and work pressure. The learning and reading time of the group more than 60 years old is the longest because older people have more free time to learn and read. By learning and reading, many older people get away with loneliness and improve the quality of their lives. Similar to the previous conclusions, the survey data also show that although the learning and reading time on weekend is longer than that

Table 6. Percentage of learning and reading time of Beijing residents of different income in their free time in 2011 (%).

Different annual income	Learning and reading time		
	Average daily	Weekday	Weekend
<0.99	3.00	6.70	7.24
1.00–1.99	2.22	4.27	6.23
2.00–2.99	3.27	6.11	10.15
3.00–4.99	2.95	7.34	6.59
5.00–9.99	2.95	6.52	6.52
10.00–19.99	2.59	5.99	6.24
>20	3.64	8.98	8.43

Table 7. Beijing residents' learning and reading time of different ages in their free time in 2011 (Unit: hours: minutes).

	Learning and reading time		
Different ages	Average daily	Weekday	Weekend
Less than 19 years old	00:23	00:06	01:04
20–24	00:20	00:14	00:34
25–29	00:18	00:14	00:27
30–39	00:15	00:10	00:28
40–49	00:18	00:14	00:27
50–59	00:18	00:13	00:30
More than 60 years old	00:26	00:25	00:30

Table 8. Percentage of learning and reading time of Beijing residents of different ages in their free time in 2011 (%).

	Learning and reading time		
Different ages	Average daily	Weekday	Weekend
Less than 19 years old	3.39	3.50	12.63
20–24	3.16	7.63	7.55
25–29	3.05	8.25	6.42
30–39	2.73	6.63	7.01
40–49	3.08	7.56	6.78
50–59	2.58	4.98	6.85
More than 60 years old	2.93	6.04	6.35

on weekday regardless of ages. The difference between income groups is bigger on weekend.

The analysis of the indicator of "percentage of learning and reading time in free time", shows that the indicator of Beijing residents have the same variation with the indicator of "the learning and reading time". The maximum value of the indicator on weekday appears in the group of between 25 to 29 years old, who are newly graduated young people with many things to learn, so they spend more time learning and reading. The maximum value of the indicator on weekend appears in the group of less than 19 years old, who are mostly primary and secondary students. Under pressure of entering the higher school, they have longer learning and reading time.

3 CONCLUSIONS

The above analysis shows that the learning and reading time of Beijing residents has great relationship with their educational levels, gender, family income, age, and so on. First, in terms of education levels, there is a positive correlation between the learning and reading time of Beijing residents and their educational levels. With the improvement of educational levels, the learning and reading time increases. Second, in terms of gender, due to the impact of the traditions of "men dominate outdoor world, while women indoors", the learning and reading time of Beijing male is longer than that of Beijing female. Third, in terms of family income, people of the highest and lowest income have the longest learning and reading time, while others' is less. Fourth, in terms of age, the learning and reading time of the group more than 60 years old and less than 19 years old is longer than others'. Because of the pressure from family and society, the learning and reading time of the group between 30 to 39 years old is the shortest.

REFERENCES

Lei. 1999. Who are the busiest urban residents. *Front* (1):30–32.

Lianbin Qing. 1990. A preliminary study of Metropolitan workers life time allocation and utilization. *Sociological Studies* (1):92–102.

Qiyan Wang. 1997. Study of life time allocation of Beijing residents. *Management World* (4):199–204.

Qiyan Wang. 2007. Beijing Citizen's Living Changes in 20 Years. *Social Science of Beijing* (5):22–26.

Engineering Management and Industrial Engineering – Leung (Ed.)
© 2015 Taylor & Francis Group, London, ISBN 978-1-138-02773-2

A preliminary discussion on reforms of engineering fluid mechanics teaching

X.J. Yang
College of Vehicles and Energy of Yanshan University, Qinhuangdao, China

ABSTRACT: Engineering fluid mechanics is an important fundamental course for petroleum engineering majors. After analyzing the basis and background of reforms in education, this paper elaborates the need for reforms of the course from three perspectives—the theoretical teaching, practical teaching, and the assessment. Based on the specific situation of the major of petroleum, the paper proposes suggestions on the teaching of the course in order to achieve better teaching results.

Keywords: engineering fluid mechanics; teaching reform; practical teaching system reform; assessment reform

1 INTRODUCTION

As the oil consumption grows, the oil industry is developing at an unprecedented rate in China[1]. In this regard, petroleum engineering and technical personnel are required to master solid and advanced expertise. Petroleum engineering has its own distinctive characteristics with its goal to train personnel who will engage in the field of petroleum engineering for oil and gas production, reservoir, and so on. As the phenomena of fluid flow commonly exist during the process of oil and gas exploration, fluid mechanics has always been a basic course for petroleum engineering majors[2]. The course, with theory, experiment and calculation as its three pillars, has formed a relatively stable teaching system in the petroleum universities in China and has been playing a pivotal role in oil-related higher education[3].

2 REFORM BASIS AND THE BACKGROUND

2.1 *Oil industry in national economy*

Rapid economic development of China's brings about growing demand for energy. As oil is an important energy source, the oil industry plays an important role in national economic development. Related to all aspects of people's livelihood, the oil industry is also known as the "lifeblood of the national economy". However, oil exploration is becoming more difficult. The several main oil fields have in general entered a declining stage. Although stable yielding is not easy to maintain, there is still potential for further development. As a basic industry, petroleum industry needs a good number of technical talents. Graduates in petroleum have great job opportunities in oil companies. Regarding the shortage of talents, many oil companies even book graduates with petroleum universities in advance.

2.2 *Reform basis*

Regarding the need for talents in the oil industry, efforts should be made to improve the students' knowledge structure to adapt to the new situation of economic construction and social development in China. Hence, the curriculum should be adjusted according to the new requirements. Rather than covering too many knowledge points, it should be intensive and concise with high starting point and new contents. Specifically, the curriculum should: 1) pay attention to the interconnections between knowledge points and the coherence between courses; 2) reduce duplication with other courses and enhance the integration of courses; 3) raise the starting point and increase the amount of information in classroom teaching. By increasing the intensity, the course is supposed to cover a wide range of knowledge within limited course hours and to achieve a balance between quality and efficiency.

In short, the guiding principle for curriculum reform of engineering fluid mechanics is to "restructure knowledge, lay solid foundation, reflect latest development, and integrate with other disciplines". Restructuring of the teaching contents should on one hand value the basic common knowledge among courses and on the other hand highlight applications. Being comprehensive and intensive, the course should also be informative, interesting and pioneering.

3 TEACHING REFORM OF ENGINEERING FLUID MECHANICS

3.1 *Theoretical teaching system reform*

3.1.1 *Textbook reform*

Engineering fluid mechanics is an important basic course for petroleum engineering majors. For this reason, textbook selection is crucial in the reform of the course. The textbook should take the basic concepts as the basis, adopt computational analysis as the methods, and focus on application. After comprehensive investigation on the existing textbooks at home and abroad, we selected the textbook entitled *Engineering Fluid Mechanics* which is edited by Yuan Enxi and published by Petroleum Industry Press. Clear in focus, the book values both theory and practice. It not only reflects the characteristics of petroleum engineering, but also follows the latest development in fluid mechanics.

3.1.2 *Teaching content reform*

The comprehensive knowledge system of engineering fluid mechanics sometimes makes the teachers stick too much to the knowledge per se and leave the students little space for free thinking, which inhibits their creativity and imagination. Therefore, when organizing the teaching content, teachers should make appropriate adjustments, break the traditional framework, and pay attention to the application of knowledge. Laying stress on the ideas and methods in solving problems, they should cultivate students' basic qualities and abilities for problem-solving. In our teaching practice, we omit the sections on obsolete topics such as fluid inertia, density, thermal expansion properties, and hydrostatic pressure computing. For sections on orifice flow, pipe, open channel, we delete the repeated contents and summarize this part of knowledge into two general categories—the internal flow and the external flow. In addition to the introduction to typical fluid types, we add relevant content on fluid measurement, flow visualization and numerical simulation of fluid dynamics.

With the development of science and technology, cross-discipline studies have brought about a number of new interdisciplinary topics, which represent the new development and directions of research. Examples include studies on non-Newtonian fluid, which has the characteristics of both substances and fluid, seepage issues concerning coastal engineering and offshore oil exploration, the phenomenon of multiphase flow involving sediment, mixed gas, burning or cavitation. In this regard, new research areas closely related to the applications of Engineering Fluid Mechanics should be appropriately introduced to the course so that the students could better understand the characteristics and essence of the subject, broaden their horizons, and enhance their adaptability to various situations. By doing so, we are trying to achieve a transition from the traditional "knowledge-based talent education" to "innovative talents education".

3.1.3 *Teaching methods and approaches reform*

Considering the respective strengths, we propose that teaching media should be retained and combined together so that they can give full play to their advantages and complement each other. As mathematics weighs much in fluid mechanics, by clearly writing each symbol on the blackboard, demonstrating the key formula derivation on site, and explaining the idea and meaning, teachers could have an effective communication with the students and guide them to have intensive thinking. It not only enables students to memorize the relevant knowledge points, but also helps develop their abilities of logical reasoning and abstract thinking. However, literal explanation could hardly help student establish a clear picture of so many concepts in engineering fluid mechanics. Multimedia resources such as videos and animation at this time could facilitate teaching and learning.

The introduction lesson is an important lesson, which can never be cut short even if the whole teaching hours are deducted. A good introduction to the course enables the students to have a clear idea about the subject including its objectives, significance, and the learning methods, and so on. Explanation of the importance of the course in their future career may enhance the students' enthusiasm in the course[6]. A main aim of the introduction lesson is to make the students realize that the course is not difficult to learn. Many of the theories are related to higher mathematics that they have learned in Year One or even to the knowledge they learned in high school. In this lesson, a lot of examples can be cited for illustration such as the story about Dayu's controlling of water, the Dujiang Dam Project and hourglasses as well as examples from various fields including aerospace, air, marine, energy, environmental protection, chemical and petroleum, and so on. As a result, students will have a better understanding of the wide applications of engineering fluid mechanics. Following a chronological order, the theoretical development of the subject might be introduced, from which students will get to know that the water spinning wheel in the Song Dynasty was invented four or five hundred years earlier than that of Britain; the complex canal gate built in the Northern Song Dynasty was about three hundred years earlier than similar ship locks in the Netherlands.

We should also improve classroom efficiency and encourage students to have creative thinking. Following the cognitive rules, teachers are supposed to guide and encourage students to ask questions in class, express opinions and participate in discussion. The teaching content should be both knowledgeable and interesting. Examples from real life and concepts students are familiar with may encourage them to share their own opinion and carry out class discussions. As many students are fond of football, we can use Beckham's banana kick as an example to illustrate the Bernoulli Principle[6]. When the football player kicks the ball, his foot only touches a small part of the ball and "shovels" the ball up. The force makes the ball rotate. When it rotates in the air and moves forward in high speed, the ball is in fact in a rigid body motion consisting the rigid body translation, fixed-point motion, and fixed axis rotation. For any point on a rigid body in motion, its speed equals to the vector sum of the speed of the basis point and the rotation rate of the point. As the speed of the ball varies on two sides, the air flow rates on both sides of the ball are also different. According to the Bernoulli Principle, the ball is subjected to a lateral pressure differential, which makes the ball deviates to one side while moving forward. In this case, the football is simultaneously involved in two linear motions, so it moves along a curved arc. In the same vein, the Bernoulli Principle could also explain questions such as why there are pits on golf balls, why two ships have to keep a distance, why the aircraft could fly, and why geese herringbone can fly.

Classroom assignments and written assignments combined. As an important part of the learning process, doing assignments is a training process to use the basic principles and equations in fluid mechanics to analysis and solve practical problems[7]. Classroom assignments primarily concern about conceptual comprehension. After-class assignments are in the form of group discussion. Each group will have a chance to make presentation in class. By doing so, students on the one hand show more interest in learning and have more confidence to speak in front of a large audience. On the other hand, it deepens students' understanding of complex notions in the course. Another type of assignments is written assignments, which are mainly assigned as after-school homework. When working independently, students may expose problems in their learning. Teachers, by correcting the exercises, could know students' learning including their understanding of basic concepts, application of formula, and so on. For those exercises from which students show a lot of problems, the teacher could set up an exercise class. Students with innovative ideas are given opportunities to state their understanding of the exercise so that the other students could also broaden their mind and have a more comprehensive and thorough understanding of the issue under discussion. Students who did the exercise incorrectly are also invited to elaborate their ideas and have a peer discussion with the other students. At the same time, teachers can also learn about difficulties for students so as to adjust their teaching pace and content.

3.2 *Experiment teaching system reform*

Experiment teaching is another important component of engineering fluid mechanics. It is by doing experiments that students obtain perceptual knowledge and verify theories and numerical calculation results. Hence, experiments are an important way which helps students have in-depth understanding of abstract notions, promotes proliferation thinking, and triggers their learning interest.

3.2.1 *Experiment design*

To enable students to have a firm grasp of what they learned. After learning each chapter, students are asked to design their own experiments so that they can verify the theorems and equations that they newly learned.

3.2.2 *Experimental class*

Every student is required to do experiments in class and complete the report independently. Any problem encountered during the experiment and their solutions should be reflected in the report. By doing so, students could increase their hands-on abilities to observe and analyze problems.

3.2.3 *Computer applications*

In order to strengthen their programming capacity, we guide students to work on projects by computers such as the calculation concerning the balance of resistance value in small pipe network.

4 ASSESSMENT REFORM

Exam is a means to check teaching and learning. In the past, assessment in engineering mechanics is solely made via written examinations, which are not conducive to the overall development of students especially their practical ability in doing experiments. For quality education, assessment on students' quality and capacity is also necessary apart from testing their professional knowledge. From such a capacity-valued reform perspective, a variety of testing forms might be included for assessment:

4.1 *Closed book exam*

Closed book exam focuses on knowledge assessment, taking up 40% of the total score. The final exam is in the form of closed book exam including questions on the basic concepts and notions, essay questions, calculation, and so on.

4.2 *Open book exam*

Open book exam aims to make assessment of students' hands-on abilities, taking up 30% of the total score. They include classroom quizzes (formula), experimental design, computer programming, and so on.

4.3 *Others*

This part concerns about students' attitude toward the course, taking up 30% of the total score. It includes attendance, classroom performance (presentation and discussion), assignment and experiment.

The combination of the above three parts makes a formative assessment including a variety of testing forms. This test method enables students to actively involve themselves in learning and conforms to the requirements of modern quality education.

5 CONCLUSION

In the teaching process, the teacher, is to optimize the integration of different teaching resources, arrange the order of teaching content, and carefully design class and experiment. Various teaching methods could mobilize students' enthusiasm in the course and encourage them to conduct autonomous learning. Intrinsic motivation assures good learning results.

In the experimental class, what the students have learned does not simply lie in the accuracy of the experiment data or the complete depiction of a curve. What is more significant is that by working on comprehensive experiments, students gain the ability to identify problems in the experiment and find ways to solve them. In addition, students could have a better understanding of what they learned and have a preliminary understanding of experimental studies. In other words, experiments increase students' ability to gain knowledge, apply knowledge and have scientific thinking, which is beneficial for their learning and future work[8].

Reform of the course of engineering fluid mechanics is a continuous process for improvement. With the rapid development of high technology and the interaction between disciplines, the top priority for education is to cultivate our students to possess broad knowledge, pioneering spirit and innovation ability so that they can meet the large demand for all-round engineering talents in the market.

REFERENCES

[1] Yuan Enxi. Engineering Fluid Mechanics [M]. Beijing: Petroleum Industry Press, 2005.

[2] Sui Xiuxiang. Discussion on a New Model of Experiment Class of Fluid Mechanics in Petroleum Engineering [J]. *Laboratory Research and Exploration*. 2012, 31 (11): 111–113.

[3] Wen Jianlong. Teaching Reform of Experiments in Hydrodynamic [J]. *Jiangsu University Journal (Higher Education Edition)*. 2002, 24 (4): 84–86.

[4] Hou Shuangyin. On "Solid Foundation, Extensive Expertise, Strong Capability, High Quality" [J]. *Journal of Hebei Agricultural University*. 2003, 5 (3): 5–7.

[5] Zhang Changjun. On "Extensive Expertise" Training [J]. *Petroleum Education*. 1997, 60 (5): 8–40.

[6] Wu Yihua & Xie Hongyong. A Preliminary Study of the Teaching Methods and Approaches in Fluid Mechanics [J]. *Shaanxi Education (Higher Education Edition)*. 2009, (8): 65.

[7] Zhao Chao. Exploring Teaching Methods of Fluid Mechanics [J]. *China Metallurgical Education*. 2010, (5): 63–64.

[8] Shi Baocheng, Lou Shan, Liao Ruiquan, et al. Some Thoughts and Practice of Curriculum Reform of Fluid Mechanics [J]. *Inner Mongolia Petrochemical*. 2008, (15): 68–70.

Engineering Management and Industrial Engineering – Leung (Ed.)
© 2015 Taylor & Francis Group, London, ISBN 978-1-138-02773-2

Training with 'theory and practice together' in 'Excellent Engineer Plan'

C.H. Lu & H.L. Liao
School of Mechanical-Electronic Engineering, Hohai University, Changzhou, China

ABSTRACT: The university training plan of Mechanical Project can help students to master the mechanical theory more easily. But the students have different ideas and their training shall be different. 'Excellent Engineer Plan' was approved by the Chinese Education Department, letting students stay in factories for a time to see the working process of mechanical engineer. The plan can provide students with opportunities to learn from some engineers practice what they have learned and adapt themselves to the working circumstance sooner after they get bachelor's degree. The students need to combine theory with practice and grasp the knowledge of mechanical project when being trained. At the same time, there are many conceptions to raise their capabilities in design. This plan is a new try to change the project education of mechanical engineering and get a good result in three years.

1 INTRODUCTION

It is well known that education decides a county's future. Many Chinese students of mechanical engineering were puzzled with how to use the theory they learned from classes. Now, 'Excellent Engineer Plan' in Hohai University, which was approved by Chinese Education Department, provides the students with opportunities to see the working process of mechanical engineer in factories for some time. A good chance was supplied to help students know the usage of theory. These students need to combine the theory with practice in all training courses in two to three years. They participate voluntarily and had ordinary (means not too bad) scores in first year in university.

The concept of 'combining theory with practice' was put forward by Yangming Wang, a very famous philosopher of Ming dynasty in China. [1]

A famous educator of UK, named John Henry, thought the real education was to foster talents and intelligence not to memorize specific knowledge points in text books. He said the initiation was not to add a little knowledge but keep absorbing and digesting a large amount knowledge we had learned and we were learning and remold our thinking. [2]

A lot of exercises made students to learn theory deeply and clearly, but they often know few facilities in productive progress. The purpose of 'Excellent Engineer Plan' is to help mechanical engineering students learn more practicable examples than others. And the students in plan can combine theory with practice after being trained. The students study as a trainee in one or two companies to finish projects before graduation, but they can turned to their lectures and engineers in companies for advice and help. It makes training flexible and the training content abundant.

The students in the plan were asked to know more knowledge from projects and designs. They visited factories each year. So they can get more experience than other students who have joined the plan. All students in the plan were divided into groups.

Following American education style, they debate questions in teams to finish their homework while few courses were given in English. In the end, students in a team are as united and fraternal as good friends. [3] When the students graduated from the universities, they can adapt to circumstance more easily. [4] 'Excellent Engineer Plan' made students knew how to analyze important points in questions. So they can solve questions faster than other students. All these students in plan got good scores and brilliant future in the end.

2 'EXCELLENT ENGINEER PLAN'

2.1 *The plan's causes*

Nowadays, it's very important to do research on education methods and thoughts for personnel training. In American universities, people are cultivated with an ability to think and analyze. They emphasize that universities are set for people to improve their knowledge level, not for the degree. [2] All the students who went aboard found they should use a few months to catch up with students born overseas. Many universities in aboard ask students to finish few projects by teamwork. Younger children in Europe or USA knew some facilities and could cut metal with machine because he followed his parents or friends to work.

But most Chinese students didn't know machine or engineering graphics until they left middle school. They must learn these in few months so as to pass the examination in overseas colleges. The pioneers felt it very difficult to study in foreign colleges at the beginning. Compared with Chinese mechanical engineer education, the overseas training method has its advantages. The foreign students seem more like engineers than Chinese because the Chinese students spent most time studying principles. If the Chinese students spend time visiting factories and doing exercises in the projects, they can become qualified engineer before they graduate.

So the 'Excellent Engineer Plan' gave students more examples in courses in study term. Students in the plan followed engineers, mastered mechanical principles from courses and knew how to finish a design or project by trainee term in factories.

2.2 *Practice in enterprises*

As a student in the 'Excellent Engineer Plan', he must grasp main points in references supply by the factories. The students should visit enterprises and stay there for about one month every two semester. This is a special settle for students in the plan and the visit time is longer than that of other students. All students in the plan can recognize how an engineer works in working site and how difficult the work is. They will realize how important the hands-on experience in the progress is. Students in the plan need to follow engineers of a factory (ordinarily, they are divided into two enterprises) to get plenty working experience in the last year. So they can finish the training in the factories in time. In the factories, the students are not just students in the plan, they often have chance to help the engineers solve problems in practical production time. In this term, the students were involved in doing few designs, and the designs should be modified in order to make them right and meet the produce requirement. Compared with other students, these students got huge progress in the practice and they respected tutors more than other students. Because they realized how the tutors grasp the key and lead a team in daily work.

2.3 *Teach with double-language*

When more and more students go aboard after graduation, they shall find that their english level can decide their salary and career development. If a person can speak and write a foreign language fluently, he shall find it easier to communicate with people. So the 'Excellent Engineer Plan' gave students chance to improve their english. One reform is that the lectures speak english in classes and ask students to answer some questions in english. Many universities in China used this method in teaching a few years ago. But there are some students cannot understand all courses, especially for some profound theories. So Chinese is used in teaching in Hohai University.

Double-language teaching helps students in the plan to enlarge their english vocabulary of project and develop english thinking. Now the Mechanical and Electrical Engineering Project has no English Project course in Hohai University, so the students are not familiar with the useful project words. The double-language teaching obliged students to grasp important project words in english. They should answer the questions in english in exams.

Considering that some people's listening english is poor, the professors can explain the principle by Chinese in time. Thus the students can understand more and adapt to foreign countries more easily when they go to foreign colleges.

3 COMPARING TO OTHER MECHANICAL ENGINEERING EDUCATION IN CHINA

3.1 *Study independently*

Mechanical engineering is a very difficult project to many people so they tend to copy when doing homework. In most projects' design, it can be found on internet pages that all students need to hand in. The professors are often satisfied with the result because the students are unwilling to think hard. Students in 'Excellent Engineer Plan' must study independently. So no one in the class copy other's homework. If any of them want to do that, he shall be criticized by other students. They all hope the plan can develop their abilities and everyone should do his best. This pressure will make a student find most answers by himself. Students in the plan would debate about theory in spare time and help the underachievers. So every student in the plan got good scores with independent spirit and made great progress in capability for work.

3.2 *The training process of 'combining theory with practice'*

Engineering education includes many projects or designs. Some projects are very practical. Students in these projects need to solve problems. The 'Excellent Engineer Plan' gives more projects than others to develop the students' abilities. Many projects are presented by enterprise engineers and can be used in the factory at once.

In China, many students in high school got used to do exercises a few times to remember the answers of questions so that they can pass the examination easily, but they do not think about the principles' usage. All these students are not qualified college students. In fact, college students don't need to design complicated thing but need to design something by themselves to prove the principles.

As engineering educators know, engineering students don't need to remember the exercise answers but need to combine theory with practice. So they can show the talent of engineering in the future. Some people remember a large number of exercise answers but cannot solve problems. They were criticized by media. In the 'Excellent Engineer Plan', the students practiced choosing different methods and paths to solve problems through projects or designs. This process made students learned how to combine the theory with practice. If a person just learns theory but has no experience, it will take him a lot of time to be an engineer. [1] Because their tasks are more than others' who have not joined the plan. Students in the plan can work easily than other students after graduation.

Students in the plan were encouraged to take part in a competition or an interesting team to train the creation thought. Many competitions asked students to use software to show the design structure, control the machine move and curve some parts with machine in competition site by themselves. So the university professors guided students in the plan and cut the parts for practicing more. Competitions include all areas but need mechanical and electrical knowledge such as robot competition and mechanism innovation competition, and so on. Students in the plan had gotten good grades in all sorts of competitions. When they won the competition, the students in plan got a great advancement indeed. Most students of them knew software well in training time and competition such as AutoCAD, UG, Solidworks and ProE, and so on. The students in plan used software to finish complicated design. They check their thoughts right through 3D technology. All the principles and technologies can be used and combined in the design process. So the students mastered knowledge through 'combining theory with practice'. And this is an ability or skill which the employers need in working.

When the students who joined the plan finished their designs in factories before graduation, they must prepare for the final test and defend their papers. Unless the students in plan answer the questions of actual production about their design asked by the professors in the end, they can get the degree. Students shall never forget the shortcomings in detail which were pointed out by teachers. They will do better after graduation in their careers.

4 CONCLUSION

'Excellent Engineer Plan' is a new education style in China's universities. Independent thought and useful training can help students to grasp main points in knowledge, and make students cooperate with others. Mechanical engineering project requires students to put theory into practice. Supported by Hohai University and the school of Mechanical-Electronic Engineering, the 'Excellent Engineer Plan' turned out to be a big success. Thus the students in plan can become good engineers in short time.

The trouble of the plan is how to let students practice the whole semester. Now, according to the project, the students should finish their courses in three years and spend the last year practicing. The plan seems to be acting with undue haste.

Finally, the result is very satisfying. 'Excellent Engineer Plan' let students train projects more and cultivate the spirit of independent study. It also let students combine theory with practice. The students who have been trained for two years all had good ability and confidence when they left the university.

REFERENCES

[1] Chunhui Lu, 2013, *Differences in University Engineering Education between China and USA*, PMSS2013, (Advanced Education Research Vol 16) P249–252.

[2] Chunhui Lu, 2011,*'Theory and Practice Together' in American Engineering Education*, (Trans Tech Publications Ltd, ICAM2011, Part 1) P538–541.

[3] Qirong Wang Edited, 2002, *Lectures in Oxford—the Energy of Spirit,* (Publishing House of Civil Aviation) Page 6–7, (In Chinese).

[4] Yubo Yin, Jianping Cai, 2010, *Research in Chinese and American Courses of CAD System*, Mechanical Courses Conference Research Papers, (Publishing House of College Education) P43–47, (In Chinese).

Engineering Management and Industrial Engineering – Leung (Ed.)
© 2015 Taylor & Francis Group, London, ISBN 978-1-138-02773-2

Research on problems and countermeasures of railway characteristics professional degree postgraduate training

X.Z. Xu
Department of Electronics Engineering, East China Jiaotong University, Nanchang, China

ABSTRACT: With the rapid development of high-speed railway and intercity railway, the demand of the railway talents has surged. This paper analyzes the contradiction between the development of high-speed railway and the demand of talents. On three aspects of students, curriculum setting and mentor configuration, discussing the professional type of postgraduate training lack enough practical reasons in detail. For these reasons, it put forward the training railway characterized by specialty of electric graduate students coping strategies. Analysis shows that only by strengthening the construction of the tutor team, optimizing the curriculum structure and a rigorous quality assurance system can we further improve the railway characteristics electrical professional degree graduate training quality to meet the needs of rapid development of the railway.

Keywords: railway characteristics; professional degree postgraduate; postgraduate training

1 INTRODUCTION

Railway is the key part of the transport system and it is a large artery of the national economy. In recent years, the high-speed railway has been developed at a fast speed. Chinese railway will reach the highest speed. The rapid development of high-speed railway construction needs huge quantities of high-quality skilled talents. With the large number of high-speed railway constantly put into operation in China, the contradiction between development and related talents demands has been an increasingly prominent problem of high-speed railway. This contradiction is mainly reflected in two aspects.

On the one hand, the scale of the existing talent is small. With the new route gradually built and put into operation, to adapt to the high-speed railway technology and equipment requirements demands contemporary talents both in quantity and quality. In recent years, the railway continued to increase along human resources development and cultivate talents to ensure the railway development to a certain extent. But compared with the railway construction scale and speed of advance, the existing talent scale is still in short supply in the state (Liu Yuenan 2014).

On the other hand, the existing personnel quality should be improved. The railway staff's overall cultural level, professional ability, professional knowledge and skill structure are not perfect enough, and a new era of high-speed railway needs to adapt to the professional management of new requirements of the high-speed railway. With the rapid development of the railway, the railway development needs to meet more complex talents of multi-field and multi-level, to further strengthen the leadership and management talents team (She Yuanfu et al. 2013).

Therefore, conducting research on the railway electrical professional degree graduate training is conducive to the sustainable development of railway industry and the progress of railway personnel resources.

2 THE CULTIVATION STATUS ANALYSIS OF THE RAILWAY CHARACTERISTICS PROFESSIONAL DEGREE POSTGRADUATE

Overall, our school electrical postgraduate training has made obvious progress in the aspects of the concept, system, mechanism and measures, the scale of enrollment and the quality of students have been steadily improved. Because compared with professional degree graduate research and academic degree students in training programs, curriculum and training methods have no obvious difference. So it led to a specialized type of postgraduate training lack of enough practice. The causes of the problem are located in many aspects. The following is a preliminary analysis of these reasons.

2.1 *The sources of the students*

East China Jiao-tong University is one of the main engineering universities. It is a teaching and research university with the characteristics of transportation. The most charming of the college of electrical for candidates is the railway characteristics. Since the master point recruitment, the number of students entering themselves for an examination of the electrical study is increasing year by year, but the number of on-line is limited. Take 2013 for example, the enrollment plan is 70 people, the number of students entering for an examination is 340 people, and the number of on-line is 14. Adjustment is the main source of full-time postgraduates in our hospital. Especially the professional degree candidates, the number of applicants is relatively small. Therefore, it should be completed in the recruitment plan through adjustment. "985 Project" and "211 Project" college students have accounted for only 1% in admission of candidates, the second stage college students accounted for 40%, the third stage college students accounted for 50%, college students accounted for 9%, students' practical application ability is uneven. In addition, the majors of students have a very wide distribution, including electronic information, physical, mechanical, software and other professional students, and some are even students of public security management and business management professional, the railway electrical engineering students are rare (Sun Jian 2012). Therefore, my school's outstanding students are less, their original professional and the character of our institute professional relevance are not high, and the overall quality is not ideal.

2.2 *The curriculum setting*

Graduate course teaching is an important aspect of the master cultivation. It reflects the school's teaching thoughts, directly affects the school graduate education quality and level. It is an important way to master professional solid knowledge of foundation theory and professional. It helps to promote postgraduate teaching quality. Since restoring the graduate admission, it has formed a relatively perfect training mechanism of graduate students after 20 years. It has established and adapted to a certain goal of the curriculum, to guarantee the postgraduate education quality.

Professional degree graduate education has achieved a rapid progress in recent years. Nevertheless, as a new training mode, it is also not standard and not perfect. In terms of curriculum, the core of basic theoretical courses is the same for academic graduate. Except for the professional practice of elective courses, training system of other learning link reference for the master of engineering, no prominent professional experimental courses and practice courses in the engineering curriculum. Inadequate curriculum will affect the quality of training of graduate student. The shortages are shown in the following aspects:

1. Graduate student curriculum content tends to be the undergraduate course that leads to lack of features.
   From the horizontal point of view, the postgraduate course content should be on scientific research, social practice and ability training, to link and complement each other talent of "knowledge, ability and quality" upgrading and integration. In the vertical perspective, the postgraduate course content should be content with the undergraduate courses on cohesion, then gradually deepening. However, the public basic courses accounted for about 40% of the existing huge repetition with undergraduate course content, and even some professional foundation courses and professional courses with the same to the undergraduate stage. It does not simply highlight the postgraduate education in the curriculum content features (Yan Dan 2012).
2. Postgraduate course content is outdated, research is not enough.
   The majority of the course content includes the past teaching content. It is simply not reflected some new knowledge and research results in the subject field. Even in the major courses, rarely involves the hot spot, the focus of professional disciplines and the latest developments. At the same time, a part of the course due to lack of crossing and penetration of basic disciplines supporting and related subjects, curriculum content always does not reflect the frontiers of science and technology development.
3. Graduate courses proportion structure unreasonable and the lack of flexibility in basic courses and professional courses.
   In a long term, the setting of most postgraduate courses led to the student with single knowledge structure and professional knowledge is too narrow. This is not conducive to the cultivation of postgraduates' innovation ability and the new academic thought. In addition, the lower proportion of elective courses is hard to face the difference between the students (Yu Shuangcheng et al. 2014). Students according to their own situation freely choose the courses, resulting in adopting perfunctory attitude. This will lose countless students' creative passion.

2.3 *Tutor configuration*

Full-time graduate professional degree cultivation by using "the double teacher" system, but there are similarly some problems in the implementation stage.

The first problem is school counselors positioning. The tutor is the fundamental guarantee for the quality of graduate students. The advantages and disadvantages of the tutor are not only its ability, the more important is to determine whether the teacher's ability match the postgraduate training type. At present, our country instructor is belonging to the "academic" teacher in university. They are mostly in the field of higher education and scientific research work for several years (Zhang Jie 2011). They have professional theory knowledge system, pay attention to the basic theory research. They lack of the necessary relative experience. So they cannot solve the problem of practice strength in the process of guiding students. It is hard to realize the training goal.

The second problem is the responsibility of enterprise supervisor. The most expert tutors are with senior professional titles. They are involved in the research of enterprise specific technology, management and market issues. They have rich practical experience. And in the current professional education of graduate students, enterprise mentors are generally each enterprise leadership or technical backbone. They do not have too much energy to guide students to work, the existence of the phenomenon of the front guide.

The last problem is the communication between tutor and mentor school enterprise. Double tutorial system will carry on the universities and enterprises resources integration and sharing in innovation ability, thus making the theory and practice into a whole, highlighting the collective cultural advantages. It is conducive for the professional degree graduate students to broaden their knowledge, improve the quality of working. And the process of professional degree graduate cultivation, between the enterprise tutor and school tutor, school communication with the enterprise is more difficult to achieve, unable to play a complementary role of the two, and it is not conducive to the guidance of students.

In the course of teaching, due to full-time professional degree teachers of graduate course teach in the current. Especially, the young teachers, many of them are directly from the master to doctor. They got a solid theoretical knowledge, but lack of experience of engineering practice. They highlight the case analysis and practical research on the classroom teaching. The teaching content cannot be combined with the theory of its application.

3 MEASURES OF THE IMPROVING RAILWAY CHARACTERISTICS PROFESSIONAL DEGREE POSTGRADUATE CULTIVATION QUALITY

3.1 *Establishing incentive mechanism and strengthening the responsibility of the instructor*

Mentor in graduate education has played a decisive role in the process, especially the practical professional mentor. Postgraduate is not tacit cooperation with mentor, double tutorial system cannot be carried forward, which can't get a big push. Electric railway characteristics, therefore, one of the master class units should strengthen their own schools and scientific research institutions, the construction of the contingent of tutors, teachers must play their creativity, stimulate their enthusiasm and initiative (Zhao Xichen et al. 2012). Especially each university has carried out the incentive for teachers. First of all, to attach importance to vocational teacher, make its type and academic theory tutor to be consistent in the management and treatment, so as to improve their status, give full play to their subjective initiative; second, to pay attention to the teacher performance appraisal work of each year, the inspection should be in a timely manner to carry out the exchange of experience after conferences and other activities, rewarding excellent tutors in accordance with the relevant provisions. Through each year or each semester teacher appraisal work, prompting each teacher's learning, constantly updating academic frontiers of knowledge, promoting innovation consciousness, so that the teacher clearly own responsibility, enhance their sense of responsibility and sense of mission.

3.2 *The construction of the supervisor team between schools and enterprises*

To actively promote the construction of the double tutorial system, establishing and improving the double tutorial system in the related job responsibilities and standards of the normative documents. First, the supervisor is from the school approved by the qualified tutor teachers. Second, teachers can be from the railway enterprises with senior titles of technical personnel. In order to ensure the quality of graduate education, schools should make teacher a regular face-to-face communication mechanism, in order to timely understand and master the graduate student learning, scientific research and progress of the work, find problems to solve in time. Various forms need to be adopted to strengthen teachers' team construction.

In addition, the cultivation of the professional degree in colleges and universities should actively establish close communication and cooperation with the enterprise, forming a relatively stable professional degree education teachers' system, the "double tutorial system" deserves it, rather than a mere formality.

3.3 *Optimization of course structure and changing the teaching content and way*

Professional degree graduate student should be aimed at the training goal, mainly choosing good platform of the current industry environment, advanced tools, meet the demand of the industry's latest curriculum setting, teaching content way and from the angle of the application and improvement, to cultivate students to master basic and professional knowledge, and to be able to use the knowledge to solve problems in the practical work. The principle of professional, comprehensive and applied goes through in professional skills, pay attention to the close union of theoretical knowledge and practical ability, pay attention to the combination of professional quality and comprehensive ability to use, finally realizing the education target (Zhu Min 2014).

3.4 *Perfecting quality assurance system and improving the self assessment mechanism*

Formulate from recruitment, training and degree-conferring, etc. the whole process of management rules and regulations, standardizing the professional master's each link in the process of training, making the professional degree graduate education management institutionalized, standardized and scientific, is to guarantee and improve the quality of professional degree graduate education condition. Building a professional degree postgraduate education quality guarantee system is an important basis. Regularly or graduate schools of assorted railway electrical teaching evaluation, this also helps promote the institutions to improve school conditions, and improve teaching quality. National education department should be taking sufficient investigation, to determine the graduate for a period of time in the future market for electric needs the number of people. It is used to determine the various specialized recruitment of students scale, prevent the graduate to find a job with difficult problems. Railway electrical characteristics of colleges and universities should establish a Chinese university graduate student evaluation system, to regulate colleges and universities of the railway electrical graduate professional school evaluation and supervision, thus effectively urged universities to increase their education level (Guo Xiuzhen 2012). An electrical graduate student teaching evaluation system should be established to guide the graduate courses of benign development.

4 CONCLUSIONS

With the cultivation of the professional master's positive change on thought concept, abandon the traditional heavy theory, light practice, light heavy academic graduate education and applied graduate education thoughts. Characteristics of railway electrical engineering master's degree in education is to go along with the rapid development of the railway industry in China, the social demand and a new mode of talent training, has opened up new ways to cultivate high-level and compound talents. In the process of the cultivation of the professional graduate, making them adapt to the needs of society, the feature of applied and professional knowledge is the important guarantee for sustainable, efficient and healthy development.

ACKNOWLEDGEMENTS

This research was supported by Jiangxi Province Education Department, P. R. China under grant number JXYJG-2012-049.

REFERENCES

Guo Xiuzhen. 2012. How to solve the lack of two-way training graduate in accounting, *Agricultural Accounting Chinese*, 8(1): 14–16.

Liu Yuenan. 2014. Research on existing problems and countermeasures of graduate education in Chinese universities, *Jiangsu Higher Education*, (2): 121–122.

She Yuanfu & Chen Jingchun. 2013. Cultivation of creative people and reform of graduate education system, *Journal of Graduate Education*, 13(2): 7–10.

Sun Jian. 2012. Study on curriculum structure and its improvement in graduate education, *China's Higher Education Research*, (3): 36–39.

Yan Dan. 2012. An inquiry into the problems in China's professional degree graduate education and the countermeasures, *Science and Education Corpus*, (2): 21–22.

Yu Shuangcheng, Zhang Ping & Luo Hong. 2014. Analysis of the formation mechanism of innovation ability for postgraduates, *China Higher Medical Education*, (2): 112–114.

Zhang Jie. 2011. Analysis of the postgraduate cultivation method in the colleges with professional features, *Journal of Jiangxi Blue Sky University*, 6(4): 15–17.

Zhao Xichen & Xia Jiting. 2012. Professional Master Degree Postgraduate Education: Development Course and Trend, *Journal of Shandong Yingcai University*, 8(1): 4–10.

Zhu Min. 2014. Reflection and practice of propelling internationalization of research university, *Higher Engineering Education Research*, (1): 86–91.

e-business and e-commerce

Engineering Management and Industrial Engineering – Leung (Ed.)
© 2015 Taylor & Francis Group, London, ISBN 978-1-138-02773-2

Research on overseas financing legal issues of electronic payment business

Q. Ding & X. Zheng
Law School, Beijing Jiaotong University, Beijing, China

ABSTRACT: With the rapid development of e-commerce, e-payment industry springs up at the historic moment. As an emerging industry, e-payment industry has been plagued by difficulty of financing. Electronic payment companies forced to adopt "curve" way to achieve overseas financing, and contractual control mode is the best way to "curve" listing. However, due to the lack of laws and regulations for contractual control mode, in actual operation, it has been wandering in the legal and regulatory "gray zone". So it is necessary to accelerate legislative process and make the laws on regulation of contractual control mode from legal perspective. In addition, it is necessary to further open up domestic and international capital markets to resolve the difficult of financing and to strengthen supervision of business return investment to ensure the legitimacy of foreign exchange funds, so that the function and value of contractual control mode could be brought into full play.

Keywords: electronic payment companies; overseas financing; Alipay incident; contractual control mode; legal issues

1 INTRODUCTION

Since entering the period of Twelfth Five-Year Plan, e-commerce industry has been in a stage of rapid development, in which the industrial scale in rapid expansion and various types of e-commerce services companies, especially electronic payment companies, are continually emerging, which greatly promotes the development of Chinese e-commerce industry. 2013 Annual China e-commerce market data monitoring report released by China Electronic Commerce Research Center showed that: in 2013, the trading volume of China e-commerce market reached 10.2 trillion, compared with 8.5 trillion in 2012, an increase of 29.9% [1]. These achievements, thanks to those domestic companies, have raised a huge amount of development funds overseas through contractual control mode. However, in 2011, the outbreak of "Alipay incident" and the release of internal report of China Securities Regulatory Commission made the contractual control mode from the edge of laws to the cusp.

2 "ALIPAY INCIDENT" AND THE RETHINKING ABOUT THE DILEMMA OF ELECTRONIC PAYMENT BUSINESS OVERSEAS FINANCING

Alipay (China) Network Technology Co., Ltd (referred to as "Alipay") was an independent third-party payment platform and a wholly owned subsidiary of Alibaba Group before the equity transfer. The ownership structure of Alibaba Group was: Yahoo owned 39% stake, Softbank owned 29.3% stake, and Jack Ma had the remaining 31.7% stake [2]. So Alipay was a foreign-controlled company. In June 2009 and August 2010, Alibaba Group transferred the entire equity interest to Zhejiang Alibaba E-Commerce Co., Ltd (referred to as "Zhejiang Alibaba") that a pure domestic company mastered by Jack Ma. After these two equity transfers, Alipay has become a pure domestic company. There are six control protocols between Alibaba (China) and Zhejiang Alibaba, including the loan agreement, the subscribe agreement, the agent agreement, the equity pledge agreement, the Chinese market business cooperation agreement, and exclusive technical service agreement [3]. Alipay carried out these two equity transfers because it wanted to obtain a domestic payment business license as soon as possible. In 2011, the People's Bank of China required company license to make literary protestations with regard to the companies whether exist foreign capital actual control by holdings, agreements or other arrangements. Jack Ma wanted Alipay to acquire the license as soon as possible, so he unilaterally tore up the control protocol between Alibaba (China) and Zhejiang Alibaba without obtaining the consent from Softbank and Yahoo, leading to intense contradictions among them, and then he immediately started negotiations related to compensation. This event is the "Alipay incident".

After the "Alipay incident" occurred, the media accused Jack Ma of "contrary to the spirit of contrac" and criticized him because he pushed the companies that use contractual control mode for overseas financing to the edge of life, even to death [4]. In fact, the contractual control mode is a product of "the oppressing policy" and also is the "crystallization of wisdom" from enterprise using the legal loopholes [4]. However, that Chinese electronic payment business use contractual control mode has itself objective reasons. The gradual generalization of contractual control mode and the financing difficulties of electronic payment companies are inseparable.

Chinese Company Law, Securities Law, Management Approach of Initial Public Offering and Listing and other laws all made strict rules about companies listing domestically. Only when companies meet the standard requirements and strictly follow up the approval process, could they achieve the purpose of public financing. According to the Securities Law, Special Provisions of the State Council on the Overseas Share Offering and Listing, Notice for the Issues on Overseas Listed Companies and other regulations, domestic companies to list overseas directly or indirectly, must be approved by the China Securities Regulatory Commission. Domestic companies listed overseas directly need to satisfy the net assets no less than 400 million RMB and no less than 60 million RMB of after-tax profits in the last year, and they should have potential for growth; according to the reasonable calculation of expected earnings, the amount of funding is no less than $50 million and each refinancing also need to be required for approval. In 2006, Provisions on Foreign Investors Merger Domestic Companies (No. 10) clearly stipulated that special purpose vehicle (referred to as "SPV") overseas listing need to be approved by China Securities Regulatory Commission and should make provisions for the establishment of overseas SPV, cross-border exchange of stock, domestic companies valuations, foreign capital mergers and acquisitions, financing funds repatriation and other contents. For this reason, contractual control mode was carefully designed and became a big weapon for the electronic payment business overseas financing.

3 THE ANALYSES OF CONTRACTUAL CONTROL MODE LEGAL SYSTEM OF ELECTRONIC PAYMENT BUSINESS OVERSEAS FINANCING

Contractual control mode is a deformation of the listed overseas indirectly. In practice, domestic shareholders must set up an SPV abroad at first, and then set up a subsidiary by the company in China. Overseas SPV does not have the corresponding shares of the target company directly, but domestic subsidiary and target company will sign a series of agreements to achieve the real control of the target company and obtain maximum interests from the target company. Through the contractual arrangement, overseas SPV could acquire a comprehensive and effective control of domestic company and ultimately the overseas SPV could successfully list overseas. Although the communities have different opinions on contractual control mode, the mode plays a great role in promoting the rapid development of Chinese enterprises, especially e-payment companies.

3.1 *The legislative and regulatory status about contractual control mode*

Throughout China's law, only two legal provisions expressly mention the term of "contractual control". First, in October 2005, the State Administration of Foreign Exchange promulgated Notice about Domestic Residents through Overseas SPV Financing and Return Investment Management Issues Related to Foreign Exchange. In Article I, the "return investment" is defined as "return investment" mentioned in this notice referring to domestic residents to carry out direct investment activities through the SPV within the territory, including but not limited to the following ways: to purchase or replace the Chinese equities in a domestic company, to establish foreign-invested enterprises in China and to purchase or contractual control the assets within the territory by the foreign invested enterprises, to purchase domestic assets by contracts and to invest to foreign-invested enterprises or increase the capital of domestic companies through the assets."

Second, in August 2011, the Ministry of Commerce issued Regulations of Security Review System about Foreign Investors Merger Domestic Companies. The Article IX states: "That foreign investors merger domestic companies should be determined from the substances and actual impacts that whether beyond the scope of security review of the transactions; foreign investors could not evade the security review of mergers and acquisitions in any ways, including but not limited to: on behalf of the holders, trusts, multi-level reinvestment, leasing, loans, contractual control, foreign trade, etc."

3.2 *The legal issues triggered by contractual control mode*

Due to the fact that the legal status of contractual control mode in our country has been in a fuzzy state, the legal status of the regulatory policy has been in a "vacuum". As a result, in the actual operation process, it inevitably leads to all sorts of legal issues.

3.2.1 *Evade legal supervisions*

This is reflected in the following aspects: first, avoid the domestic and foreign listing regulation. For businesses, funding is always the biggest challenge throughout the development process. In order to alleviate the plight of corporate financing, the stock market emerged as the time required. But in practice, for Chinese laws and regulations make strict restrictions on enterprises domestically listed and overseas directly listed, it is difficult to raise money by these two ways. Contractual control mode can successfully evade legal restrictions, because the laws and regulations are not clearly put the mode into their jurisdiction.

Second, avoid regulation of foreign access. Catalogue for the Guidance of Foreign Investment Industries classified the foreign access industries into four types—"encouraged", "permitted", "restricted", "prohibited", so for some industries, foreign capital is restricted, even prohibited. This leads to the fact that the electronic payment industry in China almost is controlled by Chinese holding, but this emerging industry is in great need of financial support and the most interested by foreign investors. The appearance of contractual control mode just could avoid the restrictions on foreign access, so that foreign capital could enter the restricted industries in this way.

Third, evade regulations on mergers and acquisitions with related party. In 2006, Provisions on Foreign Investors Merger Domestic Companies distinguished between the "SPV return investment" and "general foreign mergers and acquisitions". The approval level of the former is raised to the Department of Commerce, which greatly increases the difficulty of mergers and acquisitions and reduces the likelihood of success about mergers and acquisitions with related party. But in the Provisions on Foreign Investors Merger Domestic Companies, the description of the mergers and acquisition only refers to equity acquisitions and asset acquisitions. If it is understood in a narrow sense, contractual control mode does not belong to the scope of the Provisions on Foreign Investors Merger Domestic Companies, so the behavior that foreign enterprises merger domestic companies through contractual control mode doesn't have to be submitted to the Department of Commerce for examination and approval.

3.2.2 *Cause huge loss of tax revenue*

Currently, the overseas SPVs are set up in tax havens. They enjoy significant preferential tax policies, exemption from income tax, capital gains tax and other high tax burden, and the domestic companies also implement transfer pricing, dividends to shareholders, connected transaction and other means to transfer parts of the assets and profits to the offshore in order to reduce domestic tax base, while foreign operating income and overseas listed income could also be retained in the offshore through the layers of shareholding structure, which would lead to a huge loss of national revenue.

3.2.3 *Threat to national security*

Most enterprises using the contractual control mode belong to "restricted" or "prohibited" foreign access industries, and these industries have a crucial impact on national economy, people's livelihood, national security and smooth running of the economy. If we ignore the supervision of contractual control mode, it is likely to be exploited by some foreign organizations with ulterior motives. Ultimately, the result will be a serious threat to Chinese information security. Recently, according to data released by Alipay, at the end of 2013, the number of real-name users is nearly up to 300 million, [5] if the information of clients is leaked and abused by foreign government, then the consequences would be disastrous.

4 THE IMPROVEMENT SUGGESTIONS FOR CONTRACTUAL CONTROL MODE LEGAL SYSTEM OF ELECTRONIC PAYMENT BUSINESS OVERSEAS FINANCING

That electronic payment companies adopt contractual control mode for overseas financing is inevitable, so our legislative branch should speed up the legislative as soon as possible. The contractual control mode should be incorporated into the legal perspective. The improvement suggestions of the contractual control mode are as follows.

4.1 *Speed up the legislative process and make the laws on regulation of contractual control mode*

First, our country should make higher level of legal norms as soon as possible. Make specific provisions on the implication, the identified standards, the applicable conditions and the contents of contractual control mode from the legal perspective for the purpose of filling the vacancies in our current legislation. Second, "discrimination" regulatory principles should be clearly stated in the law, and take different regulatory approaches for different types of enterprises. The law should strictly control state-owned enterprises to use contractual control mode for overseas financing from the protection of national assets and national economic security; while the law should adopt a tolerant attitude to private enterprises, it must regulate and guide the conduction of enterprises to cross-border listing and financing by expressly acknowledging the existence of contractual control mode. Third, the law should clearly stipulate registration system. The enterprise using contractual control mode must be registered, and control protocols used by unregistered enterprise are invalid. On the one hand, it is convenient to supervise enterprises and control protocols for regulators; on the other hand, regulators accepting control protocols registration indicate that the control protocols have been recognized by regulatory authorities, increasing the confidence of foreign investors.

4.2 *Open up domestic and international capital markets to alleviate the plight of corporate financing difficulties*

The supervision of the contractual control mode should stand on the ground of corporate financing, premised on the safety of the national economy, constantly improving the multi-level and multi-stage domestic capital market system to provide a more smooth domestic environment for corporate financing. Chinese securities market can learn from international practices, thus gradually reducing the rigidity index of domestic companies listing and financing to improve the domestic financing conditions for enterprises and to provide a relaxed, fair and secure financing platform. At the same time, it also should moderately liberate restrictions on companies overseas listed. In some specific regulatory measures, it does not need to set policy barriers deliberately, excessively increasing the financing costs of overseas listed. Therefore, it should take the efficiency of corporate financing and stability of securities markets into account, thus carrying out monitoring and management of enterprise overseas listed in an appropriate manner.

4.3 *Strengthen the supervision of business return investment to ensure the legitimacy of foreign exchange funds*

Return investment brings massive capital investment inflows and outflows, which most likely causes the failure of capital controls and international balance of payments. Therefore, foreign exchange management departments should strengthen the monitoring and management of return investment and reasonably guide the return investment behavior to protect the compliance of cross-border capital flows [6]. Therefore, we could introduce the "transparency" review of foreign capital in the practice. "Transparency" in legal field means: a rule, law or legal process should be open for the general public and can be easily seen, found and get, just as people can easily and effortlessly see things through clean windows [7]. The relevant authorities should on the basis of the original notarization system further require overseas notarization institutions to disclose the situation of actual controller, and carry out substantive examination on the source of foreign funds. The source of foreign funds unable to explain or difficult to explain under suspicious circumstances should be required to provide appropriate information and required to disclose the situation of the actual controllers, or should be disallowed to make return investment.

5 CONCLUSIONS

Contractual control mode is the primary method of overseas financing of electronic payment companies, which is a good solution to satisfy the investment demands for the foreign investors and the financing needs for domestic companies under the restrictions of domestic regulatory policy and industrial policy. The outbreak of the "Alipay incident" has exposed many legal issues of contractual control mode. Under the background of economic globalization, the law has to pay more efforts to realize its value. Only by improving the financing legal system for electronic payment companies could the enterprises that have been or will be in contractual control mode for financing obtain better development.

ACKNOWLEDGEMENTS

This paper is supported by "the Fundamental Research Funds for the Central Universities" and Beijing Philosophy and Social Science Planning Project—"Research on Beijing legal system of Sustainable Utilization of rural land".

REFERENCES

[1] Agency Survey: The Volume of Chinese E-commerce Market Transaction Reached over 10 Trillion in 2013. 2014. Xinhua Web Site. http://news.xinhuanet.com/2014-03/19/c_119849576.htm.

[2] Zheng. R.F. 2014. Study on the Contracts and the Default Risk of VIE Mode. Xiamen. Amoy University Master Dissertation: 6.

[3] Feng. Y. 2011. VIE Mode Review—Based on Analysis of Alipay Incident. Changsha. *Entrepreneur World* (7):187–189.

[4] Lao. Y.B. 2012. Legal Risks and Solutions on Contractual Control Mode in Red Chip. Shanghai. Fudan University Master Dissertation: 1&9.

[5] Alipay Became Global Largest Provider for Mobile Payments. 2014. Phoenix New Media. http://finance.ifeng.com/a/20140220/11699226_0.shtml.

[6] Zheng. J.T. 2013. The Problems and Recommendations of Foreign Exchange Management and Statistical Monitoring. Baoding. *Hebei Finance* (2).

[7] Zhang. X.J. 2007. Research on WTO Transparency Principle. Beijing. *Tsinghua Law Review* (9).

Engineering Management and Industrial Engineering – Leung (Ed.)
© 2015 Taylor & Francis Group, London, ISBN 978-1-138-02773-2

How does WeChat marketing influence consumers' sharing intention

X.P. Wang
College of Information Engineering, Hangzhou Dianzi University, Hangzhou, Zhejiang, China

X.D. Wang
College of Management, Hangzhou Dianzi University, Hangzhou, Zhejiang, China

ABSTRACT: This paper gives a summarization of previous literature about the influential factors on online shopping intention of consumers on the basis of a rapidly rising of WeChat users and gradual development of WeChat marketing in China. Then it develops an empirical research considering the characteristics of WeChat marketing. In this paper, the model is based on TAM model and AISAS model, taking college WeChat users of our country's as study samples and employing structural equation modeling to validate research model. The analysis supports most research hypotheses and verifies the effectiveness of our model. At the end of the paper, we provide scientific management advices so as to introduce something for the development of WeChat marketing in our country.

1 INTRODUCTION

According to data reported by Data Center Of China Internet (DCCI), the number of mobile phone users in China will reach 720 million by the end of 2013, exceeding the number of PC users (716 million), which means that China has really entered the mobile internet age.[1] At present, people are getting more and more accustomed to accessing Internet with mobile phones. Mobile internet is gradually becoming a major force of future Internet.

WeChat is a free application program (APP) designed by Tencent in January 27th, 2011 that can provide instant messaging service for smartphones, which is originated from a foreign well-known cellphone APP, KiK Messenger. It is a widely used social tool in mobile internet field and its features of mobility, sociability and informedness have provided entry point for studies on enterprises' WeChat marketing.

WeChat marketing, namely, a kind of network marketing via WeChat, needs to be realized through mobile terminals. Its mode of information transmission lays particular stress on fragment information. It is a kind of compensatory marketing. At present, there are not many direct studies on WeChat and WeChat marketing in domestic academic circles. In terms of theoretical studies, the transmission modes of WeChat are analyzed, explained and forecasted mainly based on relevant theories of communication sciences; in terms of empirical studies, the cases during WeChat users' application are mainly expounded in forms of individual cases. Studies concerning WeChat marketing mainly focus on affirming the potential of WeChat marketing, summarizing its marketing characteristics, concluding its marketing approaches, analyzing its marketing advantages and disadvantages, etc., as well as analyzing the significance and actual effect of WeChat in marketing and business operation according to economics related theories.

Considering the new characteristics of current network consumers, enterprises are no longer the only subject to carry out marketing campaigns. Instead, consumers can also be the initiator, participant and spreader of network marketing. Hence, it is of great significance for enterprises' marketing decisions to study consumers' sharing behaviors. Based on Technology Acceptance Model (TAM) and AISAS model, this study will explore the impact factors of WeChat marketing and establish influence model of WeChat marketing on consumers' sharing intention from the perspective of marketing information itself. Through data collection from questionnaire surveys, the study uses the interrelation between the analysis variables of SPSS17.0 and AMOS17.0 to verify the hypothesis and model. On this basis, the study has reached conclusions from these theories and proposed suggestions on development strategies of WeChat marketing for enterprises.

2 LITERATURE REVIEW

2.1 *WeChat and WeChat marketing*

At present, there is no accurate academic definition for WeChat. By integrating all kinds of literature materials, WeChat can be described as follows: WeChat is a free APP designed by Tencent that can provide instant messaging service for smartphones, support cross-communication operator platform and cross-operating system platform, and is available for functions of IM, SNS (Social Network Services) and LBS (Location Based Service).

According to WeChat development company's figure: WeChat had been officially approved on November 18, 2010, successively launched iphone version, Android version and Symbian version in January 2010. On March 29, 2012, WeChat registered users reached 100 million; then in July 2013, WeChat registered users reached 400 million. Version 1.0 focuses on multimedia function and the low cost; Version 2.0 added voice intercom function; Version 3.0 achieved to Scan QR code; Version 4.0 added the function of photo album and launched the mobile social circle of friends.[2] On August 18, 2012, WeChat public platform was officially online; enterprises and organizations could register their official account, which gave the birth to the WeChat marketing.

At present, there are not many direct studies on WeChat marketing in domestic academic circles. Taking "WeChat" as a key word, a total of 145 literatures are found on China Network Knowledge Infrastructure (CNKI), with their subjects mainly distributed in fields of journalism & media, information economics & postal service, higher education and business economics. These literatures can be divided into three categories. The first one is theoretical studies where the transmission modes of WeChat are analyzed, explained and forecasted mainly based on relevant theories of communication sciences. The second one is empirical studies where the cases during WeChat users' application are expounded in forms of individual cases according to practical experience of individual user's application of WeChat. The third one is discussion on WeChat marketing where the significance and actual effect of WeChat in marketing and business operation are analyzed according to economics related theories.

It can be seen that studies on WeChat marketing in academic circles are relatively scattered, mostly focusing on the features, advantages and disadvantages of WeChat marketing, while studies on the influence of WeChat marketing on users' behavioral intention from the perspective of users' are scarce. In view of WeChat's functional characteristics and its marketing characteristics realized with the help of mobile internet, this study will focus on the "sharing intention" of users' behavioral intention with particular stress on the influence of WeChat marketing factors on users' sharing intention.

2.2 *New consumer and AISAS model*

With the popularization of mobile internet and mobile application, consumers gradually become the subject of propagation—consumers can not only get information from the internet, but also act as the subject to post messages and spread information so as to share information with more consumers. Lewis and Brider think that consumers in Internet era has become "new consumer" with new behavior characteristics of "independence, individuality, attention to participating and well-informed consumption information", owned more discourse power and undergone structural changes in their behaviors and attitudes.[3] With the statistics, Vkerlin and Kanner concluded that consumers in Internet era search commodity information much more frequently than ever before, share plenty of contents that may affect people's purchasing behaviors through Web2.0 platform and are able to influence others' consumption behaviors.[4]

Dentsu Inc. has reconstituted AIDAM consumer behavior model and puts forward AISAS model. According to AISAS model, in Internet era, after knowing the information of commodities or services and purchasing them, consumers will take the initiative to share and consequently influence other consumers, experiencing the following five links: Attention, Interest, Search, Action and Share.[5] This model treats information search and information share as two vital links because the two links are consumers' initiative action which cannot be controlled by traditional marketing tools. Thus it can be seen that mobile internet has hastened the reformation of network marketing and will constantly motivate users' strong consumption and sharing intention with this. On the whole, compared with AIDAM model, AISAS model can better explain consumers' behavioral model in interpersonal interaction in Internet era, which has been universally acknowledged in academic circles. In view of new characteristics of current network consumers, attention should be paid to the potential marketing value of users' sharing behavior. It is of great significance for enterprises' network marketing to study WeChat users' sharing intention.

2.3 *Study on consumer behavioral intention based on TAM model*

TAM was proposed by Davis in 1989, which is the basic theory for current technical studies on users' acceptance or rejection of information in system application. Its core idea is that consumers' usage behaviors for commodities or services are caused by their intention, which is determined by their attitudes towards the action. TF.[6] Over the past two decades, the explanatory power of

TAM model and the effectiveness of scale have constantly been verified on condition of different research objects and application environments, and the model itself has constantly been improved. The consumers actually have double peculiarities of commodity purchaser's and network user's while they are shopping online. Analyzing from the technical level, the process consumers receiving online shopping model has certain similarities with the process of receiving information system. Hence it is feasible to apply TAM model to study consumers' online shopping.[7] Besides, Pavlou's situational experiment, empirical analyses of Hung-Pin Shih, Donna and Kim's have all verified TAM model's effectiveness in explaining and forecasting consumers'' participation willingness for online shopping.[8]

The concept of behavioral intention comes from attitude theory and it is usually discussed with attitude theory in psychology and consumer behavior. Earlier scholars held that attitude was the reaction of individual's emotion or mood. Later, they added cognitive component into the construction of attitude because they believed that emotion was based on cognition. Peplau and Tayliniard have respectively explained cognition, emotion and behavior disposition. They proposed that cognition was people's knowledge and belief for the subject matter of attitude; emotion was people's feeling towards the subject matter of attitude; behavior was people's action or behavior disposition towards the subject matter of attitude.[9] Engel, Blaekwell & Miniard proposed that cognitive factor and emotional factor were treated as the determinative factors of attitude and behavior disposition was determined by attitude. The users' attitude towards certain subject matter will determine their behavior intention and this intention will further influence their final action.[10]

Through collating the literatures about behavioral intention, it can be seen that the scholars focus their studies of behavioral intention on purchasing intention and that behavioral intention roots in attitude theory. The author agrees with the view that behavioral intention is influenced by attitudes and attitude can be divided into cognitive attitude and emotional attitude. Thus, this study treats cognitive attitude and emotional attitude as the intermediate variable and studies the influence of WeChat marketing factors on consumers' behavioral intention.

To sum up, TAM model has been widely used in research fields of consumers' online shopping behavior intention and has achieved many important theoretical results. AISAS model is generally accepted as a theory available for explaining network consumers' behavioral intention at present. WeChat marketing, as an important branch of network marketing, has much in common with online shopping. Hence, it is feasible to study WeChat marketing based on TAM model and AISAS model. Besides, considering new characteristics of current consumers, functional characteristics of WeChat and its marketing characteristics realized with the help of mobile internet, this study will focus on the sharing intention of users' behavioral intention with particular stress on the influence of WeChat marketing factors on users' sharing intention.

3 RESEARCH MODEL AND HYPOTHESES

3.1 *Variable definition*

To extract factors of influence of WeChat marketing so as to provide the mentality and basis for research model and questionnaire design, this survey takes the method of in-depth interviews. The interviews include online interview and off-line interview, which respectively has ten participants, all of whom are WeChat users. The two interviews were conducted separately, mainly worked to solve two problems: one is to explore the factors of WeChat marketing that influence consumers' intention to share, on the basis of marketing information itself; the other is to provide analyzing basis by which to develop the scale of WeChat marketing.

Starting from the discussion of users' understanding of WeChat marketing, the interviews then successively discuss the WeChat marketing methods that are impressive to users, brands and products concerned by users, their attitudes toward WeChat marketing as well as their experience of online shopping. Also, the interviews mainly discuss users' purchasing experience and sharing behavior carried out through WeChat marketing. During the interviews, the users orderly and freely shared their opinions about WeChat marketing and specifically analyzed some individual cases of WeChat marketing.

According to the outcome of the interviews, this survey has generalized five factors of WeChat marketing, namely information quality, push frequency, amount of information, innovation and content. Also, the survey has defined the five factors as follows. Information quality refers to the literal readability and logicality and the pictorial definition and artistry of pushed information. Push frequency refers to the daily number of pushing time, concerning not the exact daily time, but the frequency of information pushing perceived by consumers. Amount of information depicts the quantity of words and pictures of pushed information, focusing on the length of time consumers need to read the information. Innovation is the novelty of pushed information in design, forms and content. Content is about whether the marketing information is full, accurate, believable and helpful to consumers.

3.2 *Proposition of theoretical model*

According to TAM Model, consumers' employing behavior of a product or a service is caused by intention, and intention is caused by their individual attitude towards the behavior. Then, according to the ideas of Angel, Blaekwell and Miniard, attitude is determined by cognition and emotion. Thus, this survey has generalized two intermediate variables influencing behavioral intentions of consumers: cognitive attitude and affective attitude, and then depicting the former as consumers' knowledge and conviction about the subject matter of attitude, and the latter as their feelings towards it.

Based on AISAS Model as well as the new traits of consumers, the functional characteristics of WeChat itself and marketing features come into being by means of mobile internet, this survey will focus on the sharing intention in users' behavioral intentions, mainly discuss the influence of WeChat marketing factors on users' intention to share, and define the sharing intention as WeChat users' intention to transmit related marketing information in WeChat and other means of social apps, or to voluntarily tell others about the information.

With the findings of the in-depth interviews, the survey has built a theoretical miniature of the influence of WeChat marketing factors on users' intention to share. It is shown as the Figure 1.

3.3 *Research hypothesis*

According to the purpose of research and theoretical miniature, this survey has put forward the following hypothesis:

H1: Information Quality has significant influence on consumers' cognitive attitude.

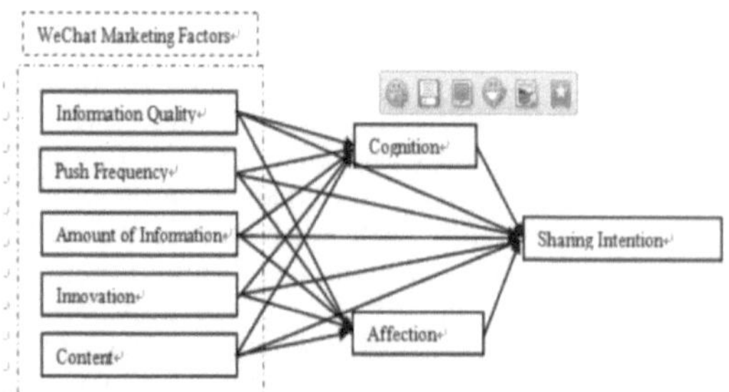

Figure 1. The model of WeChat marketing factors' influence on consumers' sharing intention.

H2: Push Frequency has significant influence on consumers' cognitive attitude.
H3: Amount of Information has significant influence on consumers' cognitive attitude.
H4: Innovation has significant influence on consumers' cognitive attitude.
H5: Content has significant influence on consumers' cognitive attitude.
H6: Information Quality has significant influence on consumers' affective attitude.
H7: Push Frequency has significant influence on consumers' affective attitude.
H8: Amount of Information has significant influence on consumers' affective attitude.
H9: Innovation has significant influence on consumers' affective attitude.
H10: Content has significant influence on consumers' affective attitude.
H11: Information Quality has significant influence on consumers' sharing intention.
H12: Push Frequency has significant influence on consumers' sharing intention.
H13: Amount of Information has significant influence on consumers' sharing intention.
H14: Innovation has significant influence on consumers' sharing intention.
H15: Content has significant influence on consumers' sharing intention.
H16: Affective attitude has significant influence on consumers' sharing intention.
H17: Cognitive attitude has significant influence on consumers' sharing intention.

4 RESEARCH AND FIGURES

4.1 *Questionnaire design and sample selection*

The formulation of initial scale mainly referred to the research findings in related areas in both China and abroad, and was appropriately amended and supplemented according to the characteristics of WeChat marketing and the findings of the in-depth interviews. Then, after repeated modifications based on the findings of localized pre-survey, the final questionnaire consists of three sections, which are the basic information of answerers, if they use WeChat or not and the body of the scale. The questionnaire utilizes the Likert Scale Level 5, and to ensure the effectiveness of the samples, the beginning of the questionnaire hints that, "If you are WeChat user, please answer the following questions."

By now, college students have become the most representative group who use WeChat in China, and are therefore an important group of target consumers for WeChat marketing. According to the researches made by such as Kuehn, most marketing-related researches could acquire reliable findings if taking students as samples. Therefore, this survey mainly targets college students. It should be illustrated that the research objects of this essay are limited to the students who use WeChat, as they are more familiar with the functions of WeChat so we could make better analysis about factors influencing users' intention to share.[11]

4.2 *Sample structure analysis*

This investigation takes 200 college students as subjects, which from 14 universities in Hangzhou. The investigation began in February, 2014 and lasted for 3 weeks. Finally, we got 184 effective questionnaires, among which there are 181 college students who were WeChat users, and the date analysis would be based on 181 questionnaires.

5 DATA ANALYSIS AND CONCLUSION

5.1 *Reliability analysis*

We use Cronbach's α to test the consistency of the questionnaire. Most scholars such as Hair proposes that if the Cronbach α are greater than 0.9, the reliability is very good, if the Cronbach α are between 0.8 and 0.9, the reliability is good, and if the Cronbach α are between 0.7 and 0.8, the reliability is acceptable.[12] According to the analysis results, all the factors' Cronbach's α are above 0.7, which means that the reliability is acceptable.

5.2 *Validity analysis*

Validity analysis, mainly consists of content validity and structure validity, is used for measuring whether research scale could accurately reflect the purpose and requirement of the research. All the items in the questionnaire of this article are based on the carried out in-depth interviews. Therefore, it is safe to consider the questionnaire meet the requirements of content validity.

Secondly, the article inspects whether the variants are suitable for factor analysis. Referring to the KMO measure index, if the index of KMO is below 0.5, the sample could be considered as suitable for factor analysis. According to the analysis results, the KMO indexes of both WeChat marketing factor and attitudinal variable are above 0.6, and the significance probability of Bartlett sphericity test is less than 0.01. Thus it meets the general research standard, and suitable for factor analysis.

Next, factor analysis is done respectively to WeChat marketing factor and attitudinal variables by SPSS17.0, testing the validity of the questionnaire through calculating factor loading. Principal Component Analysis is employed in factor extraction, and Varimax is employed in factor rotation. General social science study considers a factor loading quantity to be effective if it is above 0.4. According to the analysis results, all the factor loading quantities of 21 measurement indexes of 7 variables are above 0.5, thus the validity of this questionnaire meets the research requirements. In addition, the overall resolutions of all variable measurements to research variables are above 60%, satisfying the 30% standard that general variable measurements should explain research variable measurements. Therefore, the research questionnaire could prove to be high in structure validity.

5.3 *Correlation analysis*

Correlation analysis is a statistical method to analysis the close degree and the change trend between different variables. In this paper, we choose Pearson correlation coefficient to value the correlation bewteen the variables.

According to the analysis results, on 0.01 level, all five WeChat Marketing factors are correlated significantly with Cognitive attitude, Affective attitude and sharing intention, in addition, Cognitive attitude and Affective attitude are correlated significantly with sharing intention.

5.4 *Hypothesis model testing*

A model consisting of a group of independent variables and one or more dependent variables usually resorts to making a relation for each dependent variable to show its relationship with other dependent and independent variables. In statistics, measurable variables are frequently perplexed by measuring errors, such as errors in measuring independent variables will cause errors in estimating the parameters to regression model. Cause-and-effect relationship does not always come with correlativity and regression relationship. The cause-and-effect relationship could be tested by SEM. This article will test important variables and their hypothetical relationships with SME.

The SME structured by this survey consists of 8 variables. 5 of the 8 variables are forecasting latent variables, and they are the 5 factors of corporate WeChat marketing: quality of information, pushing frequency, amount of information, innovation of information and content of information. 3 of the 8 variables are resulting latent variables, namely cognitive attitude, emotional attitude and sharing intention (willingness to share). The SME is shown in Figure 2.

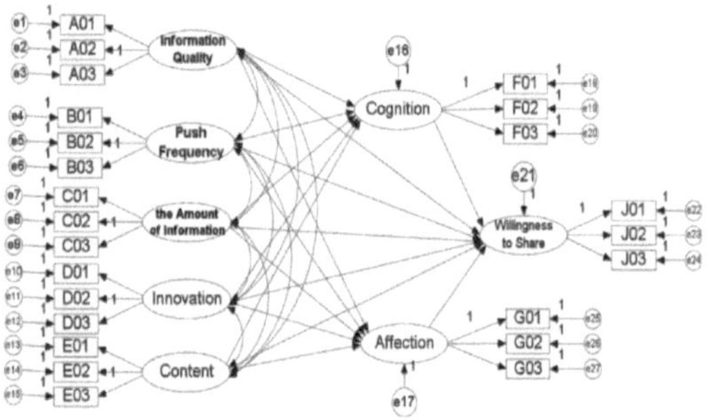

Figure 2. Structural equation model.

5.4.1 *Fit analysis of the structure equation*
AMOS has lots of fit index, according to the experience of predecessors and research practice, this paper would take 7 indexes which are CMIN/DFNFI, GFI, RFI, IFI, CFI, RMSEA to evaluate the Model.

According to the analysis results, CMIN/DF is 1.577 and is less than 5, which means the good fit of the model. Also, all the value of GFI, RFI, IFI, CFI are close to 1, which means the fit is good. The value of RMSEA is 0.057, which between 0.05 and 0.1, also indicates the good fit. It turned out that our model is reasonable and has explanatory power.

5.4.2 *Verification results of research hypothesis*
Results of research hypothesis are tested under the circumstances that the model's degree of fitting is all right. It is generally thought that when the absolute value of C.R. (critical ratio) (the standard critical radio that distinguishes whether regression coefficient has prominent difference from zero) is equal to or greater than 2, the numerical value of regression coefficient could be considered as distinctively different from zero above the 0.05 level of significance.

Through the assessment of the constructed model, hypothesis 2, 4 and 5 are verified, and hypothesis 1 and 3 are rejected for they are under the level of significance. In other words, their quality and quantity of information only have non-significant influence on consumers' attitudinal cognition. The reason might be that, as the pace of life in current society is accelerated, and consumers are increasingly used to accept instant and fragmented information through mobile internet, when they receive a large amount of information of high-quality, they are not always willing to absorb and understand the information with a lot of time. Hypothesis 6, 7, 8, 9 and 10 are also verifies, namely all the five micro-marketing factors have significant influence on consumers' emotional attitude. To a degree hypothesis 16 and 17 conform to the classical TAM, which means cognitive attitude and emotional attitude have a significant influence on consumers' intention to share. While hypothesis 11, 12 and 14 are verified, hypothesis 13 and 15 are rejected for being under the level of significance, or their quality and quantity of information only have non-significant influence on consumers' intention to share.

It should be noted that, the 4 hypotheses are rejected may also be a result of incompleteness of the research. So deeper discussion is still to be carried out in further research.

6 SUMMARY AND SUGGESTIONS

For modern corporations, WeChat has very important value of marketing. This article focuses on the market background of gradual development of WeChat marketing and explores the major factors of WeChat marketing that influence consumers' intention to share on the basis of TAM generally used in the area of technology acceptance and AISAS Model generally accepted in the area of consuming behavior in the circumstance of e-commerce. In the research, 5 factors of WeChat marketing are introduced so as to better reflect the true story of how WeChat marketing influences consuming behavior. The analysis results verify most hypotheses, and prove the explaining effectiveness of research models. The empirical research about the intention to share in consuming behavior has enriched the theoretical achievements in the new area of consuming behavior, acquired conclusion with practical significance. The research is conducive for corporations to market better with the help of the behavior of consumers themselves and guides them to initiate and transmit corporate WeChat marketing.

The research findings of this article suggest that stimulating the consumers' intention to share and making them initiators, participants and propagators of marketing are extremely important for the success of corporate marketing.

REFERENCES

[1] China Internet development report of China Internet network information center [R]. 2012.
[2] Jin Cao, Zuona Lv. An analysis of new medium in the angle of popular culture: based on WeChat [M]. Southeast Communication, 2012.9:14–16.
[3] Lewis D, Bridger D. The soul of the new consumer: Authenticity what we buy and why in the new ecomomy [M]. London: Nicholas Brealey Publishing, 2000:204–208.
[4] Verkin D, Kanner B. Watch this, listen up, click here: Inside the 300 billion dollar business behind the media you constantly consume [M]. New Jersey: John Wiley & Sons, 2007:25–32.

[5] Kobayashi Y. A study of engagement in Japan [J]. Aoyama Journal of Business, 2009, 43(4): 39–60.
[6] Davis FD. Perceived usefulness, perceived ease of use, and user acceptance of information technology [J]. MIS Quarterly, 1989, 13 (3): 319–340.
[7] Aron O' Cass, Tion Fenech. Web Retailing Adoption: Exploring the Nature of Internet Users Web Retailing Behavior [J]. Journal of Retailing and Consumer Service, 2003, 10 (2): 81–94.
[8] Peijing Zhou, Lun Hou. An Analysis of Factors Affecting the Intention of Consumers about Micro-blogging Marketing: Based on TAM and IDT Model [M]. Managerialist, 2011. 12:12–40.
[9] Peplau & Tayliniard Social Psychololy [M]. 7th ed. Englewood Cliffs NJ: Prentice Hall, 1991.
[10] Engel, Blaekwell& Miniard. Consumer Behavior [M]. The Dryden Press, Chieago, 1986.
[11] Kuehn, K.R, R.P. Khandekar and C.R. Scott. The Effects of Marginality and Reward on Matrix Conflict [J]. Project Management Journal, 1996, 27(3).
[12] Hair, J.F., Anderson, R.E., etal. Multivariate data analysis: With readings [M]. New York: Macmillan, 1992.

Engineering Management and Industrial Engineering – Leung (Ed.)

An inventory problem considering multiply suppliers in an e-commerce company

C.J. Guo & B.H. Gao
Graduate School at Shenzhen, Tsinghua University, China

ABSTRACT: e-commerce companies are different from traditional selling companies. The Inventory Problem (IP) is much more complex in an e-commerce company for there are dozens or even hundreds of suppliers providing products for the company. Meanwhile, considering that customers will return goods as they are unsatisfied with them, suppliers will retrieve the return goods from warehouse. We formulate an IP model to arrange the order point and retrieve point for the multiply suppliers. By using the IP model, we set a case study of "JuMei YouPin" to test the practicability of it.

Keywords: e-commerce; inventory; order quantity; retrieve

1 INTRODUCTION

With the development of the internet, the number of e-commerce companies has increased rapidly in China. One of the important problems confronting decision makers in these modern organizations is to control and to maintain the inventories. However, this problem becomes much more complex than that in traditional companies. For example, "JuMei YouPin" is an e-commerce company which sells make-ups in China. Unlike other offline companies which always have only one supplier to provide just one brand products, it contains dozens of suppliers to provide hundreds of brand products. When these suppliers transport products to warehouses of JuMei located in different areas of China randomly, some non-urgent products often enter warehouse in advance and occupy the space, leading to some urgent products failing to enter warehouse in time. As a result it delays the sales. Under this circumstance, how to arrange each supplier's time of entering warehouse and quantity should be paid much more attention.

The remaining of this paper is organized as follows. In the following section, we briefly review the literature in IP and the contribution of this paper. In section 3, we dedicate to the conceptualization and formulation of the model and the solution procedure in section 4. Finally, section 5 contains some concluding remarks.

2 LITERATURE REVIEW

There are so many researchers who have attempted to study the IP. The Economic Order Quantity (EOQ) model of Harris (1913) is the foundation of modern-day inventory models. Within the IP literatures, a great number of efforts have been devoted to developing lot sizing order models to develop the EOQ model. Karlin (1958) was the first to address this issue. He made assumptions underlying the structure of the inventory cost components and presented three single stage newsvendor models to characterize the optimal ordering policy under random supply. In a later paper, Lee and Rosenblatt (1986) studied a joint lot sizing and inspection policy for an EOQ model with a fixed fraction of defective products. Gerchak et al. (1988) analyzed production problem in a single period in which the production process was characterized by uncertain demand. Inderfurth (2004) determined an optimal production policy for a uniformly distributed demand and yield rate. We suggest reader consult Wright and Mehrez (1998) for a detailed review of the literatures dealing with the relationship between quality and inventory, and Bill (2005) for an origins review of the formula of EOQ.

However, to our best knowledge, most existing models are concerned about IP which do not fit to be applied in e-commerce companies like JuMei YouPin as these companies should made order decisions with multiply suppliers instead of only one supplier in studies of the past.

The contributions of this paper to the IP is twofold: (1) Compared to the existing EOQ model, the most distinctive feature of ours is to arrange the order point and quantity for multiply suppliers at the same time. (2) Concerning the reverse logistic, generally, warehouses store the returned goods from customers until suppliers take them back. In our model, the retrieve point and quantity for multiply suppliers also will be arranged.

3 MODEL FORMULATION

3.1 *Problem description*

The network of our IP model is shown in Figure 1. It involves the forward and reverse logistic.

In order to formulate a mathematical model for this problem, we make the following assumptions:

1. One week is a cycle. Each day's working time of the warehouse is broken up into periods. The duration of the period is according to the number of suppliers.
2. In each period, warehouse can handle only one supplier's entering warehouse and only one supplier's retrieving goods.
3. At the beginning of each cycle, the company must determine how many units of each supplier's products should be delivered to each warehouse in each period. Supply ability of each supplier is unlimited.
4. The warehouses have limited storage capacity. A per-unit holding cost is incurred on ending inventory of each period.
5. Each day's demand must be met on time from inventory, if not, there will incur a unit time delay punishment.
6. At the end of each cycle, the remaining inventory of each warehouse must reach a certain quantities to ensure the normal sale at the beginning of the next cycle.

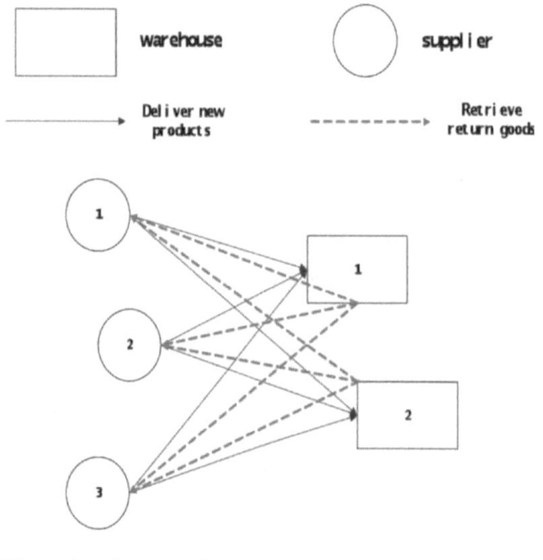

Figure 1. IP network structure.

The following notation will be used to describe the model:

i, I: index and set of suppliers,
j, J: index and set of warehouses,
d, D: index and set of days in each cycle,
t, T: index and set of periods in each day.

Parameters:

c_i^j unit distance transport costs from supplier i to warehouse j;
d_i^j distance between supplier i to warehouse j;
k_i Fix costs of each vehicle belongs to supplier i;
K_i Fix costs of supplier i for once transportation (including deliver or retrieve goods);
n_{id} The number of vehicles to be used of supplier i in day d;
P_j Capacity of warehouse j;
R_i Capacity of each vehicle belongs to supplier i;
h_i Unit storage cost in supplier i;
s^j Unit storage cost of warehouse j;
$\tilde{s}^j$ Unit storage cost of warehouse j for return goods;
f_i Unit delay punishment for products belong to supplier i for unit time;
g_{id}^j The need of warehouse j in day d for supplier i;
N_i^j Minimum quantity of products belong to supplier i in warehouse j at the end of each cycle;
θ Rejection rate;
v Speed of the vehicle.

Intermediate variables:

G_{id}^j Quantity of products belongs to supplier i should be meet (including current need and previous need that cannot be meet in time) of warehouse j in day d;
M_{id}^j Quantity of products belongs to supplier i to be meet in reality of warehouse j in day d;
a_{id}^j Inventory of products belongs to supplier i of warehouse j in day d;

Decision variables:

a_{id}^j Quantity of products delivered to warehouse j of supplier i in day d;
b_{id}^j Quantity of returned goods belong to supplier i of warehouse j in day d;

$$x_{idt}^j = \begin{cases} 1, & \text{if supplier i deliver products to} \\ & \text{warehouse j at period t in day d} \\ 0, & \text{otherwise} \end{cases}$$

$$y_{idt}^j = \begin{cases} 1, & \text{if supplier i retrieve goods from} \\ & \text{warehouse j at period t in day d} \\ 0, & \text{otherwise} \end{cases}$$

$$X_{id}^j = \begin{cases} 1, & \text{if supplier i deliver products to} \\ & \text{warehouse j in day d} \\ 0, & \text{otherwise} \end{cases}$$

$$Y_{id}^{j}=\begin{cases}1, & \text{if supplier i retrieve goods from} \\ & \text{warehouse j in day d} \\ 0, & \text{otherwise}\end{cases}$$

3.2 *Model*

Subject to:

$$Q_{i(d+1)}^{j}=Q_{id}^{j}-M_{id}^{j}+a_{id}^{j} \quad \forall\ I,\mathrm{J},D \tag{1}$$

$$G_{i(\mathrm{d}+1)}^{j}=G_{id}^{j}-M_{id}^{j}+g_{i(\mathrm{d}+1)}^{j} \quad \forall\ I,\mathrm{J,D} \tag{2}$$

$$M_{id}^{j}\le g_{id}^{j} \quad \forall\ I,\mathrm{J,D} \tag{3}$$

$$G_{i1}^{j}=g_{i1}^{j} \quad \forall\ I,\mathrm{J} \tag{4}$$

$$Q_{i8}^{j}\ge N_{i}^{j} \quad \forall\ I,\mathrm{J,D} \tag{5}$$

$$a_{id}^{j}\le n_{id}R_{i}X_{id}^{j} \quad \forall\ I,\mathrm{J,D} \tag{6}$$

$$b_{id}^{j}\le n_{id}R_{i}Y_{id}^{j} \quad \forall\ I,\mathrm{J,D} \tag{7}$$

$$\sum_{i}a_{id}^{j}+\sum_{i}Q_{id}^{j}\le P_{j} \quad \forall\ \mathrm{J,D} \tag{8}$$

$$v\times(24\cdot X_{id}^{j}+(t-1)\cdot x_{idt}^{j})\ge d_{i}^{j} \quad \forall\ I,\mathrm{J,D,T} \tag{9}$$

$$\sum_{d}b_{id}^{j}=\theta\sum_{d}M_{id}^{j} \quad \forall\ I,\mathrm{J} \tag{10}$$

$$\sum_{t}x_{idt}^{j}=1 \quad \forall\ I,\mathrm{J,D} \tag{11}$$

$$\sum_{i}x_{idt}^{j}=1 \quad \forall\ \mathrm{J,D,T} \tag{12}$$

$$x_{idt}^{j}\le X_{id}^{j} \quad \forall\ I,\mathrm{J,D,T} \tag{13}$$

$$\sum_{t}y_{idt}^{j}=1 \quad \forall\ I,\mathrm{J,D} \tag{14}$$

$$\sum_{i}y_{idt}^{j}=1 \quad \forall\ \mathrm{J,D,T} \tag{15}$$

$$y_{idt}^{j}\le Y_{id}^{j} \quad \forall\ I,\mathrm{J,D,T} \tag{16}$$

$$x_{idt}^{j}-y_{idt}^{j}\le m\cdot(X_{id}^{j}-Y_{id}^{j})^{2} \quad \forall\ I,\mathrm{J,D,T} \tag{17}$$

m is a sufficiently large number

$$x_{idt}^{j},y_{idt}^{j},X_{id}^{j},Y_{id}^{j}\in\{0,1\} \tag{18}$$

The objective function minimizes the sum of costs including transportation cost, fixed cost of vehicles, holding cost and delay punishment. The first four terms of the above objective function give the total transportation costs incurred, while the fifth to seventh give the sum of holding and costs. Finally, the last term give the delay punishment. Eqs. (1)~(5) are the balance equation for each day's needs and storage. Eqs. (6)~(7) guarantee that the quantity of transported each time cannot exceed the capacity of vehicles. Eq. (8) ensures that the inventory cannot exceed the capacity of warehouse. Eq. (9) ensures that there must be enough time for suppliers to arrive the warehouse before the entering warehouse point. Eq. (10) prohibits the quantity relationship between return and needs. Eqs. (11)~(17) ensures that in each period the number of suppliers at each warehouse cannot exceed one and each supplier cannot occupy two or more periods in each day. Finally, constraint (18) enforces the binary restriction on the x, y, X and Y decision variables.

4 MODEL SOLUTION

4.1 *Data collection*

For application of the proposed model, we set a case study of an e-commerce company "JuMei YouPin". The company has four warehouses located in TianJin, ChengDu, ShangHai and GuangZhou. We choose ten suppliers from the company and divide each day into eight periods.

4.2 *Solution approach*

The IP model is a zero-one Mixed Integer-nonlinear Programming (MIP) problem. We find that each warehouse is mutual independence. We propose a solution procedure to divide the model into sub-models according to the warehouses to solve this problem. However, the sub-model is also an MIP problem, so we use ILOG Cplex to solve each sub-model. The solution procedure is coded in JAVA. The overall solution procedure is as follows (Fig. 2).

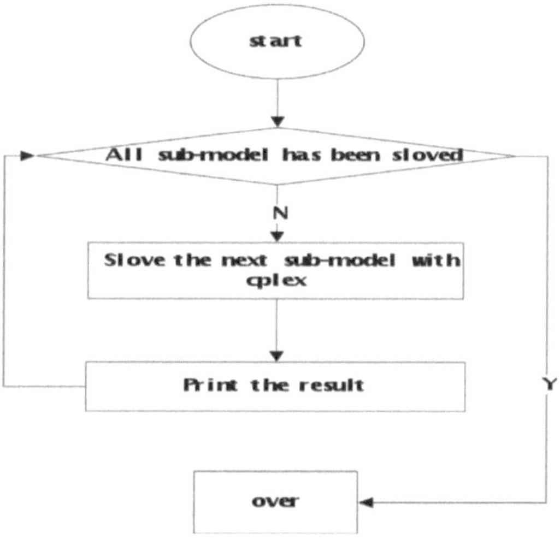

Figure 2. Overall solution procedure.

Table 1. Warehouse of TianJin.

	Mon.	Tue.	Wed.	Thu.	Fri.	Sat.	Sun.
8:00–9:00		4(R) 2500,1300	5(R) 5000,800		2 2500	4 3600	
9:00–10:00		6(R) 2800,1200	1(R) 4800,300				
10:00–11:00							
11:00–12:00		2(R) 1800,300			3(R) 4500,500		6 3000
14:00–15:00							
15:00–16:00							8 2500
16:00–17:00		8(R) 2300,800		9(R) 5800,400			
17:00–18:00		7(R) 2000,500			10(R) 4700,200	7 2700	

*Supplier delivers new products and retrieves return goods when they back.

Table 2. Warehouse of ChengDu.

	Mon.	Tue.	Wed.	Thu.	Fri.	Sat.	Sun.
8:00–9:00		4(R) 3000,800	5(R) 4200,200			4 2500	
9:00–10:00			6(R) 4600,200	1(R) 5300,220			
10:00–11:00			2(R) 3900,180				
11:00–12:00					3(R) 3700,150		
14:00–15:00							
15:00–16:00							8 1800
16:00–17:00		8(R) 2000,800		9(R) 4500,200			
17:00–18:00		7(R) 3900,300			10(R) 4000,150		

Table 3. Warehouse of ShangHai.

	Mon.	Tue.	Wed.	Thu.	Fri.	Sat.	Sun.
8:00–9:00		4(R) 4000,900	2(R) 4800,250	3(R) 4500,180		4 2500	
9:00–10:00		1(R) 2900,800				1 2500	
10:00–11:00		5(R) 2000,800		6(R) 4500,180		5 2800	
11:00–12:00		8(R) 3800,200	9(R) 4000,180				
14:00–15:00							
15:00–16:00							
16:00–17:00							
17:00–18:00		7(R) 1800,200			10(R) 4800,150	7 2000	

Table 4. Warehouse of GuangZhou.

	Mon.	Tue.	Wed.	Thu.	Fri.	Sat.	Sun.
Supplier							
Quantity of new products							
8:00–9:00		4(R)	1(R)	5(R)		4	
		3000,1000	3800,250	4800,300		2200	
9:00–10:00		2(R)	3(R)			2	
		3300,300	5000,330			2800	
10:00–11:00		9(R)		6(R)			
		4800,320		4500,280			
11:00–12:00							
14:00–15:00							
15:00–16:00			8(R)				
			4300,280				
16:00–17:00							
17:00–18:00		10(R)		7(R)			
		4800,300		3500,200			

4.3 *Results*

The result of the model shows in Tables 1–4.

5 SUMMARY

Data collection is always difficult, especially when there are more than several suppliers in an e-commerce company. With the above results, we are confident to say that our model is useful for e-commerce companies.

However, with the increasing number of suppliers, the solving time of Cplex increases sharply. In the future, it is necessary to propose a heuristic solution to solve the model in a more efficient way.

REFERENCES

Bill Roach, 2005. Origins of the Economic Order Quantity Formula. School of Business, Washburn University.

Gerchak, Y., Vickson, R.G., Parlar, M., 1988. Periodic review production models with variable yield and uncertain demand. IIE Transactions 20 (2), 144–150.

Harris, F.W., 1913. What quantity to make at once. In: The Library of Factory Management. Operation and Costs, The Factory Management Series, vol. 5 A.W. Shaw Co., Chicago, pp. 47–52.

Inderfurth, K., 2004. Analytical solution for a single-period production-inventory problem with uniformly distributed yield and demand. Central European Journal of Operations Research 12 (2), 117–127.

Karlin, S., 1958. One stage inventory models with uncertainty. In: Karlin, S., Scarf, H.(Eds.), Studies in the Mathematical Theory of Inventory and Production. Stanford University Press, Stanford, CA.

Rosenblatt, M.J., Lee, H.L., 1986. Economic production cycles with imperfect production processes. IIE Transactions 18 (1), 48–55.

Wright, C.K., Mehrez, A., 1998. An overview of representative research of the relationship between quality and inventory. Omega 26 (1), 29–47.

Engineering Management and Industrial Engineering – Leung (Ed.)
© 2015 Taylor & Francis Group, London, ISBN 978-1-138-02773-2

How does organization structure influence CWB in the e-commerce times: From the perspective of employee psychology climate

W.B. Liu
University of Electronic Science and Technology of China, Chengdu, China

J. Tang
Fujian Normal University, Fuzhou, China

ABSTRACT: This paper constructs a mediating effect model to analyze the relationship between organization structure and employee CWB from the perspective of individual psychological climate in the e-commerce times. We used hierarchical regression method to analyse the data base on 366 questionnaires from 29 firms, and we found both complexity and centralization had a significant positive effect on CWB. Coordination mechanism had a significant negative effect on CWB. Meanwhile, employee's work autonomy and job insecurity psychological climate had separate significant negative and positive effect on CWB respectively.

1 INTRODUCTION

Daft (2003) affirms the organization structure design and transform is meaningful, because it can influence the organization performance through organizing employee's behavior. However, in the process of exploring the organization structure's outcome variables, the influence of organization structure on individual psychology, attitude and behavior is still not very clear (Tata & Prasad, 2004). Especially, very few researches pay close attention to employee's Counterproductive Work Behavior (CWB) in the e-commerce times (Ones & Viswesvaran, 2012).

2 DATA AND METHODS

2.1 *Theoretical model*

This paper analyses the relationship between organization structure and employee CWB from the perspective of individual psychological climate in the e-commerce times, we construct a mediating effect model, the following Figure 1 shows the relationships established in the hypotheses proposed above.

2.2 *Data collection and variable measurement*

Self-administered questionnaires were randomly distributed to 600 full-time employees working in 29 firms. In the end, 366 completed questionnaires were returned within three months. In all variables, t-tests show no significant differences between these two groups of firms. Finally, Most of the respondents worked in trading, IT, foodstuff, transportation, and oil and energy corporations. Their company scale range from less than 50 to more than 5000 employees. Sixty-one per cent of the respondents were women. The age of the respondents are between 21 and 46 (M = 29.6 years) and the average tenure of their present organization is 5.2 years, with a range from 0.8 to 19.5 years.

2.3 *Reliability and validity analysis*

Cronbach's alpha coefficient shows adequate values (see diagonal in Table 1). All analyses provide reasonable evidence that the measures used are reliable and valid. LISREL 6.0, which evaluates concept reliability, convergent validity and unidimensionality, served to perform this analysis. To assess scale reliability, the study computes the composite reliability estimate, which has a value of

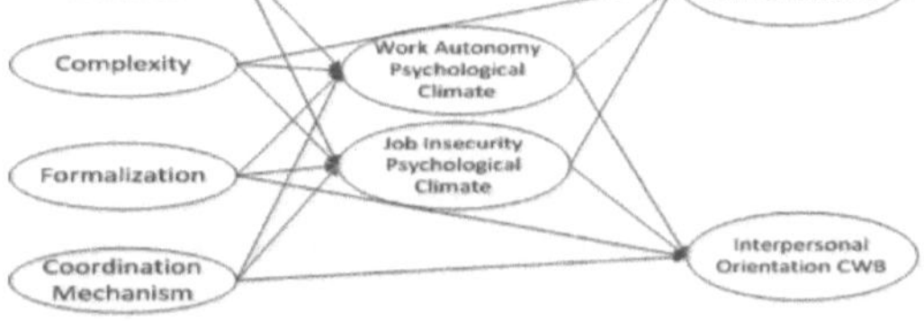

Figure 1. Mediation Model 1.

Table 1. Means, standard deviations, correlation and Cronbach's alphas.

	1	2	3	4	5	6	7	8
Mean	4.2	4.6	3.8	3.6	3.2	4.1	3.9	3.5
S.D	0.87	0.72	0.76	0.91	0.83	0.77	0.93	0.86
Centralization	***0.76***							
Complexity	0.32*	***0.82***						
Formalization	0.28*	0.23*	***0.79***					
Coordination	–0.17	–0.18*	–0.16*	***0.83***				
Work autonomy	–0.23**	–0.27*	–0.26*	0.22**	***0.77***			
Job insecurity	0.25*	0.22*	0.32	–0.31	–0.19**	***0.78***		
Organizational-CWB	0.36*	0.21**	0.36**	–0.26*	–0.18**	0.17**	***0.89***	
Interpersonal-CWB	0.17**	0.29**	0.22*	–0.19*	–0.12**	0.13**	0.28*	***0.87***

0.761. This value is greater than the minimum recommended level of 0.7 (Bagozzi & Yi, 1988). The significance of the factor loadings serves to examine convergent validity. All the loadings of the measurement items on the hypothesized construct are significant ($p < 0.05$) in providing evidence of convergent validity (Anderson and Gerbing, 1988). The results obtained in the confirmatory factor analysis indicate that there exist convergent validity and unidimensionality. The chi-square statistic ($\chi^2 = 12.16$, df = 9) is significant and the fit indices approach the preferred threshold of 0.90 (NFI = 0.92; NNFI = 0.96; AGFI = 0.95; GFI = 0.97). Further, the small magnitude of residuals (RMR = 0.08; RMSEA = 0.05) also indicates appropriate model data fit.

3 EMPIRICAL RESULT

3.1 *Descriptive statistics analysis*

Means, standard deviations, and Cronbach's alphas for all study variables are presented in Table 2, the Pearson correlation coefficient among major independent and dependent variables are also shown in Table 1.

3.2 *Regression analysis*

Firstly, we analyze the relations between the four organization structure variables and the two categories of CWB, and the relations between the four organizational structure variables and the two categories of psychological climate (work autonomy and job insecurity). Table 2 shows that centralization was significantly positive related to organizational-CWB ($r = 0.108$, $P < 0.01$), and interpersonal-CWB ($r = 0.087$, $P < 0.05$); complexity was significantly positive related to organizational-CWB ($r = 0.096$, $P < 0.05$), and interpersonal-CWB ($r = 0.103$, $P < 0.01$); coordination was significantly negative related to organizational-CWB ($r = -0.116$, $P < 0.01$), and interpersonal-CWB ($r = -0.182$, $P < 0.01$). But, the value of regression coefficient for formalization and CWB is non-significant. At the same time, centralization was significantly negative related to work autonomy psychological climate ($r = -0.203$, $P < 0.01$), complexity was significantly negative related to work autonomy psychological climate ($r = -0.096$, $P < 0.05$). Formalization was significantly negative related to work autonomy psychological climate ($r = -0.102$, $P < 0.05$), coordination was significantly positive related to work autonomy psychological climate ($r = 0.128$, $P < 0.01$). Finally, centralization was significantly positive related to job insecurity psychological climate ($r = 0.182$, $P < 0.01$). Complexity was significantly positive related to job insecurity psychological climate ($r = 0.103$, $P < 0.01$). Coordination mechanism was significantly negative related to job insecurity psychological climate ($r = -0.152$, $P < 0.01$). But, the value of regression coefficient for formalization and job insecurity psychological climate is non-significant.

Secondly, we looked at the relationships between psychological climate and the two categories of CWB. Table 3 shows that the R^2 coefficient increases significantly when the regression model includes the psychological climate (Model 2) in comparison with Model 1. Additionally, work autonomy psychological climate was significantly negative related to organizational-CWB ($r = -0.212$, $P < 0.01$), and interpersonal-CWB ($r = -0.104$, $P < 0.01$). At the same time, job insecurity psychological climate was significantly positive related to organizational-CWB ($r = 0.181$, $P < 0.01$), and interpersonal-CWB ($r = 0.196$, $P < 0.01$).

3.3 *Mediate effect analysis*

The process of three-step regression analyses and the statistical results are presented in Table 4.

Table 2. Results drawn from the regression analyses of CWB and psychological.

	Organizational-CWB		Interpersonal-CWB	
	Model 1	Model 2	Model 1	Model 2
Gender	0.107*	0.210	0.187*	0.320*
Age	0.124	0.231	0.126	0.212
Educational level	–0.176*	–0.097*	–0.097*	–0.313*
Marital status	0.204	0.116	0.116	0.166
Work tenure	0.107*	0.136*	0.136*	0.132*
Number of staff	0.116	0.125	0.125	0.156
Centralization		0.108**		0.087*
Complexity		0.126*		0.138*
Formalization		0.098		–0.132
Coordination mechanism		–0.116**		–0.182**
R^2	0.156	0.321	0.132	0.278
F	30.226***	36.611***	29.873***	32.226***
ΔR^2	0.065	0.165	0.050	0.146
	Work autonomy psychological climate		Job insecurity psychological climate	
	Model 1	Model 2	Model 1	Model 3
Gender	–0.106	–0.103	0.131	0.146
Age	0.218*	0.162*	0.223	0.208
Educational level	0.124*	0.156**	–0.212**	–0.217**
Marital status	–0.231	–0.139	0.124	0.136
Work tenure	0.156	0.176	–0.213	–0.202*
Number of staff	–0.230*	–0.201*	0.178*	0.196*
Centralization		–0.203**		0.182**
Complexity		–0.096*		0.103**
Formalization		–0.102*		0.161
Coordination mechanism		0.128**		–0.152**
R^2	0.162	0.302	0.182	0.312
F	32.241***	36.316***	29.386***	30.122***
ΔR^2	0.059	0.140	0.093	0.130

Table 3. Results drawn from the regression analyses of CWB.

	Organizational-CWB		Interpersonal-CWB	
	Model 1	Model 2	Model 1	Model 2
Gender	0.107*	0.218	0.187*	0.218*
Age	0.124	0.221	0.126	0.176
Educational level	–0.176*	–0.112*	–0.097*	–0.218*
Marital status	0.204	0.125	0.116	0.161
Work tenure	0.107*	0.126*	0.136*	0.172*
Number of staff	0.116	0.115	0.125	0.162
Work autonomy psychological climate		–0.212**		–0.104**
Job insecurity psychological climate		0.181**		0.196**
R^2	0.156	0.309	0.132	0.312
F	30.226***	32.121***	29.873***	35.166***
ΔR^2	0.065	0.153	0.050	0.180

Table 4. Three-step regression analyses for the mediation model.

	Step 1		Step 2		Step 3
	Org-CWB	Work autonomy	Job insecurity	Org-CWB	Org-CWB
Gender	0.210	–0.103	0.146	0.218	0.161
Age	0.231	0.162*	0.208	0.221	0.206
Educational level	–0.097*	0.156**	–0.217**	–0.112*	–0.109*
Marital status	0.116	–0.139	0.136	0.125	0.211
Work tenure	0.136*	0.176	–0.202*	0.126*	0.116*
Number of staff	0.125	–0.201*	0.196*	0.115	0.102
Centralization	0.108**	–0.203**	0.182**	–	0.068*
Complexity	0.126*	–0.096*	0.103**	–	0.112*
Formalization	0.098	–0.102*	0.161	–	0.108
Coordination mechanism	–0.116**	0.128**	–0.152**	–	–0.136
Work autonomy	–	–	–	–0.212**	–0.208**
Job insecurity	–	–	–	0.181**	0.196**
R^2	0.321	0.302	0.312	0.309	0.336
F	36.611***	36.316***	30.122***	32.121***	36.766***

	Step 1		Step 2		Step 3
	Inter-CWB	Work autonomy	Job insecurity	Inter-CWB	Inter-CWB
Gender	0.320*	–0.103	0.146	0.218*	0.201*
Age	0.212	0.162*	0.208	0.176	0.165
Educational level	–0.313*	0.156**	–0.217**	–0.218*	-206*
Marital status	0.166	–0.139	0.136	0.161	0.156
Work tenure	0.132*	0.176	–0.202*	0.172*	0.192*
Number of staff	0.156	–0.201*	0.196*	0.162	0.136
Centralization	0.087*	–0.203**	0.182**	–	0.112
Complexity	0.138*	–0.096*	0.103**	–	0.115*
Formalization	–0.132	–0.102*	0.161	–	–0.107
Coordination mechanism	–0.182**	0.128**	–0.152**		–0.142**
Work autonomy	–	–	–	–0.104**	–0.116*
Job insecurity e	–	–	–	0.196**	0.182**
R^2	0.278	0.302	0.312	0.312	0.316
F	32.226***	36.316***	30.122***	35.166***	36.127***

As shown in Table 4, when we put organizational structure and psychological climate variables into the regression model together, we found: (1) work autonomy psychological climate was significantly negative related to organizational-CWB ($r = -0.208$, $P < 0.01$), and interpersonal-CWB ($r = -0.116$, $P < 0.01$). At the time, the relationship between organizational structure and CWB were not significant or decreased with the coefficient, except the formalization. Thus, we partly confirm H4. (2) job insecurity psychological climate was significant positive related to organizational-CWB ($r = 0.196$, $P < 0.01$), and interpersonal-CWB ($r = 0.182$, $P < 0.01$). At the same time, the relationship between organizational structure and CWB were not significant or decreased with the coefficient, except the formalization.

4 CONCLUSION AND DISCUSSION

4.1 *Management implications*

In practice, people can perceive a wide range of behaviors as wrong, inappropriate, unfair, or unethical. Further, under certain contextual conditions breaches of normative conventions will be perceived as moral transgressions, and will undergo a shift from preference to universal absolute. We proposed and found that an important individual cognition factor predicting of counterproductive work behavior is the psychological climate used to influence others, specifically for work autonomy psychological climate and job insecurity psychological.

Firstly, we found work autonomy psychological climate was significant negative related to both organizational-CWB and interpersonal-CWB. Freedom of movement, freedom to establish and execute plans for task accomplishment, and freedom from immediate supervision were important for giving people a sense of marginal control over their lives at work. No doubt this sense of control was psychologically beneficial in its own right (Averill, 1973; Fisher, 1984). But these dimensions of work autonomy were psychologically consequential for other reasons as well. In the workplace context they were meaningful as indicators of organizational structure. Many people valued and sought work autonomy because of its meaning in these respects and what it concomitantly meant about persons who had it.

We believe that freedom of movement, for example, was important to many people as an indicator of the degree of challenge a job entailed and of the responsibility the person need to take in the job. This was especially true for maintenance mechanics, who often compared themselves on this basis to stationary production workers. Mechanics commonly claimed that while immobility might be tolerable to those "mentally suited for repetitive work". For them, it would be intolerable because of their needs for more challenging work and work which demanded mobility. It was also implied that freedom of movement was something that could be granted only to trustworthy workers who didn't need to be watched continuously. The latter sentiment was not uniquely held by maintenance mechanics, but was also expressed by a maintenance clerk, a machine oiler, several parts expeditors, and a number of secretaries, all of whom enjoyed considerable freedom of movement in their jobs relative to their status peer. At the same time, freedom of establishing one's own plans for task accomplishment was important as an indicator of how much initiative and responsibility a job demanded, and an indicator of how much trust the person in the job was warranted.

Work autonomy psychological is having more freedom to move around at will, to plan one's own tasks, or to work alone. In the workplace culture it is a badge of status, an indicator of a job's skill and responsibility demands, and, perhaps most important, a reward for reliable and competent performance. The implication of this is that work autonomy psychological can't be understood solely as 'latitude' within limits imposed by bosses, work rules, company policies, and so forth. It means much more than this to those who possess it and those who strive for it. In sum, work autonomy will help the manager to restraint employees' CWB efficiently.

Secondly, we found job insecurity psychological climate was significant positive related to both organizational-CWB and interpersonal-CWB. The findings of this study also make it clear the job insecurity psychological climate encompasses much more than simply fear of losing one's job next week or next month. Instead, the prospect of demotion, deterioration in working conditions, or even the long-term prospect of eventual job loss is also associated with decreased well-being and work commitment. In fact, because it takes the employee a long time to perceive job insecurity and really loss their job. So when the job insecurity psychological climate occurs, they will compensate their possible psychological and material damage from various CWB. From this perspective, we argued that concluding that job insecurity does indeed constitute a major psychological stressor for employee. However, it is necessary to take into account that the 'objective' indices of insecurity did not directly influence mental health and had only a slight impact on their work commitment and psychological contract. Instead, it was the individual's subjective appraisal of risk that negatively affected well-being and work effort.

4.2 *Research limitations*

It is necessary to explain that we conducted our research in China, a setting that deserves more research attention. The trend of organizational restructuring—in the forms of corporate downsizing, mergers and acquisitions, plant closures, and workforce reorganizations—has expanded to the global workforce (Probst & Lawler, 2006). In particular, the Chinese economy has undergone significantly reforms in the past two decades, shifting from an "iron rice bowl" policy of lifetime employment to a market-oriented setting. The economic reforms have had significantly impacts on employment security among Chinese workers. Furthermore, China offers a particularly appropriate context to test claims that people from collectivist cultures react more negatively to the threat of job loss than do individualists, because of their greater emphasis on group ties, higher need for affiliation, and preferences for security (Probst & Lawler, 2006).

However, these findings also need to be considered in light of several limitations. Our design is correlation, so we cannot draw causal conclusions. Thus, in order to draw causal conclusions, the further studies also should use a cross-lagged panel longitudinal design to measure both constructs at both times and examine the dynamic relationship between organizational structure, psychological climate and counterproductive work behavior. For example, it might be interesting to examine the conditions in which respondents reappraise their psychological climate and regulate their CWB, suggested by Gross's (1998) model of emotional regulation.

Another limitation is the self-reported nature of our measures, which may contribute to common method bias. However, we tried to minimize this bias by using multiple sources so that the organizational structure was confirmed by supervisors. Our use of a time-lagged design also should reduce the method bias in our findings. In addition, we calculated the variance explained by the method factor following Williams, Cote, and Buckley (1989). Results show that the variance explained was 6.8 percent, which was much lower than the 25 percent observed in William et al. (1989), suggesting that common method variance is not a pervasive problem in this study. Additional studies that assess organizational practices more objectively and test counterproductive work behavior using different sources of data would be desirable.

ACKNOWLEDGMENT

This research was supported by The National Social Science Fund under Grant No. 14CGL018.

REFERENCES

[1] Anderson, J.C., & Gerbing, D.W. (1988). Structural equation modeling in practice: a review and recommended two-step approach. Psychology Bullet, 103(3):411–423.

[2] Averill, J.R. (1973). Personal control over aversive stimuli and its relationship to stress. Psychological Bulletin, 80(2):286–303.

[3] Bagozzi, R.P., & Yi, Y. (1988). On the evaluation of structural equation models. Journal of Academic Marketing Science, 16:74–94.

[4] Fisher, S. (1984). Stress and the Perception of Control. Hillsdale, NJ: Lawrence Erlbaum.

[5] Gross, J.J. (1998). The emerging field of emotion regulation: An integrative review. Review of General Psychology, 2(2):271–299.

[6] Marshall, J., Smith, S., & Buxton, S. (2009). Learning organizations and organizational learning: what have we learned? Management Services, 53(2):36–44.

[7] Ones, D.S., & Viswesvaran, C. (2012). Integrity Test Predict Counterproductive Work Behavior and Job Performance Well: Comment on Van Iddekinge, Roth, Raymark, and Odle-Dusseau. Journal of Applied Psychology, 97(3):537–542.

[8] Probst, T.M., & Lawler, J. (2006). Cultural values as moderators of the outcomes of job insecurity: The role of individualism and collectivism. Applied Psychology, 55(1):234–254.

[9] Tata, J., & Prasad, S. (2004). Team Self-management, Organizational Structure, and Judgments of Team Effectiveness. Journal of Managerial Issues, 16(2):248–265.

[10] Williams, L.J., Cote, J.A., & Buckley, M.R. (1989). Lack of method variance in self-reported affect and perceptions at work: Reality or artifact? Journal of Applied Psychology, 74(3):462–468.

Engineering Management and Industrial Engineering – Leung (Ed.)
© 2015 Taylor & Francis Group, London, ISBN 978-1-138-02773-2

Author index